Recent Advances in the Development of Thin Films

Recent Advances in the Development of Thin Films

Editors

Zohra Benzarti
Ali Khalfallah

Basel • Beijing • Wuhan • Barcelona • Belgrade • Novi Sad • Cluj • Manchester

Editors

Zohra Benzarti
CEMMPRE, Department of
Mechanical Engineering
University of Coimbra
Coimbra
Portugal

Ali Khalfallah
CEMMPRE, Department of
Mechanical Engineering
University of Coimbra
Coimbra
Portugal

Editorial Office
MDPI AG
Grosspeteranlage 5
4052 Basel, Switzerland

This is a reprint of articles from the Special Issue published online in the open access journal *Coatings* (ISSN 2079-6412) (available at: https://www.mdpi.com/journal/coatings/special_issues/Thin_Film).

For citation purposes, cite each article independently as indicated on the article page online and as indicated below:

Lastname, A.A.; Lastname, B.B. Article Title. *Journal Name* **Year**, *Volume Number*, Page Range.

ISBN 978-3-7258-1855-6 (Hbk)
ISBN 978-3-7258-1856-3 (PDF)
doi.org/10.3390/books978-3-7258-1856-3

© 2024 by the authors. Articles in this book are Open Access and distributed under the Creative Commons Attribution (CC BY) license. The book as a whole is distributed by MDPI under the terms and conditions of the Creative Commons Attribution-NonCommercial-NoDerivs (CC BY-NC-ND) license.

Contents

About the Editors vii

Zohra Benzarti and Ali Khalfallah
Recent Advances in the Development of Thin Films
Reprinted from: *Coatings* 2024, 14, 878, doi:10.3390/coatings14070878 1

Daniel Cristea, Ioana-Laura Velicu, Luis Cunha, Nuno Barradas, Eduardo Alves and Valentin Craciun
Tantalum-Titanium Oxynitride Thin Films Deposited by DC Reactive Magnetron Co-Sputtering: Mechanical, Optical, and Electrical Characterization
Reprinted from: *Coatings* 2022, 12, 36, doi:10.3390/coatings12010036 5

Moussa Grafoute, Kouamé Boko Joël-Igor N'Djoré, Carine Petitjean, Jean François Pierson and Christophe Rousselot
Influence of Oxygen Flow Rate on the Properties of FeO_XN_Y Films Obtained by Magnetron Sputtering at High Nitrogen Pressure
Reprinted from: *Coatings* 2022, 12, 1050, doi:10.3390/coatings12081050 21

Olena Okhay, Paula M. Vilarinho and Alexander Tkach
Structure, Microstructure, and Dielectric Response of Polycrystalline $Sr_{1-x}Zn_xTiO_3$ Thin Films
Reprinted from: *Coatings* 2023, 13, 165, doi:10.3390/coatings13010165 36

Roger Bujaldón, Alba Cuadrado, Dmytro Volyniuk, Juozas V. Grazulevicius, Joaquim Puigdollers and Dolores Velasco
Role of the Alkylation Patterning in the Performance of OTFTs: The Case of Thiophene-Functionalized Triindoles
Reprinted from: *Coatings* 2023, 13, 896, doi:10.3390/coatings13050896 44

Ali Khalfallah and Zohra Benzarti
Mechanical Properties and Creep Behavior of Undoped and Mg-Doped GaN Thin Films Grown by Metal–Organic Chemical Vapor Deposition
Reprinted from: *Coatings* 2023, 13, 1111, doi:10.3390/coatings13061111 57

Jiedong Deng, Feng Jiang, Xuming Zha, Tao Zhang, Hongfei Yao, Dongwei Zhu, et al.
Impact Resistance of CVD Multi-Coatings with Designed Layers
Reprinted from: *Coatings* 2023, 13, 815, doi:10.3390/coatings13050815 71

Chaoqun Ma, Donghong Peng, Xuanyao Bai, Shuangqiang Liu and Le Luo
A Review of Optical Fiber Sensing Technology Based on Thin Film and Fabry–Perot Cavity
Reprinted from: *Coatings* 2023, 13, 1277, doi:10.3390/coatings13071277 83

Jianxiao Bian, Yingtang Sun, Jinchang Guo, Xin Liu and Yang Liu
Enhancing the Performance and Stability of Perovskite Solar Cells via Morpholinium Tetrafluoroborate Additive Engineering: Insights and Implications
Reprinted from: *Coatings* 2023, 13, 1528, doi:10.3390/coatings13091528 111

Ali Khalfallah, Amine Khalfallah and Zohra Benzarti
Identification of Elastoplastic Constitutive Model of GaN Thin Films Using Instrumented Nanoindentation and Machine Learning Technique
Reprinted from: *Coatings* 2024, 14, 683, doi:10.3390/coatings14060683 124

Elena Dmitriyeva, Igor Lebedev, Ekaterina Bondar, Anastasia Fedosimova, Abzal Temiraliev, Danatbek Murzalinov, et al.
The Influence of Lyophobicity and Lyophilicity of Film-Forming Systems on the Properties of Tin Oxide Films
Reprinted from: *Coatings* **2023**, *13*, 1990, doi:10.3390/coatings13121990 . . . . **140**

Tao He, Dexin Wang, Yu Xu and Jing Zhang
The Facile Construction of Anatase Titanium Dioxide Single Crystal Sheet-Connected Film with Observable Strong White Photoluminescence
Reprinted from: *Coatings* **2024**, *14*, 292, doi:10.3390/coatings14030292 . . . . **156**

Artem Shiryaev, Konstantin Rozanov, Vladimir Kostishin, Dmitry Petrov, Sergey Maklakov, Arthur Dolmatov and Igor Isaev
Retrieving the Intrinsic Microwave Permittivity and Permeability of Ni-Zn Ferrites
Reprinted from: *Coatings* **2023**, *13*, 1599, doi:10.3390/coatings13091599 . . . . **164**

Arezou Khezerlou, Hajar Zolfaghari, Samira Forghani, Reza Abedi-Firoozjah, Mahmood Alizadeh Sani, Babak Negahdari, et al.
Combining Non-Thermal Processing Techniques with Edible Coating Materials: An Innovative Approach to Food Preservation
Reprinted from: *Coatings* **2023**, *13*, 830, doi:10.3390/coatings13050830 . . . . **178**

About the Editors

Zohra Benzarti

Zohra Benzarti is a Professor (Associate) in the Department of Physics at the University of Sfax, Faculty of Sciences of Sfax (USF-FSS), in Tunisia. She received her PhD degree in the Physics of Materials (Thin Films Semiconductors) from the University of Tunis El-Manar, Tunisia, in 2006. In 2016, she obtained her Habilitation (HDR) in the Physics of Thin Films from the University of Monastir, Tunisia. Prof. Benzarti joined the CICECCO at the University of Aveiro in Portugal in 2022. In 2023, she moved to the Centre of Mechanical Engineering Materials and Processes (CEMMPRE) and joined the Materials and Processes Group-B, in the Department of Mechanical Engineering of the Faculty of Sciences and Technology of the University of Coimbra (FCTUC) in Portugal. Since the early 2000s, Prof. Benzarti has been working on the fabrication of III-nitride thin films using metal–organic vapor deposition (MOCVD) and molecular-beam epitaxy (MBE) techniques and their multiphysics characterization. Her research interests include the correlation between the nanomechanical and physical properties of thin-film nitride semiconductors. She has also worked on the development of multifunctional materials, including the elaboration of nanomaterials and piezoelectric nano-composites, ZnO doped with transition metals and rare earths, hybrid and ceramic perovskites for light sensors, photovoltaic cells, White LEDs, and energy storage devices. Currently, she is exploring copper corrosion, decorative deposition on polymer substrates, and Li-ion batteries. Prof. Benzarti is the author and co-author of over 50 publications, including books, book chapters, papers in international journals, and presentations at international conferences. She is indexed in the Google Scholar database with over 703 citations and an h-index of 15. Prof. Benzarti has been invited to join the editorial board, serve as a guest editor, and peer review for several prestigious journals.

Ali Khalfallah

Ali Khalfallah (Prof. Dr.) is a Full Professor in the Department of Mechanical Engineering of the Higher Institute of Applied Sciences and Technology at the University of Sousse in Tunisia (DME-UST). He was the Head of the Department of Mechanical Engineering from 2017 to 2019. He received his PhD degree from the University of Tunis El-Manar (Tunisia) in 2004. In 2015, he obtained the diploma of the "Habilitation à Diriger des Recherches - HDR" in Mechanical Engineering from the University of Sousse in Tunisia. In 2019, he joined the Centre of Mechanical Engineering Materials and Processes (CEMMPRE) in the Department of Mechanical Engineering of the Faculty of Sciences and Technology of the University of Coimbra (FCTUC) in Portugal. Prof. Khalfallah's research activities focus on various range of topics, including the multidisciplinary modeling of the mechanical behavior of materials and the identification of thin-film properties, using a combination of computational physics, artificial intelligence, and multi-scale simulations, with a recent exploration into tribology.

He is author and co-author of 65+ publications (i.e., book, book chapters, papers in international journals), an invited speaker at international conferences, and the owner of an issue patent. In the Google Scholar indexing database, the publications of Prof. Khalfallah received more than 449 citations and h-index of 12.

Prof. Khalfallah has been invited to join the editorial board of several international journals and serve as a reviewer for many international journals and conferences. He received the outstanding reviewers' awards (2023-IOP).

Editorial

Recent Advances in the Development of Thin Films

Zohra Benzarti and Ali Khalfallah *

CEMMPRE, Department of Mechanical Engineering, University of Coimbra, Rua Luís Reis Santos, 3030-788 Coimbra, Portugal; zohra.benzarti@dem.uc.pt
* Correspondence: ali.khalfallah@dem.uc.pt; Tel.: +351-933411080

Thin films and coatings are an integral part of modern technology, with applications including solar cells [1–3], cutting-edge electronics [4], cutting tools [5,6] and even food preservation [7]. The development of thin films is a dynamic and rapidly evolving field of research, driven by the need for their improved performance, durability, and functionality. Tailoring thin-film materials with specific properties, which can then be utilized across a variety of industries [8–10], is facilitated by the monitoring of modern coating processes [11–13] and the application of multiphysics characterization techniques. The advanced knowledge gained from these techniques enable the manufacturing of sound components for potential application in a broad range of fields [9,14], further expanding the reach and impact of thin-film technology.

This Special Issue presents a selection of high-quality papers reflecting the recent advances in the development of thin films. It comprises thirteen contributions, including two reviews and eleven research articles, covering several aspects of thin films that highlight the importance of this field and the ongoing progress towards potential and diverse applications across various industry sectors.

One key area of focus in this Special Issue is the deposition and characterization of thin films, with contributions exploring novel deposition techniques and their impact on film properties. Cristea et al. (Contribution 1) investigated the deposition of tantalum–titanium oxynitride thin films using DC reactive magnetron co-sputtering, and they found that controlling the Ti and Ta content results in a tunable elemental composition and structure, with higher Ti content leading to increased oxygen content and influencing optical and electrical properties. Grafoute et al. (Contribution 2) also contributed to this area of research by examining the influence of oxygen flow rate on the properties of Fe-O-N films deposited by magnetron sputtering, and they found that higher oxygen content results in well-crystallized films with good semiconductor properties, which is in agreement with previous research [15,16].

The synthesis and characterization of thin films with specific structural and electrical properties are also covered in this Special Issue. Okhay et al. (Contribution 3) reported the successful preparation of monophasic $Sr_{1-x}Zn_xTiO_3$ thin films with high Zn content using a sol–gel technique. They found that morpholinium tetrafluoroborate additive engineering in perovskite solar cells significantly improves performance and stability, yielding a power conversion efficiency of 23.83% and retaining 92% of the initial PCE after 2000 h. Bujaldón et al. (Contribution 4) synthesized thiophene-containing triindole derivatives and investigated the impact of alkylation patterning on their performance as organic semiconductors. They found that derivatives with longer N-alkyl chains resulted in thin films with improved crystallinity and semiconductor properties.

The assessment of the mechanical and electrical properties of thin films is another key area of research. Khalfallah et al. (Contribution 5) studied the impact of Mg doping on the mechanical behavior of GaN thin films. They found significant improvements in hardness and Young's modulus and an effect on creep behavior due to increased dislocation density and observed a transition from n-type to p-type conductivity. Deng et al. (Contribution 6)

Citation: Benzarti, Z.; Khalfallah, A. Recent Advances in the Development of Thin Films. *Coatings* **2024**, *14*, 878. https://doi.org/10.3390/coatings14070878

Received: 27 June 2024
Accepted: 10 July 2024
Published: 12 July 2024

Copyright: © 2024 by the authors. Licensee MDPI, Basel, Switzerland. This article is an open access article distributed under the terms and conditions of the Creative Commons Attribution (CC BY) license (https://creativecommons.org/licenses/by/4.0/).

contributed to this area of research by evaluating the impact resistance of CVD multi-coatings with designed layers, providing insights into the dynamic milling process and the effect of layer thickness on coated cutting tools. They found that thicker coatings and coatings without a TiN surface layer showed a decrease in impact resistance.

Thin films are also applied in sensing and solar cell technologies [9,17–20]. Ma et al. (Contribution 7) [21] reviewed optical fiber sensing technology based on thin films and Fabry–Perot cavities, highlighting the versatility and applicability of this technique. Bian et al. (Contribution 8) enhanced the performance of perovskite solar cells using morpholinium tetrafluoroborate as an additive, achieving notable improvements in power conversion efficiency and long-term stability, addressing challenges related to defect presence [22].

Additionally, Khalfallah et al. (Contribution 9) [23] proposed an inverse analysis approach to identify the elastoplastic parameters of GaN thin films, combining instrumented nanoindentation with finite element simulations and an artificial neural network model. This approach offers a computationally efficient and accurate method for identifying the elastoplastic parameters of thin films, leveraging the combined power of nanoindentation testing, finite element simulations, and ANNs. Moreover, it improves our understanding of the complex behavior of thin films under stress [24].

Several of the papers presented in this Special Issue explore the properties of thin films. Dmitriyeva et al. (Contribution 10) investigated the influence of lyophobicity and lyophilicity on the characteristics of tin oxide films, observing that the addition of a fluorinating agent does not increase specific conductivity, and confirming the incorporation of fluorine ions through X-ray diffraction analysis. He et al. (Contribution 11) [25] presented a novel method for the construction of anatase titanium dioxide single-crystal sheet-connected films, resulting in strong white photoluminescence comparable to commercial fluorescent lamp coatings, achieved through the incorporation of oxygen defects.

Shiryaev et al. (Contribution 12) explored the applicability of mixing rules to predict the permittivity and permeability of composite materials containing Ni-Zn ferrites. They revealed differences in the domain structures and demagnetizing fields between particles and bulk ferrite, and found a discrepancy between the measured and retrieved permeability values. The authors attributed this discrepancy to the difference between the domain structures and demagnetizing fields of particles and bulk ferrite.

In their review, Khezerlou et al. (Contribution 13) [26] highlighting the potential of combining edible coating materials with non-thermal processing technologies to enhance food preservation [27]. They reported that edible coatings made from food-grade structuring ingredients, like proteins, polysaccharides, and lipids, can be fortified with functional additives to improve food quality, safety, and shelf life by reducing ripening, gas exchange, and microbial decay.

In summary, this Special Issue showcases a diverse range of advancements in the development of thin films, highlighting their importance across various sectors. From novel deposition techniques to enhanced characterization methods and applications, this issue underscores the ongoing innovation in this field. The contributions herein not only reflect the current state of the art, but also provide a glimpse into the future of thin-film technologies, paving the way for their further advancement and application in industry and technology.

This Special Issue is a testament to the hard work and dedication of the authors who contributed their insightful research. We extend our sincere gratitude to each author for their valuable contributions, to the peer reviewers for their dedication in ensuring the quality of the published papers, and to everyone else involved in this process. We believe this reprint will provide a comprehensive overview of the latest advancements in the development of thin films and inspire further exploration and innovation.

The contributions to this Special Issue are listed as follows:

Contribution 1—Cristea, D.; Velicu, I.L.; Cunha, L.; Barradas, N.; Alves, E.; Craciun, V. Tantalum-Titanium Oxynitride Thin Films Deposited by DC Reactive Magnetron Co-Sputtering: Mechanical, Optical, and Electrical Characterization. *Coatings* **2022**, *12*, 36.

Contribution 2—Grafoute, M.; N'Djoré, K.B.J.I.; Petitjean, C.; Pierson, J.F.; Rousselot, C. Influence of Oxygen Flow Rate on the Properties of FeOXNY Films Obtained by Magnetron Sputtering at High Nitrogen Pressure. *Coatings* **2022**, *12*, 1050.

Contribution 3—Okhay, O.; Vilarinho, P.M.; Tkach, A. Structure, Microstructure, and Dielectric Response of Polycrystalline $Sr_{1-x}Zn_xTiO_3$ Thin Films. *Coatings* **2023**, *13*, 165.

Contribution 4—Bujaldón, R.; Cuadrado, A.; Volyniuk, D.; Grazulevicius, J.V.; Puigdollers, J.; Velasco, D. Role of the Alkylation Patterning in the Performance of OTFTs: The Case of Thiophene-Functionalized Triindoles. *Coatings* **2023**, *13*, 896.

Contribution 5—Khalfallah, A.; Benzarti, Z. Mechanical Properties and Creep Behavior of Undoped and Mg-Doped GaN Thin Films Grown by Metal–Organic Chemical Vapor Deposition. *Coatings* **2023**, *13*, 1111.

Contribution 6—Deng, J.; Jiang, F.; Zha, X.; Zhang, T.; Yao, H.; Zhu, D.; Zhu, H.; Xie, H.; Wang, F.; Wu, X.; Yan, L. Impact Resistance of CVD Multi-Coatings with Designed Layers. *Coatings* **2023**, *13*, 815.

Contribution 7—Ma, C.; Peng, D.; Bai, X.; Liu, S.; Luo, L. A Review of Optical Fiber Sensing Technology Based on Thin Film and Fabry–Perot Cavity. *Coatings* **2023**, *13*, 1277.

Contribution 8—Bian, J.; Sun, Y.; Guo, J.; Liu, X.; Liu, Y. Enhancing the Performance and Stability of Perovskite Solar Cells via Morpholinium Tetrafluoroborate Additive Engineering: Insights and Implications. *Coatings* **2023**, *13*, 1528.

Contribution 9—Khalfallah, A.; Khalfallah, A.; Benzarti, Z. Identification of the Elastoplastic Constitutive Model of GaN Thin Films Using Instrumented Nanoindentation and Machine Learning Technique. *Coatings*, **2024**, *14*, 683.

Contribution 10—Dmitriyeva, E.; Lebedev, I.; Bondar, E.; Fedosimova, A.; Temiraliev, A.; Murzalinov, D.; Ibraimova, S.; Nurbaev, B.; Ele-mesov, K.; Baitimbetova, B. The Influence of Lyophobicity and Lyophilicity of Film-Forming Systems on the Properties of Tin Oxide Films. *Coatings* **2023**, *13*, 1990.

Contribution 11—He, T.; Wang, D.; Xu, Y.; Zhang, J. The Facile Construction of Anatase Titanium Dioxide Single Crystal Sheet-Connected Film with Observable Strong White Photoluminescence. *Coatings* **2024**, *14*, 292.

Contribution 12—Shiryaev, A.; Rozanov, K.; Kostishin, V.; Petrov, D.; Maklakov, S.; Dolmatov, A.; Isaev, I. Retrieving the Intrinsic Microwave Permittivity and Permeability of Ni-Zn Ferrites. *Coatings* **2023**, *13*, 1599.

Contribution 13—Khezerlou, A.; Zolfaghari, H.; Forghani, S.; Abedi-Firoozjah, R.; Alizadeh Sani, M.; Negahdari, B.; Jalalvand, M.; Ehsani, A.; McClements, D.J. Combining Non-Thermal Processing Techniques with Edible Coating Materials: An Innovative Approach to Food Preservation. *Coatings* **2023**, *13*, 830.

Funding: This research received no external funding.

Conflicts of Interest: The authors declare no conflicts of interest.

References

1. Hossain, M.I.; Mansour, S. A Critical Overview of Thin Films Coating Technologies for Energy Applications. *Cogent Eng.* **2023**, *10*, 2179467. [CrossRef]
2. Wang, F.; Wu, J. Applications in Solar Thin Films. In *Modern Ion Plating Technology*; Elsevier: Amsterdam, The Netherlands, 2023; pp. 321–340. ISBN 978-0-323-90833-7.
3. Song, N.; Deng, S. Thin Film Deposition Technologies and Application in Photovoltaics. In *Thin Films—Deposition Methods and Applications*; Yang, D., Ed.; IntechOpen: London, UK, 2023; ISBN 978-1-80356-455-5.
4. Kwon, H.-J. Applications of Thin Films in Microelectronics. *Electronics* **2022**, *11*, 931. [CrossRef]
5. Kim, W.R.; Heo, S.; Kim, J.-H.; Park, I.-W.; Chung, W. Multi-Functional Cr–Al–Ti–Si–N Nanocomposite Films Deposited on WC-Co Substrate for Cutting Tools. *J. Nanosci. Nanotechnol.* **2020**, *20*, 4390–4393. [CrossRef] [PubMed]

6. Yasuoka, M. Study on the Tribological Behavior of Cutting Tools Coated with Films. *Int. J. Mod. Phys. B* **2011**, *25*, 4261–4264. [CrossRef]
7. Carrascosa, C.; Raheem, D.; Ramos, F.; Saraiva, A.; Raposo, A. Microbial Biofilms in the Food Industry—A Comprehensive Review. *Int. J. Environ. Res. Public Health* **2021**, *18*, 2014. [CrossRef] [PubMed]
8. Kim, J.S.; Park, Y.M.; Bae, M.K.; Kim, C.W.; Kim, D.W.; Shin, D.C.; Kim, T.G. Cutting Performance of Tungsten Carbide Tools Coated with Diamond Thin Films after Etching for Various Times. *Mod. Phys. Lett. B* **2018**, *32*, 1850236. [CrossRef]
9. Bhattacharyya, N.; Bhattacharyya, S.; Ghosh, K.; Pal, S.; Jana, A.; Mukherjee, S. Fabrication of Sensor Technology Using Thin Films for Biosensing, Agricultural and Environmental Applications. In *Comprehensive Materials Processing*; Elsevier: Amsterdam, The Netherlands, 2024; pp. 88–99. ISBN 978-0-323-96021-2.
10. Iqbal, A.; Naqvi, S.A.R.; Sherazi, T.A.; Asif, M.; Shahzad, S.A. Thin Films as an Emerging Platform for Drug Delivery. In *Novel Platforms for Drug Delivery Applications*; Elsevier: Amsterdam, The Netherlands, 2023; pp. 459–489. ISBN 978-0-323-91376-8.
11. Christensen, T.M. Overview of Thin Film Growth. In *Understanding Surface and Thin Film Science*; CRC Press: Boca Raton, FL, USA, 2022; pp. 151–177. ISBN 978-0-429-19454-2.
12. Ghazal, H.; Sohail, N. Sputtering Deposition. In *Thin Films—Deposition Methods and Applications*; Yang, D., Ed.; IntechOpen: London, UK, 2023; ISBN 978-1-80356-455-5.
13. Benzarti, Z.; Sekrafi, T.; Khalfallah, A.; Bougrioua, Z.; Vignaud, D.; Evaristo, M.; Cavaleiro, A. Growth temperature effect on physical and mechanical properties of nitrogen rich InN epilayers. *J. Alloys Compd.* **2021**, *885*, 160951. [CrossRef]
14. Song, Y. Introduction: Progress of Thin Films and Coatings. In *Inorganic and Organic Thin Films*; Song, Y., Ed.; Wiley: Hoboken, NJ, USA, 2021; pp. 1–58. ISBN 978-3-527-34497-0.
15. Petitjean, C.; Grafouté, M.; Pierson, J.F.; Rousselot, C.; Banakh, O. Structural, Optical and Electrical Properties of Reactively Sputtered Iron Oxynitride Films. *J. Phys. D Appl. Phys.* **2006**, *39*, 1894–1898. [CrossRef]
16. Grafouté, M.; Petitjean, C.; Rousselot, C.; Pierson, J.F.; Grenèche, J.M. Structural Properties of Iron Oxynitride Films Obtained by Reactive Magnetron Sputtering. *J. Phys. Condens. Matter* **2007**, *19*, 226207. [CrossRef]
17. Kiran, K. Synthesis and Characterization of CZTS and $Cu_{2-x}Ag_xZnSnS_4$ Thin Films for the Application of Solar Cells. *Sci. Vis.* **2024**, *24*, 1–11. [CrossRef]
18. Silva, D.; Monteiro, C.S.; Silva, S.O.; Frazão, O.; Pinto, J.V.; Raposo, M.; Ribeiro, P.A.; Sério, S. Sputtering Deposition of TiO_2 Thin Film Coatings for Fiber Optic Sensors. *Photonics* **2022**, *9*, 342. [CrossRef]
19. Elanjeitsenni, V.P.; Vadivu, K.S.; Prasanth, B.M. A Review on Thin Films, Conducting Polymers as Sensor Devices. *Mater. Res. Express* **2022**, *9*, 022001. [CrossRef]
20. Khmissi, H.; Azeza, B.; Bouzidi, M.; Al-Rashidi, Z. Investigation of an Antireflective Coating System for Solar Cells Based on Thin Film Multilayers. *Eng. Technol. Appl. Sci. Res.* **2024**, *14*, 14374–14379. [CrossRef]
21. Ma, C.; Peng, D.; Bai, X.; Liu, S.; Luo, L. A Review of Optical Fiber Sensing Technology Based on Thin Film and Fabry–Perot Cavity. *Coatings* **2023**, *13*, 1277. [CrossRef]
22. Koné, K.E.; Bouich, A.; Soro, D.; Soucase, B.M. Surface Engineering of Zinc Oxide Thin as an Electron Transport Layer for Perovskite Solar Cells. *Opt. Quant. Electron.* **2023**, *55*, 574. [CrossRef]
23. Khalfallah, A.; Khalfallah, A.; Benzarti, Z. Identification of Elastoplastic Constitutive Model of GaN Thin Films Using Instrumented Nanoindentation and Machine Learning Technique. *Coatings* **2024**, *14*, 683. [CrossRef]
24. Barkachary, B.M.; Joshi, S.N. Numerical Simulation and Experimental Validation of Nanoindentation of Silicon Using Finite Element Method. In *Advances in Computational Methods in Manufacturing*; Narayanan, R.G., Joshi, S.N., Dixit, U.S., Eds.; Lecture Notes on Multidisciplinary Industrial Engineering; Springer: Singapore, 2019; pp. 861–875. ISBN 978-981-329-071-6.
25. He, T.; Wang, D.; Xu, Y.; Zhang, J. The Facile Construction of Anatase Titanium Dioxide Single Crystal Sheet-Connected Film with Observable Strong White Photoluminescence. *Coatings* **2024**, *14*, 292. [CrossRef]
26. Khezerlou, A.; Zolfaghari, H.; Forghani, S.; Abedi-Firoozjah, R.; Alizadeh Sani, M.; Negahdari, B.; Jalalvand, M.; Ehsani, A.; McClements, D.J. Combining Non-Thermal Processing Techniques with Edible Coating Materials: An Innovative Approach to Food Preservation. *Coatings* **2023**, *13*, 830. [CrossRef]
27. Mathew, S.; Radhakrishnan, E.K. Edible Nanocoatings and Films for Preservation of Food Matrices. In *Nano-Innovations in Food Packaging*; Apple Academic Press: Boca Raton, FL, USA, 2022; pp. 217–246. ISBN 978-1-00-327742-2.

Disclaimer/Publisher's Note: The statements, opinions and data contained in all publications are solely those of the individual author(s) and contributor(s) and not of MDPI and/or the editor(s). MDPI and/or the editor(s) disclaim responsibility for any injury to people or property resulting from any ideas, methods, instructions or products referred to in the content.

Article

Tantalum-Titanium Oxynitride Thin Films Deposited by DC Reactive Magnetron Co-Sputtering: Mechanical, Optical, and Electrical Characterization

Daniel Cristea [1], Ioana-Laura Velicu [2,*], Luis Cunha [3], Nuno Barradas [4], Eduardo Alves [5] and Valentin Craciun [6]

1 Materials Science Department, Transilvania University, 500036 Brasov, Romania; daniel.cristea@unitbv.ro
2 Faculty of Physics, Alexandru Ioan Cuza University, 700506 Iasi, Romania
3 Physics Centre of Minho and Porto Universities, 4710-057 Braga, Portugal; lcunha@fisica.uminho.pt
4 Nuclear Science and Technology Centre, Technical Superior Institute, University of Lisbon, 2695-066 Bobadela LRS, Portugal; nunoni@ctn.tecnico.ulisboa.pt
5 Institute of Plasma and Nuclear Fusion, Technical Superior Institute, University of Lisbon, 2695-066 Bobadela LRS, Portugal; ealves@ctn.tecnico.ulisboa.pt
6 Laser Department, National Institute for Laser, Plasma, and Radiation Physics, 077125 Magurele, Romania; valentin.craciun@inflpr.ro
* Correspondence: laura.velicu@uaic.ro

Abstract: The possibility to tune the elemental composition and structure of binary Me oxynitride-type compounds (Me_1Me_2ON) could lead to attractive properties for several applications. For this work, tantalum-titanium oxynitride (TaTiON) thin films were deposited by DC reactive magnetron co-sputtering, with a −50 V bias voltage applied to the substrate holder and a constant substrate temperature of 100 °C. To increase or to decrease in a controlled manner, the Ti and Ta content in the co-sputtered films, the Ti and Ta target currents were varied between 0.00 and 1.00 A, in 0.25 A steps, while keeping the sum of the currents applied to the two targets at 1.00 A. The reactive gases flow, consisting of a nitrogen and oxygen gas mixture with a constant N_2/O_2 ratio (85%/15%), was also kept constant. The single-metal oxynitrides (TaON and TiON) showed a low degree of crystallinity, while all the other co-sputtered films revealed themselves to be essentially amorphous. These two films also exhibited higher adhesion to the metallic substrate. The TaON film showed the highest hardness value (14.8 GPa) and the TiON film a much lower one (8.8 GPa), while the co-sputtered coatings exhibited intermediary values. One of the most interesting findings was the significant increase in the O content when the Ti concentration surpassed the Ta one. This significantly influenced the optical characteristic of the films, but also their electrical properties. The sheet resistivity of the co-sputtered films is strongly dependent on the O/(Ta + Ti) atomic ratio.

Keywords: ternary oxynitride; co-sputtering; hardness; adhesion; wear

Citation: Cristea, D.; Velicu, I.-L.; Cunha, L.; Barradas, N.; Alves, E.; Craciun, V. Tantalum-Titanium Oxynitride Thin Films Deposited by DC Reactive Magnetron Co-Sputtering: Mechanical, Optical, and Electrical Characterization. Coatings 2022, 12, 36. https://doi.org/10.3390/coatings12010036

Academic Editor: Philipp Vladimirovich Kiryukhantsev-Korneev

Received: 30 November 2021
Accepted: 23 December 2021
Published: 28 December 2021

Copyright: © 2021 by the authors. Licensee MDPI, Basel, Switzerland. This article is an open access article distributed under the terms and conditions of the Creative Commons Attribution (CC BY) license (https://creativecommons.org/licenses/by/4.0/).

1. Introduction

Transition metal oxynitrides are a group of ceramic materials that are characterized by the possibility of changing the ratio between the nitrogen and oxygen content, and the one between the transition metal and the non-metallic elements (Me/(O + N)), which allows for controlling a broad spectrum of properties suitable for many applications: optical properties, especially obtaining a wide range of colours; electrical properties, from conductive to insulating, variable mechanical properties, etc.

Hereinafter, a short overview on the tantalum and titanium-based oxides, nitrides, and oxynitrides (TaO, TaN, TaON, TiO, TiN, TiON) will be presented, with emphasis on their actual and potential applications.

One of the most common variants of tantalum oxides is tantalum pentoxide (Ta_2O_5). It is extensively studied due to its high dielectric constant, high refractive index and good thermal and chemical stability, properties which make it a strong candidate for microelectronics applications. Other applications of tantalum pentoxide include anti-reflection coatings deposited on the surface of solar cells, optical transmission paths (optical fibres), insulating material in devices requiring high permittivity [1], as well as in the manufacture of metal-oxide-semiconductor field effect transistors [2]. Tantalum pentoxide can also be used as protective coatings or components of storage capacitors in HDRAM (High-density random-access memory) units. Related to the example presented above, there are a few drawbacks, the most significant being related to the fact that tantalum pentoxide crystallizes at temperatures above 600 °C, with negative effects on its performance [1]. Due to its high chemical stability, tantalum pentoxide is also a promising biocompatible material, and it can be used as a catalytic material for wastewater treatment and for the decomposition of water to obtain hydrogen for energy purposes [3,4].

Transition metal nitrides are known for their remarkable properties which include high hardness, high temperature stability, chemical stability, etc. These nitrides can be used as wear-resistant coatings, protective coatings and structural elements in integrated circuits, among other applications. Tantalum nitride can be used as thin-layer resistors [5], diffusion barriers [6] or wear-resistant hard layers [7,8]. The properties of tantalum nitride thin films depend on their microstructure and composition. A search in the literature, regarding the reported tantalum nitride phases, resulted in an extensive list. Some examples are $TaN_{0.05}$, Ta_2N, $TaN_{0.8-0.9}$, and TaN [9], ε-TaN, θ-TaN and δ-TaN [10], Ta_2N, Ta_5N_6, and Ta_3N_5 [11]. The electronic properties range from good conductors, like Ta_2N, TaN or Ta_5N_6 to more semiconductor phases, like Ta_3N_5 [11].

Tantalum oxynitrides demonstrate having application potential, in general, in a much larger domain, in relation to the corresponding metallic nitrides and oxides [12–14]. In terms of electrical resistivity, it was found that this parameter varies in a very large domain. As a thin film, produced by magnetron sputtering, its electrical resistivity may vary from 5.29×10^{-4} Ω cm (for a flow of 2.5 sccm of reactive gas - 15% O_2 + 85% N_2 - and a grounded substrate holder), up to 1.93×10^6 Ωcm (for a flow of 30 sccm of reactive gas mixture and a substrate polarization voltage of −100 V) [12]. Hardness values for tantalum oxynitride coatings, varying from ~7 GPa up to ~22 GPa, depending on their composition and structure, have been previously reported [14].

Titanium oxide is one of the most studied titanium-based materials. Titanium oxide films were proposed for photocatalytic, electronic, optical, and optoelectronic applications [15–17]. Moreover, other applications are reported for TiO_2, namely: corrosion resistant coatings, anti-bacterial coatings, self-cleaning surfaces, etc. [18].

Titanium nitride has also been extensively studied. It is a versatile ceramic material, which exhibits good wear and corrosion resistant properties, being widely applied on cutting tools. Moreover, titanium nitride has biocompatible properties as well as a combination of high ductility and hardness, leading to its use in the medical implants field [19].

The properties of TiON thin films depend on the O/N and Ti/(O + N) ratios. TiON films can be used in many fields of applications: medical devices, selective solar absorbers, dielectric layers, resistive layer in memory devices, thin film resistors, and as a photocatalyst [20]. Moreover, TiON films can be also used as decorative coatings due to the colour variation when the O/N ratio is changed [21].

Since oxynitride coatings based on transition metals (MeON) can benefit from the properties of the respective oxides or nitrides of the particular metal, or can be characterized by entirely new properties, the possibility to tune the elemental composition and the structure of binary Me oxynitride-type compounds, containing two transitional metals (Me_1Me_2ON-type), could further lead to attractive properties for several applications. Hereinafter, results concerning some mechanical properties (hardness, elastic modulus, adhesion to the substrate, wear behaviour), optical properties (colour coordinates and reflectance), and electrical properties (sheet resistance) of co-sputtered TaTiON coatings will be correlated to their deposition parameters, and to their chemical composition and structure.

2. Materials and Methods

Films belonging to the TaTiON system were deposited onto several types of substrates by DC reactive co-magnetron sputtering. The selected substrates, needed for various analysis techniques, were glass slides, silicon wafers, and AISI 316L steel disks (20 mm in diameter, 1.5 mm in thickness) polished to a mirror-like finish. Before being inserted in the chamber, all the substrates were cleaned with ethanol, to remove impurities from the surface. Prior to the deposition process, the chamber was evacuated to a base pressure lower than 1×10^{-5} Pa. The substrate holder was positioned at 70 mm from the high purity targets (Ta and Ti 99.6%) with dimensions 200 mm × 100 mm × 6 mm. A suitable rotation speed of the substrate holder was employed, to avoid the deposition of multilayer structures. Before each deposition, the substrates were plasma etched, using a pulsed direct current of 0.6 A, during 500 s, in an argon atmosphere, with a partial pressure around 0.3 Pa. The gas atmosphere during the deposition process was composed of argon as working gas and a reactive mixture of nitrogen + oxygen (15% O_2 + 85% N_2). The argon flow (70 sccm) was kept constant during all depositions, as well as the N_2 + O_2 gas mixture flow (10 sccm). The films were deposited with a −50 V bias voltage applied to the substrate holder. The substrate holder was heated by the Joule effect, thermostatically controlled to maintain the temperature at 100 °C for all depositions. The temperature was measured using a thermocouple. The homogeneity of the substrate temperature was guaranteed by connecting the heater significantly before the beginning of the deposition (never less than 3 h) and by the rotation of the substrate holder during the deposition. The main variable parameter was the applied target current on each target. To increase or to decrease, in a controlled manner, the Ti and Ta content in the deposited films, the Ti and Ta targets currents were varied between 0.00 and 1.00 A, in 0.25 A steps, but keeping the sum of the currents in both targets always equal to 1.00 A. The equivalent total current density in each deposition was 50 A/m^2. These current configurations led to five different deposition sessions, each one with the duration of 3600 s. Detailed deposition parameters can be found in Table 1.

Table 1. Values of the experimental parameters used to deposit by reactive magnetron co-sputtering films belonging to the TaTiON system (t_f = thickness of the films).

Sample Code	Targets Current (A)		Targets Voltage (V)		(N_2 + O_2) Flow (sccm)	Ar Flow (sccm)	Bias (V)	t_f (nm)
	Ta	Ti	Ta	Ti				
Ta1–Ti0–ON	1	0	391	-	10	70	−50	934 ± 85
Ta0.75–Ti0.25–ON	0.75	0.25	424	360	10	70	−50	1006 ± 96
Ta0.5–Ti0.5–ON	0.50	0.50	421	383	10	70	−50	677 ± 73
Ta0.25–Ti0.75–ON	0.25	0.75	425	397	10	70	−50	417 ± 63
Ta0–Ti1–ON	0	1	-	436	10	70	−50	220 ± 44

The chemical composition of the coatings was assessed by Rutherford backscattering Spectrometry (RBS). The measurements were made at the CTN/IST Van de Graaff accelerator in the small chamber, where three detectors are installed: standard at 140°, and two pin diode detectors located symmetrical each other, both at 165° (detector 3 on same side as standard detector 2). Spectra were collected for 2.3 MeV ^{4}He$^+$, under normal incidence. The RBS data were analysed with the IBA DataFurnace NDF v9.6 h [22]. Double scattering was calculated with the algorithms given elsewhere [23], while pileup was calculated with the algorithms given in [24]. Several of the samples showed spectral features which are characteristic to the presence of surface roughness, in some cases eventually with the formation of islands or other forms of surface non-uniformity. Consequently, the simulations were done by introducing a thin Ta layer with very large thickness standard deviation, between the film and the substrate.

The structure of the films was analysed by grazing incidence and symmetrical X-ray diffraction (GIXRD and XRD) investigations, performed on the films deposited on Si substrates with an Empyrean instrument (Panalytical, Almelo, Netherlands) set to work in a parallel beam geometry with CuK$_\alpha$ radiation (λ = 1.540598 Å), in the range of 2θ = 30–100°.

Instrumented indentation measurements were performed on the films, to obtain the indentation hardness and the indentation elastic modulus, using a CSM Instruments/Anton Paar NHT^2 nanoindenter (Corcelles-Cormondrèche, Switzerland), equipped with a Berkovich geometry diamond pyramidal tip (tip radius = 100 nm), with the following protocol: 30 s loading to the desired load, 10 s pause (dwell time, to minimize the creep effect), 30 s unloading. The maximum applied load per each sample was chosen as a function of the coating thickness, to avoid the influence of the substrate on the final results. At least 20 indentations were performed on each sample. The coating thickness was measured in several locations by ball cratering, using a Calotest machine from CSM Instruments, equipped with a 30 mm steel ball and diamond slurry as the abrasive entity.

The adhesion of the coatings to the substrate was quantified by scratch tests, using a CSM Instruments/Anton Paar Microscratch Tester (Corcelles-Cormondrèche, Switzerland), equipped with a diamond Rockwell geometry tip (radius = 100 µm). On each sample, at least 5 tracks were performed, with the following protocol: progressive load from 0.03 N to 10.00 N, loading rate 5.00 N/min, and scratch length of 2 mm. The results of interest are the load responsible for the occurrence of the first cracks in the coating (Lc1), the load necessary for the initial delamination (Lc2), and the load needed for the removal of more than 50% of the coating (Lc3).

The wear behaviour was assessed on a rotational tribometer from CSM Instruments/Anton Paar (Corcelles-Cormondrèche, Switzerland), against 6 mm diameter Al$_2$O$_3$ balls. At least 3 wear tracks were carried out on each sample, with the following protocol: applied load of 1 N, linear speed of 20 cm/s, stop condition 200 m. Before each test, the samples and the friction couples were cleaned with ethanol and blow dried with compressed air.

The optical reflectance and the colour coordinates in the CIE L*a*b* space were recorded with a 3nh YS4510 portable spectrophotometer (wavelength 400–700 nm), in at least 3 regions of each sample, on the films deposited on glass slides, with a D65 standard illuminant source, a measurement aperture of ϕ 4 mm, and a 10° observation angle.

The sheet resistance was measured with an Ossila 4-point probe meter with a 1.27 mm spacing between the measuring probes, on the samples deposited on glass slides, using a linear fit for the voltage–current curves.

3. Results and Discussion

3.1. Chemical Composition and Structural Development

Figure 1 presents the thickness of the deposited films as a function of the Ti and Ta targets' applied currents. The trend coincides to that of the deposition rate, since the deposition period was set to 3600 s for each batch.

The thickness of the films decreases significantly from ~900 nm, for the sample without titanium (Ta1–Ti0–ON), down to ~200 nm, for the sample without tantalum (Ta0–Ti1–ON). It seems that a slight increase in the deposition rate happens from I_{Ti} = 0.00 A (I_{Ta} = 1.00 A) to I_{Ti} = 0.25 A (I_{Ta} = 0.75 A), but, for titanium target currents above 0.25 A (and tantalum currents below 0.75 A), the co-sputtered coatings follow a decreasing linear trend of the thickness. The sputtering yield of these two elements is rather similar, which translates to relatively similar sputtering rates for the pure elements, the one for Ta being 85 Å/s, while the deposition rate for Ti is 80 Å/s [25].

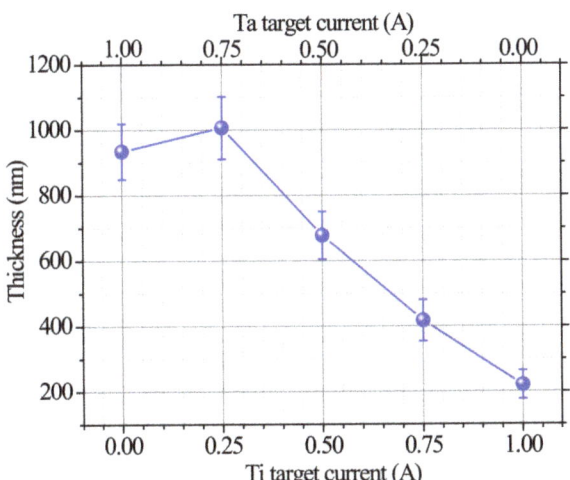

Figure 1. Coating thickness as a function of the Ti and Ta target applied currents.

The justification for the observed behaviour should be sought elsewhere. The sputtered atoms ejected into the gas phase are not in a pure thermodynamic equilibrium. In any case, the standard enthalpies of formation of the compounds, which might form during the deposition process, may help to explain the trend of the thickness (Figure 1) and the chemical composition (Figure 2). The values found in literature are: $\Delta H_f^0 \text{(TaN)} = -251 \text{ kJ·mol}^{-1}$, $\Delta H_f^0 \text{(TaO}_2\text{)} = -201 \text{ kJ·mol}^{-1}$, $\Delta H_f^0 \text{(Ta}_2\text{O}_5\text{)} = -2046 \text{ kJ·mol}^{-1}$, $\Delta H_f^0 \text{(TiN)} = -265.8 \text{ kJ·mol}^{-1}$, $\Delta H_f^0 \text{(TiO)} = -519.7 \text{ kJ·mol}^{-1}$, $\Delta H_f^0 \text{(TiO}_2\text{)} = -944 \text{ kJ·mol}^{-1}$, $\Delta H_f^0 \text{(Ti}_2\text{O}_3\text{)} = -1520.9 \text{ kJ·mol}^{-1}$, $\Delta H_f^0 \text{(Ti}_3\text{O}_5\text{)} = -2459.4 \text{ kJ·mol}^{-1}$ [26]. Based on these values, it would seem that, for similar stoichiometry compounds (e.g., TaN vs. TiN, TaO$_2$ vs. TiO$_2$, etc.), the ones based on titanium are more favoured to be formed. The higher reactivity of Ti towards N, and particularly towards O, when compared to the affinity of Ta to the same elements, is reflected in Figure 2. When the current of the Ti target increases, the non-metal/metal ratios tend to increase, particularly the O/Ti when $I(\text{Ti}) > 0.5$ A (Figure 2b). Consequently, when the amount of Ti atoms sputtered from the target is increased, due to the higher applied current, the probability of formation of compounds (especially oxides) is higher. However, this phenomenon could lead to a higher degree of target poisoning, since the compounds are not only formed on the substrate surface, but on other surfaces of the deposition chamber (including the target) as well. Generally, the sputtering rate of compounds is significantly lower than that of the elemental targets. Since the target would be working in poisoned mode, fewer elemental atoms are released through sputtering, which could explain the lower deposition rate. Another aspect that may contribute to explaining the higher deposition rate observed for Ta richer films is related to the higher chemical bond length of Ta compounds compared to analogous Ti compounds (e.g., TaN vs. TiN, TaO$_2$ vs. TiO$_2$, etc.).

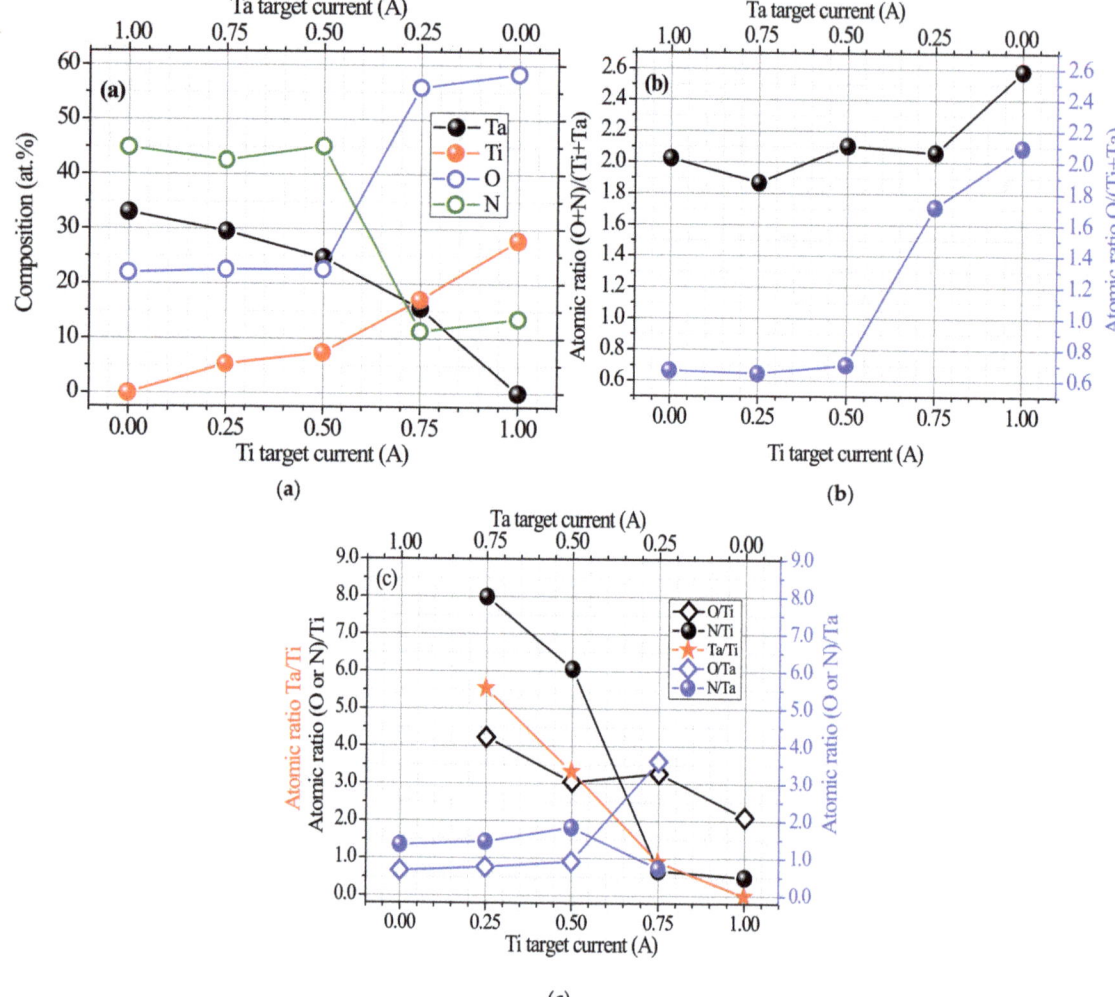

Figure 2. Chemical composition (**a**) and atomic ratios (**b**) and (**c**), as a function of Ti and Ta target applied currents.

Figure 3a,b presents grazing incidence XRD patterns of the deposited oxynitride films. According to these results, due to the presence of characteristic diffraction peaks, the single metal element oxynitrides (Ta1–Ti0–ON and Ta0–Ti1–ON, Figure 3a) exhibit a certain degree of crystallinity. On the other hand, all the other co-sputtered samples (TaTiON, Figure 3b) showed an amorphous structure, regardless of the deposition parameters.

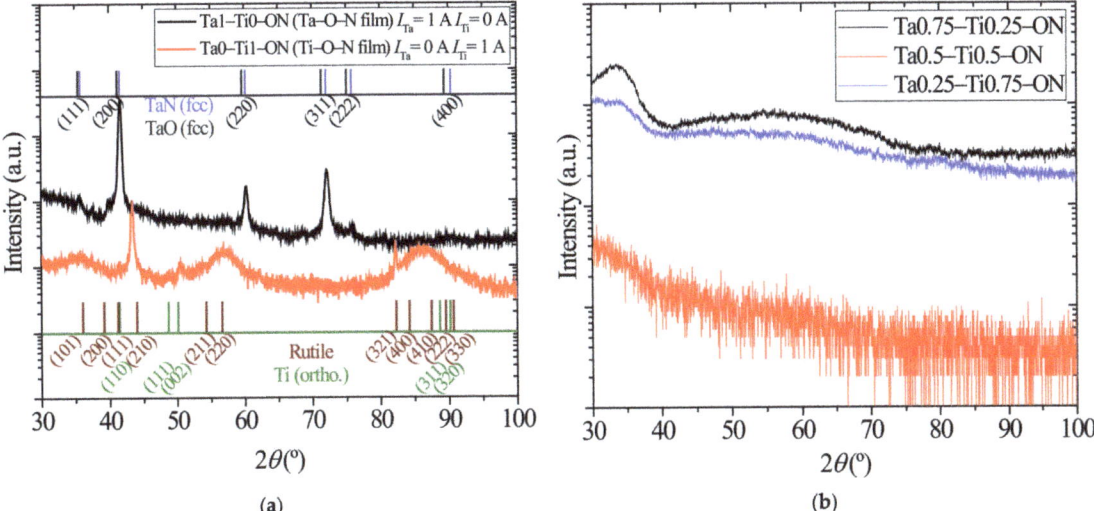

Figure 3. Grazing incidence XRD patterns of the single-sputtered coatings (**a**) and co-sputtered coatings (**b**).

It seems that the competition occurring between the metallic elements, to react with the gases and to form compounds, leads to a hindering of the crystal formation and growth. It cannot be overlooked that the relatively high atomic radius difference between Ti and Ta does not contribute to the formation of perfect crystals, particularly in the case of substitution of Ti by Ta or Ta by Ti, in the crystal lattices. The influence of the chemical composition can also be observed in relation to the phase attribution of the crystalline phases, depicted in Figure 3a. The formation of nitride crystallites was favoured during the deposition of the Ta1–Ti0–ON sample (TaON film), which can be correlated to the higher nitrogen content in the working atmosphere. In fact, when comparing the composition ratios of this sample (Figure 2c), the N/Ta ratio (~1.4) is approximately double that of the O/Ta ratio (~0.7). The XRD pattern of the TaON film exhibits significantly better-defined peaks, when compared with the TiON diffraction pattern, and reveals a polycrystalline structure. The diffraction peaks can be assigned to the face-centred cubic (fcc) phase of TaN (ICDD card 04-019-2403), probably doped with some oxygen. These diffraction peak positions are very close to those of the fcc phase of TaO (ICDD card 03-065-6750). Although the formation of Ta_2O_5 is thermodynamically favoured compared to the formation of TaN (see enthalpy of formation values), the kinetic mode prevails due to significantly higher N_2 partial pressure when compared to O_2.

Conversely, the diffraction peaks and bands corresponding to the Ta0–Ti1–ON sample (TiON film) coincide with the location of either metallic orthorhombic phase of Ti (ICDD card 00-055-0345), but more probably, based on the chemical composition of the sample, to the rutile phase of TiO_2 (ICDD card 00-021-1276). One exception can be observed in the case of the diffraction peak located at 2θ = ~43.5°, where a peak of the $TiN_{0.6}O_{0.4}$ structure could be assigned (ICDD card 00-049-1325). In the case of the TiON film, the O/Ti ratio (2.1) is almost stoichiometric to TiO_2, and it is significantly higher than the N/Ti ratio (0.5) (Figure 2c). Consequently, the formation of oxides is favoured over that of nitrides, although the O_2/N_2 ratio in the working atmosphere is less than 0.2. It seems that, in this case, the thermodynamic mode is predominant, particularly when compared to the phenomena observed for the TaON film. One can observe from the thermodynamic data that the difference between the enthalpy of formation values of the Ti oxides compared to the Ti nitrides is higher relatively to the analogous Ta oxides and nitrides. This fact has an impact on the composition (Figure 2a) and on the structural evolution (Figure 3a) of the films.

Furthermore, the samples obtained with intermediary current values (Ta0.75–Ti0.25–ON and Ta0.25–Ti0.75–ON) exhibit a broad band in the 2θ = ~30°–40° region, which could signify the formation of poorly developed crystallites in an amorphous matrix. Going into more detail, for the co-sputtered film richer in Ta (Ta/Ti = 5.5), the kinetic mode seems to dominate, inferred from the following: N/(Ta or Ti) > O/(Ta or Ti). For the film richer in Ti (Ta/Ti = 0.9), the thermodynamic mode seems to prevail, since O/(Ta or Ti) > N/(Ta or Ti). In a certain way, these results confirm the sputtering modes discussed above for the single metal oxynitride films.

The sample deposited with 0.5 A applied to each metallic target is entirely amorphous.

3.2. Mechanical Properties

The variation of both indentation hardness and elastic modulus, as a function of the applied current on each target, is shown in Figure 4. One can observe that these parameters vary in a rather large domain, from ~7 GPa up to ~15 GPa for the hardness, and from ~70 GPa up to ~220 GPa for the indentation elastic modulus. The observed trend is probably related to the higher O content promoted by the increase of the current on the Ti target current. The deviation observed in the case of Ta0–Ti1–ON sample (TiON film), especially in the case of hardness, can be a consequence of the significantly lower penetration depth of the indenter (~20 nm), due to the lower thickness of the coating (~200 nm). In situations like these, the combination of two factors may be responsible for the observed behaviour: the roughness and the indentation size effect (ISE), where the reduction of the load on the indenter and the indent size lead to an increase in the hardness as the load on the indenter decreases [27]. Using these two material characteristics (H and E), the H/E and H^3/E^2 ratios can be obtained, useful for predicting the behaviour of a material.

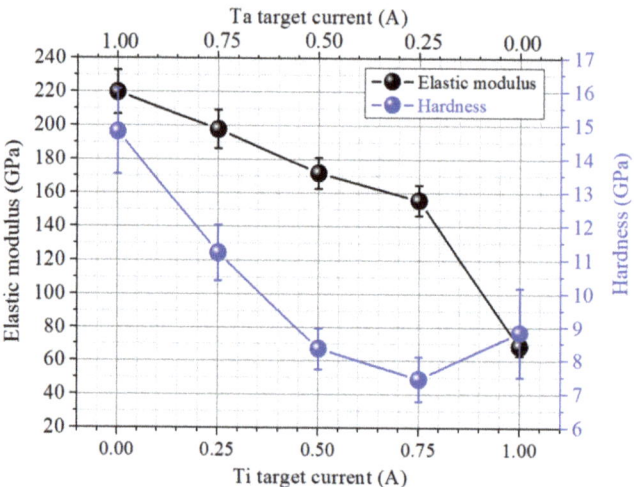

Figure 4. The variation of the indentation hardness and elastic modulus as a function of Ti and Ta target applied currents.

The variation of the H/E and H^3/E^2 ratios as function of the deposition conditions is presented in Figure 5. The H/E ratio, called elastic strain to failure, can be used to predict the resistance against wear of a material. Higher values for this ratio generally signify a better wear and failure resistance. Coatings with high hardness and rather low values of the indentation elastic modulus should exhibit improved facture toughness. Materials with high H/E ratio have a small plasticity index, meaning that their deformation under load is more likely to be elastic than plastic. The consensus regarding this H/E ratio is that, if it is larger than 0.1, the material is resistant to cracking [28]. The H^3/E^2 ratio is used to assess the resistance to plastic deformation, highlighting the material's elasticity.

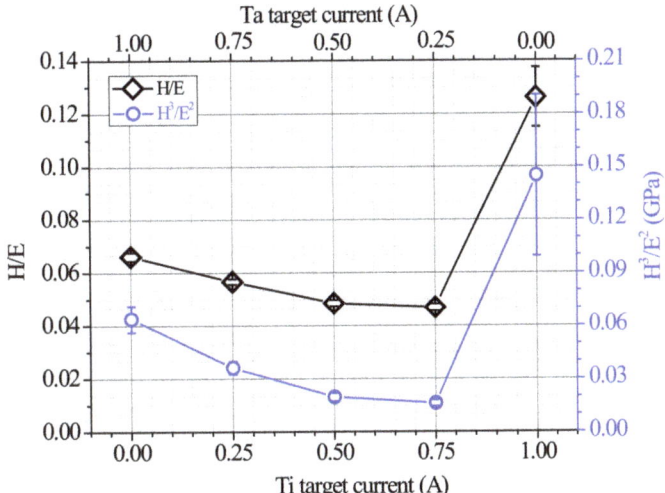

Figure 5. The variation of the H/E and H^3/E^2 ratios (H/E—elastic strain to failure ratio, H^3/E^2—indicator of a material's resistance against plastic deformation) as a function of Ti and Ta target currents.

The critical loads obtained after the scratch tests, which give an indication on the adhesion/cohesion of the substrate/coating system, exhibiting relatively different trends, compared to the observed trends of the variation of H and/or E (Figure 6).

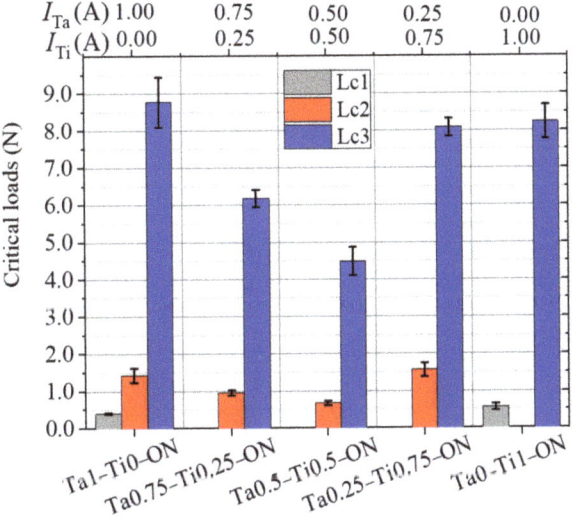

Figure 6. The variation of the scratch test adhesion critical loads: $Lc1$—first cracks; $Lc2$—first delamination; $Lc3$—total film removal.

It is worth mentioning that not all critical loads were observed on each sample. The amorphous samples did not exhibit the formation of any cracks for the selected interval of loads applied on the indenter, most probably due to their amorphous character and relatively low H/E ratio values. Once the current applied on the Ta target is gradually decreased, both the load which causes the first delamination ($Lc1$) and the one necessary for film removal ($Lc3$) decrease. However, the samples with the highest oxygen content (Ta0–Ti1–ON) and the lowest nitrogen content (Ta0.25–Ti0.75–ON), both produced with $I(\text{Ti}) \geq 0.75$ A and $I(\text{Ta}) \leq 0.25$ A, delaminate entirely due to increasingly higher loads.

Comparing the common critical load observed in all the substrate/film systems ($Lc3$) with the composition (Figure 7), a decrease of the critical load is observed when I_{Ta} decreases (I_{Ti} increases) until the targets are under the same current (0.50 A). This is probably due to the decrease of the Ta content, although the N content is more or less constant. As it was discussed previously, comparing the two metallic elements' affinity to the reactive species, tantalum atoms showed higher affinity to nitrogen than titanium atoms, forming tantalum nitride crystals (eventually with some O doping). The richest film in crystalline tantalum nitride (Ta1–Ti0–ON) is the hardest and the most adherent, and the continuous decrease of the Ta content does not contribute to the formation of this metal nitride and apparently the hardness and the adhesion are both decreasing, as a consequence. However, when $I_{\text{Ti}} > 0.50$ A, a significant increase of O content (and strong decrease of N content) was observed, and this fact seems to strongly contribute to an adhesion increase, although the hardness continues to decrease. This may be related with the relatively high ductility of these films, as the micrographs presented in Figure 8 show.

Based on the micrographs from Figure 8, the coatings seem to become more and more ductile as the oxygen content is increased (inferred from the aspect of the first delaminations—$Lc2$). The difference between the crystalline samples is most evident if one observes the shape and direction of the cracks in the coating ($Lc1$), where sample Ta1–Ti0–ON exhibits tensile cracking, while sample Ta0–Ti1–ON exhibits conformal cracking. Moreover, sample Ta0–Ti1–ON was the only configuration that did not exhibit the occurrence of intermediary delaminations, a phenomenon which might be related to the fact that this sample is the only one which exhibits a H/E ratio larger than 0.1.

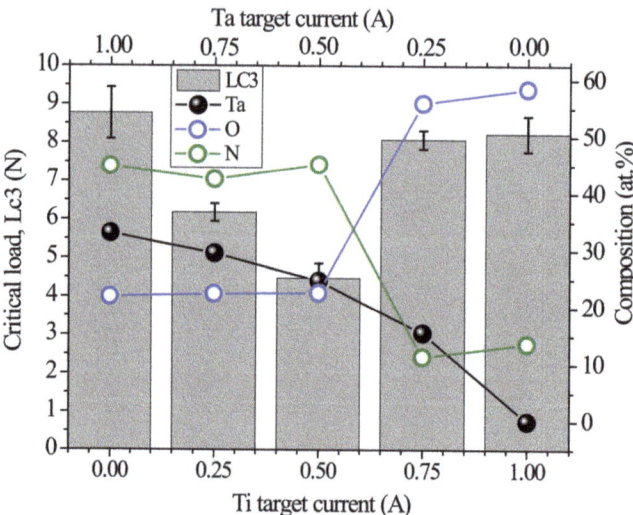

Figure 7. Correlation between critical load $Lc3$ and the composition and deposition parameters.

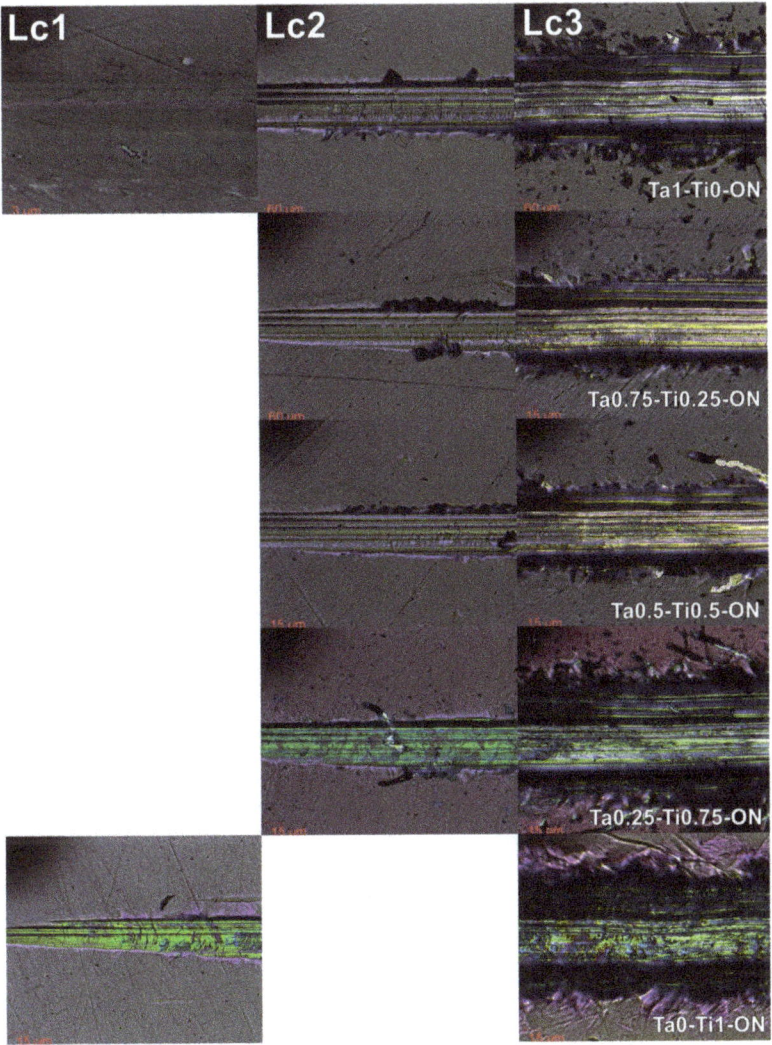

Figure 8. Representative micrographs of coating failure: *Lc*1—first cracks; *Lc*2—first delamination; *Lc*3—total film removal.

The wear behaviour was assessed on a rotational tribometer, against Al_2O_3 balls. From the variation of the friction coefficient as a function of distance, shown in Figure 9, the following observation can be made: all samples, with the exception of the bare substrate, exhibit a low friction regime, from up to ~30 m for sample Ta1–Ti0–ON, down to ~10 m for sample Ta0.5–Ti0.5–ON, followed by a high friction regime, which would signify that the substrate was breached by the Al_2O_3 ball. The better wear behaviour of sample Ta1–Ti0–ON could be linked to its higher hardness. Interestingly, even if sample Ta0–Ti1–ON has the lowest thickness, it behaves adequately during wear tests, exhibiting a low friction regime for a larger duration relative to the other samples.

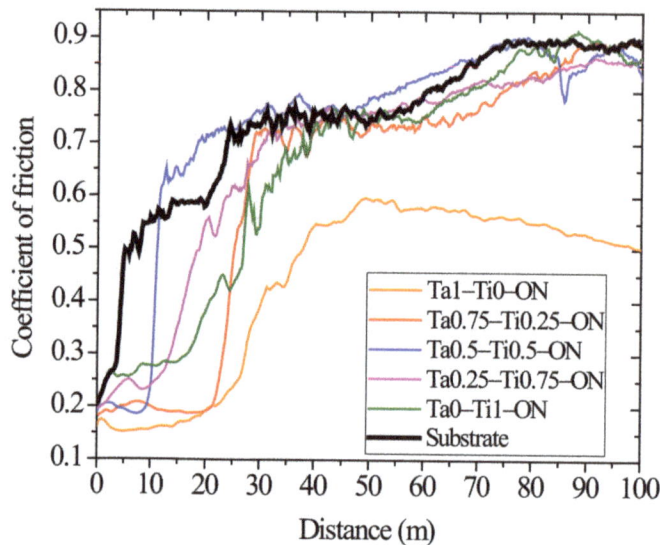

Figure 9. The variation of the friction coefficient as a function of distance.

3.3. Colour Coordinates and Reflectance

The perceived colour of the coatings deposited on steel substrates is shown in Figure 10. One can observe that two films with highest Ta-content (Ta1–Ti0–ON and Ta0.75–Ti0.25–ON) are dark-grey, while the ones with increasing Ti and O content exhibit typical interference coloration.

Figure 10. The perceived colour of the TaTiON coatings: 1—Ta1–Ti0–ON; 2—Ta0.75–Ti0.25–ON; 3—Ta0.5–Ti0.5–ON; 4—Ta0.25–Ti0.75–ON; 5—Ta0–Ti1–ON.

Since the perceived colour of a material depends on the illuminant, a proper comparison was made using the CIE L*a*b* colour coordinates system. The variation of the L*, a*, and b* coordinates from the CIE L*a*b* colour space, as a function of the deposition parameters, is shown in Figure 11. The colour parameters of the two films produced with highest Ta target current (Ta1–Ti0–ON and Ta0.75–Ti0.25–ON) are significantly different from the remaining three films (shadowed region of Figure 11). The variation of the colour parameters between these two samples is relatively small, as expected from the perceived colour of Figure 10. The lightness component (L*) is close to 0, coherently with the dark grey perceived colour. With regard to the chromatic components (a* and b*), both exhibit low chroma (a* and b* very close to 0) and both suffer a slight shift to lower values: a* decreases from 0.13 ± 0.22 to -0.39 ± 0.19, for sample Ta1–Ti0–ON (TaON film), and b* from 0.05 ± 0.83 to -2.12 ± 0.48, for sample Ta0.75–Ti0.25–ON.

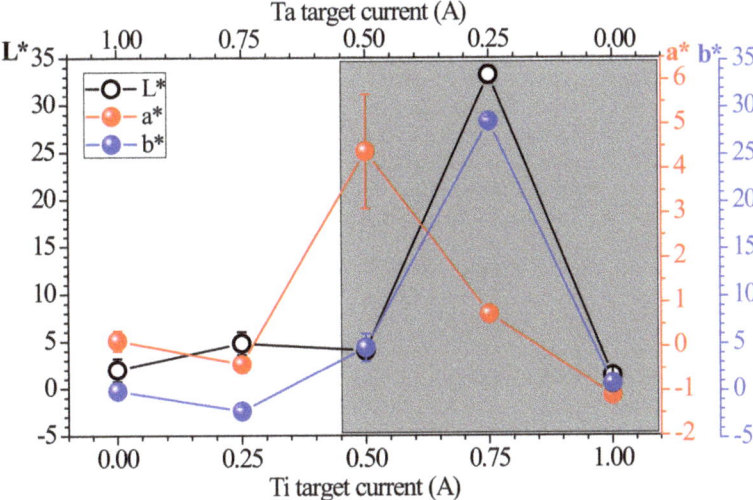

Figure 11. The variation of the colour coordinates in the CIE L*a*b* colour space.

The remaining three films exhibit large differences among them. It must be mentioned that these changes are not related only with the interaction with visible radiation with the film surface. The relatively high O content promotes the formation of oxides, which are generally transparent to visible light, as it is the case of TiO_2. Because of this transparency, the colour parameters are resultant not only from the intrinsic colour of the film but also from what it is below the surface, substrate included. These observations can be confirmed by the diffuse reflectance spectra of the samples (Figure 12a,b).

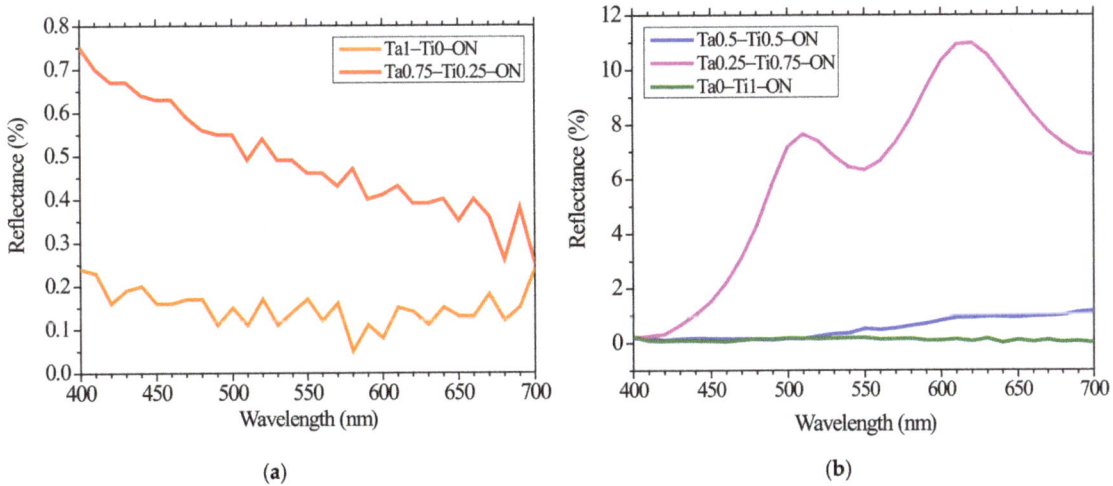

Figure 12. The variation of the reflectance as a function of the wavelength of the incident radiation: (a) Ta1–Ti0–ON and Ta0.75–Ti0.25–ON; (b) Ta0.50–Ti0.50–ON, Ta0.25–Ti0.75–ON and Ta0–Ti1–ON.

The reflectance spectra of Ta1–Ti0–ON and Ta0.75–Ti0.25–ON samples (Figure 12a) are low (dark films). The reflectance of the Ta0.25–Ti0.75–ON film is typical of a transparent material, exhibiting interference fringes coherent with its thickness (Table 1). The Ta0–Ti1–ON sample (TiON film) seems to be too thin to exhibit the interference fringes, and the film produced with the same current applied to both targets (Ta050–Ti0.50–ON) is probably in the limit of being transparent to the visible radiation, although some waviness is detected when λ > 520 nm.

3.4. Sheet Resistance

The values of the sheet resistance and of the O/metal ratio of the films produced in co-sputtering mode are depicted in Figure 13. As expected, a significant increase of the resistivity is observed when the O content of the samples increases. It would be difficult to find a better correlation between the sheet resistance and the atomic ratio O/(Ta + Ti) as the one expressed in the figure.

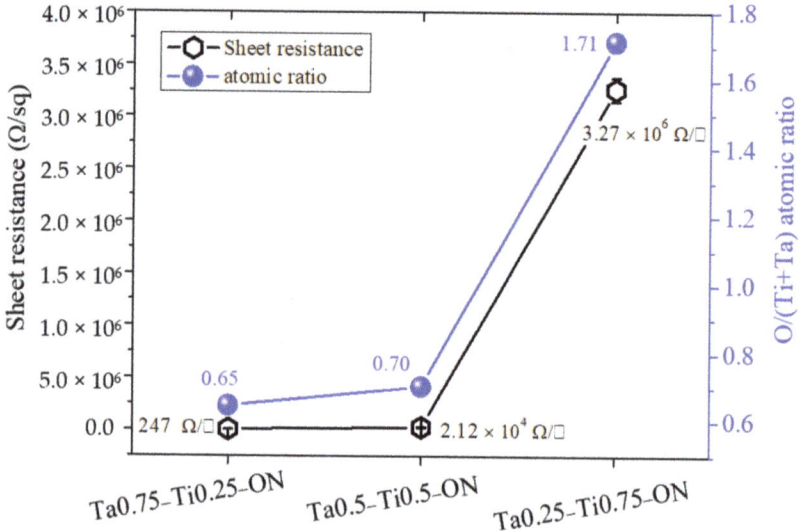

Figure 13. Sheet resistance measured by the 4-point probe method for the co-sputtered TaTiON coatings and correlation between sheet resistance and the O/(Ta + Ti) atomic ratio.

4. Conclusions

Co-sputtered tantalum-titanium oxynitride (TaTiON) thin films were deposited systematically by reactive magnetron sputtering. A titanium oxynitride (TiON) and a tantalum oxynitride (TaON) film were deposited in the same conditions of the co-sputtered films. Only these two single metal oxynitride films revealed some degree of crystallinity, higher for the TaON than the one exhibited by the TiON film. It was observed that the deposition rate is linked to the Ti content, the higher the Ti content, the lower deposition rate of the produced films. The deposition rate decreased from 2.8 Å/s, for the film with the lowest Ti content, to 0.6 Å/s for the film with the highest Ti content. The single metal oxynitride films also exhibited the highest critical load values (total film removal), both higher than 8 N. The TaON coating exhibited the highest hardest of all coatings (14.8 GPa), while the hardness of the TiON film was significantly lower (8.8 GPa). The co-sputtered coatings have intermediary properties, in terms of mechanical behaviour (hardness, adhesion critical loads). A significant increase of the O content occurred for the films where the Ti concentration was larger than the one of Ta. Consequently, the optical characteristic of the films changed significantly: the films with highest O content became transparent to visible radiation, due to the formation of oxides, particularly titanium dioxide. The sheet resistance of the co-sputtered films revealed to be strongly dependent on the O/(Ta + Ti) atomic ratio, varying in a rather large domain, from 247 ± 0.07 Ω/sq for the Ta0.75–Ti0.25–ON sample, up to 3.26 ± 0.11 × 10^6 Ω/sq for the Ta0.25–Ti0.75–ON sample.

Author Contributions: Conceptualization, D.C. and L.C.; methodology, D.C.; formal analysis, D.C., N.B., E.A. and V.C.; resources, D.C. and L.C.; data curation, D.C., I.-L.V. and L.C.; writing—original draft preparation, D.C. and L.C.; writing—review and editing, L.C. and I.-L.V.; project administration, D.C. and L.C.; funding acquisition, D.C. and L.C. All authors have read and agreed to the published version of the manuscript.

Funding: This research was funded by a grant of the Romanian Ministry of Education and Research, CNCS–UEFISCDI, project number PN-III-P1-1.1-TE-2019-1209, within PNCDI III. This work was also supported by the Portuguese Foundation for Science and Technology (FCT) in the framework of the Strategic Funding UIDB/04650/2020. V.C. acknowledges the Romanian Ministry of Research, NUCLEU Program LAPLAS VI (contract No. 16N/2019) and ELI-RO_2020_12.

Institutional Review Board Statement: Not applicable.

Informed Consent Statement: Not applicable.

Data Availability Statement: Not applicable.

Conflicts of Interest: The authors declare no conflict of interest.

Disclaimer: N.P. Barradas: Any views or opinions reflected here are those expressed by the author.

References

1. Jagadeesh Chandra, S.V.; Uthanna, S.; Mohan Rao, G. Effect of substrate temperature on the structural, optical and electrical properties of DC magnetron sputtered tantalum oxide films. *Appl. Surf. Sci.* **2008**, *254*, 1953–1960. [CrossRef]
2. Wei, A.X.; Ge, Z.X.; Zhao, X.H.; Liu, J.; Zhao, Y. Electrical and optical properties of tantalum oxide thin films prepared by reactive magnetron sputtering. *J. Alloys Compd.* **2011**, *509*, 9758–9763. [CrossRef]
3. Yang, W.M.; Liu, Y.W.; Zhang, Q.; Leng, Y.X.; Zhou, H.F.; Yang, P.; Chen, J.Y.; Huang, N. Biomedical response of tantalum oxide films deposited by DC reactive unbalanced magnetron sputtering. *Surf. Coat. Technol.* **2007**, *201*, 8062–8065. [CrossRef]
4. Jagadeesh Kumar, K.; Ravi Chandra Raju, N.; Subrahmanyam, A. Properties of pulsed reactive DC magnetron sputtered tantalum oxide (Ta_2O_5) thin films for photocatalysis. *Surf. Coat. Technol.* **2011**, *205*, S261–S264. [CrossRef]
5. Sun, X.; Kolawa, E.; Chen, J.-S.; Reid, J.S.; Nicolet, M.A. Properties of reactively sputter-deposited Ta-N thin films. *Thin Solid Films* **1993**, *236*, 347–351. [CrossRef]
6. Kim, S.K.; Cha, B.C. Deposition of tantalum nitride thin films by D.C. magnetron sputtering. *Thin Solid Films* **2005**, *475*, 202–207. [CrossRef]
7. Aryasomayajula, A.; Valleti, K.; Aryasomayajula, S.; Bhat, D.G. Pulsed DC magnetron sputtered tantalum nitride hard coatings for tribological applications. *Surf. Coat. Technol.* **2006**, *201*, 4401–4405. [CrossRef]
8. Westergard, R.; Bromark, M.; Larsson, M.; Hedenqvist, P.; Hogmark, S. Mechanical and tribological characterization of DC magnetron sputtered tantalum nitride thin films. *Surf. Coat. Technol.* **1997**, *97*, 779–784. [CrossRef]
9. Swisher, J.H.; Read, M.H. Thermodynamic properties and electrical conductivity of Ta_3N_5 and TaON. *Metall. Trans.* **1972**, *3*, 489–494. [CrossRef]
10. Liu, L.; Huang, K.; Hou, J.; Zhu, H. Structure refinement for tantalum nitrides nanocrystals with various morphologies. *Mater. Res. Bull.* **2012**, *47*, 1630–1635. [CrossRef]
11. Stampfl, C.; Freeman, A.J. Stable and metastable structures of the multiphase tantalum nitride system. *Phys. Rev. B* **2005**, *71*, 024111. [CrossRef]
12. Cristea, D.; Crisan, A.; Cretu, N.; Borges, J.; Lopes, C.; Cunha, L.; Ion, V.; Dinescu, M.; Barradas, N.P.; Alves, E.; et al. Structure dependent resistivity and dielectric characteristics of tantalum oxynitride thin films produced by magnetron sputtering. *Appl. Surf. Sci.* **2015**, *354*, 298–305. [CrossRef]
13. Cristea, D.; Crisan, A.; Munteanu, D.; Apreutesei, M.; Costa, M.F.; Cunha, L. Tantalum oxynitride thin films: Mechanical properties and wear behavior dependence on growth conditions. *Surf. Coat. Technol.* **2014**, *258*, 587–596. [CrossRef]
14. Cristea, D.; Cunha, L.; Gabor, C.; Ghiuta, I.; Croitoru, C.; Marin, A.; Velicu, L.; Besleaga, A.; Vasile, B. Tantalum Oxynitride Thin Films: Assessment of the Photocatalytic Efficiency and Antimicrobial Capacity. *Nanomaterials* **2019**, *9*, 476. [CrossRef]
15. Lou, B.S.; Chen, W.T.; Diyatmika, W.; Lu, J.H.; Chang, C.T.; Chen, P.W.; Lee, J.W. Effect of target poisoning ratios on the fabrication of titanium oxide coatings using superimposed high power impulse and medium frequency magnetron sputtering. *Surf. Coat. Technol.* **2021**, *421*, 127430. [CrossRef]
16. Kara, F.; Kurban, M.; Coskun, B. Evaluation of electronic transport and optical response of two-dimensional Fe-doped TiO_2 thin films for photodetector applications. *Optik* **2020**, *210*, 164605. [CrossRef]
17. Lu, J.H.; Chen, B.Y.; Wang, C.H. Investigation of nanostructured transparent conductive films grown by rotational-sequential-sputtering. *J. Vac. Sci. Technol. A* **2014**, *32*, 02B107. [CrossRef]
18. Rahimi, N.; Pax, R.A.; Gray, E.M. Review of functional titanium oxides. I: TiO_2 and its modifications. *Prog. Solid State Chem.* **2016**, *44*, 86–105. [CrossRef]

19. Santecchia, E.; Hamouda, A.M.S.; Musharavati, F.; Zalnezhad, E.; Cabibbo, M.; Spigarelli, S. Wear resistance investigation of titanium nitride-based coatings. *Ceram. Int.* **2015**, *41*, 10349–10379. [CrossRef]
20. Khwansungnoen, P.; Chaiyakun, S.; Rattana, T. Room temperature sputtered titanium oxynitride thin films: The influence of oxygen addition. *Thin Solid Films* **2020**, *711*, 138269. [CrossRef]
21. Vaz, F.; Cerqueira, P.; Rebouta, L.; Nascimento, S.M.C.; Alves, E.; Goudeau, P.; Rivière, J.P.; Pischow, K.; de Rijk, J. Structural, optical and mechanical properties of coloured TiN_xO_y thin films. *Thin Solid Films* **2004**, *447–448*, 449–454. [CrossRef]
22. Barradas, N.; Jeynes, C.; Webb, R.P.; Kreissig, U.; Grötzschel, R. Unambiguous automatic evaluation of multiple Ion Beam Analysis data with Simulated Annealing. *Nucl. Instrum. Methods Phys. Res. B* **1999**, *149*, 233–237. [CrossRef]
23. Barradas, N.P.; Pascual-Izarra, C. Double scattering in RBS analysis of PtSi thin films on Si. *Nucl. Instrum. Methods Phys. Res. B* **2005**, *228*, 378–382. [CrossRef]
24. Barradas, N.P.; Reis, M. Accurate calculation of pileup effects in PIXE spectra from first principles. *X-ray Spectrom.* **2006**, *35*, 232–237. [CrossRef]
25. Angstrom Sciences, Sputtering Yields. Available online: www.angstromsciences.com (accessed on 1 November 2021).
26. Speight, J.G. *Lange's Handbook of Chemistry*, 16th ed.; McGraw-Hill: New York, NY, USA, 2005; pp. 1–274.
27. Milman, Y.V.; Golubenko, A.A.; Dub, S.N. Indentation size effect in nanohardness. *Acta Mater.* **2011**, *59*, 7480–7487. [CrossRef]
28. Zhang, S. (Ed.) *Thin Films and Coatings: Toughening and Toughness Characterization*; CRC Press Taylor & Francis Group: Boca Raton, FL, USA, 2015.

Article

Influence of Oxygen Flow Rate on the Properties of FeO$_X$N$_Y$ Films Obtained by Magnetron Sputtering at High Nitrogen Pressure

Moussa Grafoute [1,*], Kouamé Boko Joël-Igor N'Djoré [1], Carine Petitjean [2], Jean François Pierson [2] and Christophe Rousselot [3,*]

1. Laboratoire de Technologie, Université Félix Houphouët Boigny, UFR SSMT 22, Abidjan BP 258, Côte d'Ivoire; joelndjore@outlook.fr
2. Institut Jean Lamour—UMR 7198, Université de Loraine, 54011 Nancy, France; carine.petitjean@univ-lorraine.fr (C.P.); jean-francois.pierson@univ-lorraine.fr (J.F.P.)
3. Département MN2S, Institut FEMTO-ST (CNRS/UFC/ENSMM/UTBM), Université de Franche Comté, 25211 Montbéliard, France
* Correspondence: gramouss@hotmail.com (M.G.); christophe.rousselot@univ-fcomte.fr (C.R.)

Abstract: Fe-O-N films were successfully deposited by magnetron sputtering of an iron target in Ar-N$_2$-O$_2$ reactive mixtures at high nitrogen partial pressure 1.11 Pa (Q(N$_2$) = 8 sccm) using a constant flow rate of argon and an oxygen flow rate Q(O$_2$) varying from 0 to 1.6 sccm. The chemical composition and the structural and microstructural nature of these films were characterized using Rutherford Backscattering Spectrometry, X-ray diffraction, and Conversion Electron Mössbauer Spectrometry, respectively. The results showed that the films deposited without oxygen are composed of a single phase of γ''-FeN, whereas the other films do not consist of pure oxides but oxidelike oxynitrides. With higher oxygen content, the films are well-crystallized in the α-Fe$_2$O$_3$ structure. At intermediate oxygen flow rate, the films are rather poorly crystallized and can be described as a mixture of oxide γ-Fe$_2$O$_3$/Fe$_3$O$_4$. In addition, the electrical behavior of the films evolved from a metallic one to a semiconductor one, which is in total agreement with other investigations. Comparatively to a previous study carried out at low nitrogen partial pressure (0.25 Pa), this behavior of films prepared at higher nitrogen partial pressure (1.11 Pa) could be caused by a catalytic effect of nitrogen on the crystallization of the hematite structure.

Keywords: reactive sputtering; Fe-O-N films; X-ray diffraction; Mössbauer spectrometry; electrical and optical properties

1. Introduction

Nowadays, the search for new materials with tunable physical properties is one of the greatest preoccupations of the scientific community. Several processes are well-established to synthesize new and/or metastable materials. Among them, the reactive sputtering process remains one of the most powerful for the deposition of high-performance thin films. Indeed, this process is widely used to deposit materials with a complex microstructure (nanocomposites, multilayers, etc.) [1], to synthesize new or metastable compounds [2] or to extend the solubility limit in binary or ternary systems [3].

In the past decades, iron oxide coatings have been extensively studied due to their applications in many industrial domains [4,5]. Since the development of transition metal oxynitride films [6,7], it appears that these compounds exhibit properties that can be tuned between those of nitrides and oxides by adjusting the nitrogen and the oxygen concentrations. Iron oxide films (Fe-O) are studied because of their optical and magnetic properties [8]. Iron nitride (Fe-N) films are studied because of their interesting magnetic properties [9] and their resistance to corrosion and wear [10]. The interest in the study of iron oxynitride films (Fe-O-N) is to elaborate a material that exhibits intermediate properties between those of iron oxide and iron nitride films. The Fe-O-N thin films have been studied for potential application to photoelectrochemical water splitting to generate hydrogen, solar application [11], magnetic [12], optical, electronic, etc. properties [13,14].

At low nitrogen (or oxygen) content, this element could be dissolved into the oxide (or nitride) network. Since Mössbauer spectrometry is a relevant method to characterize the local environment of iron atoms, the study of Fe-O-N films could bring relevant information about the structure of transition metal (oxynitrides). However, little information is available in the literature on the structure and the properties of iron oxynitrides films [15]. M. Grafoute et al. [16,17] have prepared Fe-O-N films by magnetron sputtering of an iron target in Ar-N_2-O_2 reactive mixtures at low nitrogen partial pressure (0.25 Pa) using a nitrogen flow of $Q(N_2) = 2$ sccm. However, the effect of the increase in nitrogen partial pressure or nitrogen flow rate on structural, chemical, and physical properties is not widespread in the literature.

In this work, we aim to study the influence of high nitrogen partial pressure (1.11 Pa) corresponding to $Q(N_2) = 8$ sccm on the properties of Fe-O-N coatings formed by reactive magnetron sputtering of an iron target in Ar-O_2-N_2 reactive mixtures. In addition, the correlation between structural and physicochemical properties of Fe-O-N films will be discussed. Finally, we will compare the results of this work to those obtained by M. Grafoute et al. with $Q(N_2) = 2$ sccm [16,17]. The increase in the oxygen flow rate leads to a rapid enrichment of the oxygen content that comes with a strong decrease in the nitrogen content in the films.

2. Materials and Methods

Iron oxynitride coatings of about 500 nm were deposited by magnetron sputtering of iron target in Ar-N_2-O_2 reactive mixtures using a conventional process. Reactive sputtering experiments were performed using Alliance Concept AC450 sputter equipment with a vacuum chamber volume of about 70 L. A base pressure of 10^{-5} Pa was obtained with a turbomolecular pump backed with a mechanical one. The working Ar pressure was kept constant at 0.3 Pa using Brooks mass flow rate controllers and a constant pumping speed $S = 10$ L s^{-1}. A pure (99.5%) metallic iron disc of 50 mm in diameter, which was located at 60 mm from the substrate, was dc sputtered with a constant current density of 100 A m^{-2}. The substrates were cleaned with acetone and alcohol before charging in the deposition chamber. Presputtering for 10 min in pure argon was carried out to clean the iron target. Within the conventional process, nitrogen flow rate $Q(N_2)$ value was chosen at 8 sccm. Comparisons are performed sometimes with previous films [16] deposited with gas flow parameters: $Q(N_2) = 2$ sccm. The oxygen flow rate $Q(O_2)$ was varied between 0 and 1.6 sccm. The Fe-O-N coatings were deposited without external heating. Thus, the deposition temperature was expected to be lower than 323 K.

The thickness of the films deposited on glass and (100) silicon substrates was measured with a Dektak 3030 profilometer. The deposition rates were calculated from the sputtering time.

X-ray diffraction (XRD) patterns were obtained using Phillips X'pert diffractometer with Cu Kα radiation. A wavelength of λ = 1.5405 Å was used to obtain the data at a grazing angle of 0.7°. Rutherford backscattering equipment (RBS) was used to estimate the films compositions. RBS measurements were performed with the van de Graaff accelerator using a 2 MeV He$^+$ beam and 2 MeV proton beam. Proton was used because of the increased sensibility for light elements such as nitrogen or oxygen.

Mössbauer spectrometry was also used to observe the different types of Fe environment in these Fe-O-N coatings films. Mössbauer spectra of the films deposited on silicon have been measured at room temperature (300 K) by Conversion Electron Mössbauer Spectrometry (CEMS) in a standard reflection geometry with a constant acceleration signal and a ^{57}Co source diffused into a rhodium matrix using a He/CH_4 gas flow proportional counter. The Mössbauer spectrometer, precisely the Rikon 5 (ORTEC) was calibrated using α-Fe, and the isomer shift values are given relative to that of α-Fe at room temperature. The spectra were fitted with the MOSFIT program.

The films' electrical resistivity was deduced from sheet resistance measurements by the four-point probe method using a JANDEL device at room temperature. The conductivity of the films was evaluated between 20 and 170 °C using the Van der Pauw method. The optical transmittance of approximately 500 nm-thick films deposited on glass substrates was studied by UV–visible spectroscopy in the 200–1100 nm range using a Perkin Elmer Lambda 950 spectrophotometer. Finally, the refractive index (n) and the extinction coefficient (k) were deduced from spectroscopy ellipsometry analysis at an incidence angle of 70° in a 0.75–4.5 eV energy range (1700–270 nm), with a step of 0.02 eV.

3. Results and Discussion

3.1. Chemical Composition

Table 1 presents the concentration of iron, nitrogen, and oxygen in the films. It is important to emphasize that no traces of minor elements have been detected in the films (detection limit about 1 at.%).

Table 1. Chemical composition, Fe/(O + N) and chemical formula determined by RBS measurements of the films deposited with $Q(N_2)$ = 8 sccm and varying $Q(O_2)$ from 0 to 1.6 sccm.

$Q(O_2)$ sccm	Fe (% at.)	N (% at.)	O (% at.)	$Q(O_2)/(Q(O_2) + Q(N_2))$	Fe/(O + N)	Chemical Formula	Compound Type
0	50.9	49.1	0	0	1.04	$Fe_{1.04}N$	γ''-FeN
0.4	41.2	12.9	45.9	0.05	0.70	$Fe_{2.10}O_{2.34}N_{0.66}$ or $Fe_{2.80}O_{3.12}N_{0.88}$	Maghemite-like (γ-Fe_2O_3)
0.6	40.7	8.8	50.5	0.07	0.69	$Fe_{2.06}O_{2.55}N_{0.45}$ or $Fe_{2.76}O_{3.41}N_{0.59}$	Magnetite-like (Fe_3O_4)
0.8	39.9	5.5	54.6	0.09	0.66	$Fe_{1.95}O_{2.69}N_{0.31}$	Hematite-like (α-Fe_2O_3)
1.0	39.4	6.2	54.4	0.11	0.65	$Fe_{1.99}O_{2.73}N_{0.27}$	
1.4	39.7	2.2	58.1	0.15	0.66	$Fe_{1.98}O_{2.89}N_{0.11}$	
1.6	39.2	2.2	58.5	0.17	0.65	$Fe_{1.94}O_{2.89}N_{0.11}$	

For the film deposited without oxygen, the RBS result indicates nearly the same average nitrogen and iron contents, 50%. From the data, the stoichiometry of the sample was calculated to be $Fe_{1.04}N$. Regarding the available literature about the iron nitrides [18,19], the structure of this film can be easily attributed to the iron nitride FeN with ZnS or NaCl structural type. The XRD analysis will give an answer to this assertion.

When the oxygen flow rate $Q(O_2)$ increases from 0.6 to 1.6 sccm, the nitrogen concentration decreases from 8.8 to 2.2% and the oxygen concentration increases from 50 to 58%.

It is interesting to note that the concentration of iron seems to be unaffected by the $Q(O_2)$ increase from 0.6 to 1.6 sccm. Then, from Figure 1, the ratio Fe/(O + N) for films containing oxygen, is ranged from 0.65 to 0.7.

Based on the RBS results, we have calculated the composition of the films as summarized in Table 1. The presence of both elements in these films deposited with $Q(O_2) \geq 0.4$ sccm, i.e., oxygen and nitrogen, confirms that the films are not pure oxide but rather oxidelike oxynitrides. These results of chemical composition clearly indicate the transition from iron nitride to oxidelike oxynitrides films. Regarding the films grown at $Q(O_2)$ = 0.4 sccm ($Fe_{2.1}O_{2.34}N_{0.66}$) and 0.6 sccm ($Fe_{2.06}O_{2.55}N_{0.45}$), we note that they present an overstoichiometry (Table 1) on Fe in comparison to the chemical formula type of $Fe_2(O, N)_3$. In addition, for these two films, a chemical formula type of $Fe_3(O,N)_4$ can be proposed: $Fe_{2.80}O_{3.12}N_{0.88}$ and $Fe_{2.76}O_{3.41}N_{0.59}$. Earlier, research groups [20,21] have shown that it is possible to obtain intermediate nonstoichiometric $Fe_{3-\delta}O_4$ films between Fe_3O_4 and γ-Fe_2O_3 by adjusting the NO_2 pressure during deposition by NO_2-assisted molecular beam epitaxy. The formula representation of Fe_3O_4 is $[Fe^{3+}]_{tet}[Fe_{1-3\delta}{}^{2+},Fe_{1+3\delta}{}^{3+},]_{oct}O_4$, indicating that the Fe^{2+} ions and the vacancies occupy octahedral sites and that the Fe^{3+} ions are distributed evenly over octahedral and tetrahedral sites. These results are an important indication for the possible compound type that may be formed in the films, which will be then accurately investigated by crossing with the results of both X-ray diffraction (XRD) and Conversion Electron Mössbauer spectrometry (CEMS).

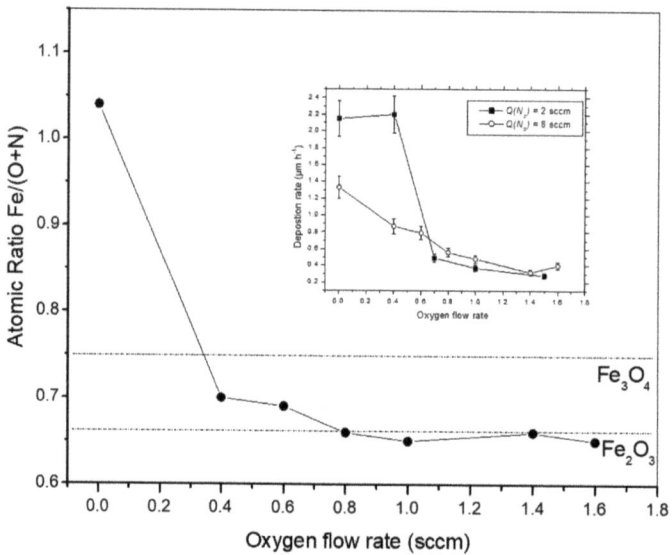

Figure 1. Influence of the oxygen flow rate on the Fe/(O + N) atomic ratio chemical composition of films deposited with $Q(N_2)$ = 8 sccm for $Q(O_2)$ varying from 0 to 1.6 sccm. In the insert is the deposition rate versus the oxygen flow rate.

3.2. XRD Analysis

Figure 2 presents the X-ray diffractograms of the films. We note a structural change from iron nitride to iron oxide ones.

Thus, for $Q(O_2)$ = 0 sccm, the diffraction peaks correspond to both γ''-FeN or γ'''-FeN. As mentioned previously, the nitrogen and iron concentrations obtained by RBS measurements are nearly equal to 50 at.% each. This result is in agreement with the XRD analysis. The structure of the γ''-FeN was determined to be ZnS-type, whereas the γ'''-FeN was NaCl-type [18,22]. In the ZnS-type structure, the nitrogen atoms are in the tetrahedral sites of the fcc iron lattice. In the NaCl-type structure, all the octahedral sites are filled by the nitrogen atoms. Many authors [23] have reported in the literature that the lattice parameter of the γ''-FeN lies between 0.428 nm and 0.433 nm. In the case of the present work, the film is well-crystallized and the lattice parameter calculated is 0.431 nm. Thus, based on this lattice parameter, this film is attributed to the γ''-FeN with the ZnS structure type. It is well-established that the lattice constant of the γ'''-FeN is around 0.45 nm. Taking into account of the presence of probable texture in the film, the diffraction intensity cannot be used to differentiate the two structures. However, it is important to emphasize that the lattice constants of the Fe-N system (γ'': 0.431nm and γ''': 0.45nm) are not so close. I. Jouanny et al. [24] have reported that the γ'''-FeN phase crystallizes in the ZnS-type structure. In their study, the γ''' may be a disordered and nonstoichiometric form of the γ'' structure. Let us note that no trace of texture is evidenced in their film and the lattice parameter is around 0.455 nm.

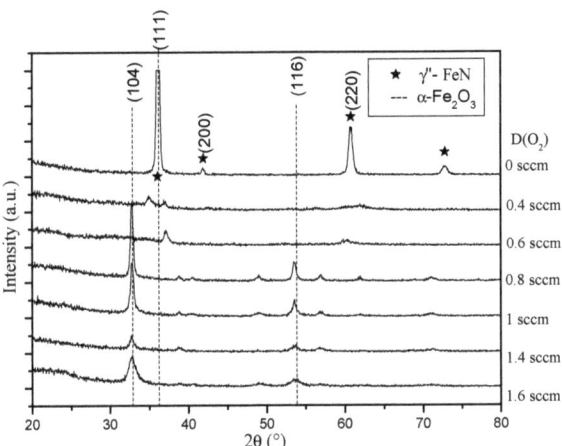

Figure 2. Influence of the oxygen flow rate on the X-ray diffractograms of films deposited with $Q(N_2) = 8$ sccm for $Q(O_2)$ varying from 0 to 1.6 sccm (JCPDS 50-1087 and 89-0596).

Regarding the film grown at $Q(O_2) = 0.4$ sccm or 0.6 sccm, we note that their structure (Figure 2) changes and is poorly crystallized. However, the diffraction peak detected after XRD analysis corresponds to common reflections of maghemite (γ-Fe$_2$O$_3$) and magnetite (Fe$_3$O$_4$). Thus, it is difficult to differentiate between the two oxides from their X-ray diffraction pattern (Figure 3b).

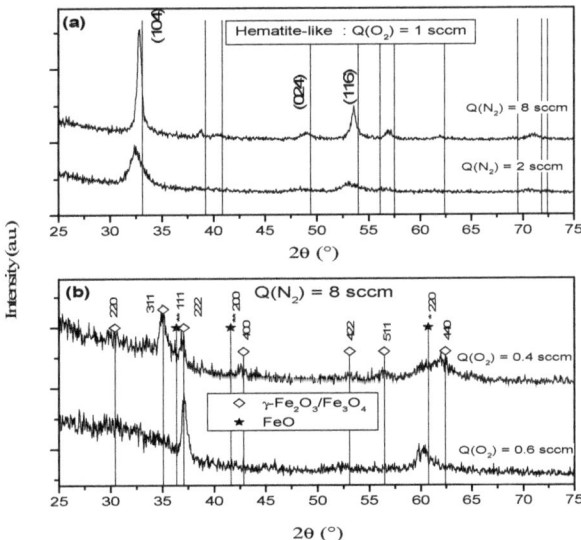

Figure 3. Comparison of (a) the X-ray diffractograms of hematite-like films deposited with $Q(N_2) = 2$ sccm and $Q(N_2) = 8$ sccm for $Q(O_2) = 1$ sccm and (b) X-ray diffractograms of γ-Fe$_2$O$_3$/Fe$_3$O$_4$ deposited with $Q(O_2) = 0.4$ sccm and $Q(O_2) = 0.6$ sccm for $Q(N_2) = 8$ sccm (JCPDS89-0596).

It is well-known that magnetite has an inverse spinel structure with Fe^{3+} ions distributed randomly between octahedral and tetrahedral sites, and Fe^{2+} ions in octahedral sites. Maghemite has a spinel structure that is similar to that of magnetite but with Fe^{2+} ions replaced by vacancies in sublattice. Then, several solutions can be proposed, for instance, the possibility of having an intermediate compound [20,21] Fe$_{3-\delta}$O$_4$ between Fe$_3$O$_4$ and

γ-Fe$_2$O$_3$, or a mixture of oxide [25,26] γ-Fe$_2$O$_3$/Fe$_3$O$_4$ or γ-Fe$_2$O$_3$/FeO, γ-Fe$_2$O$_3$/FeON, or γ-Fe$_2$O$_3$ can be assumed. As previously mentioned, these samples or oxidelike present an over- or understoichiometric on Fe in comparison to the chemical formula type of Fe$_2$(O, N)$_3$ or Fe$_3$(O, N)$_4$, respectively. Let us remember that the oxynitride film Fe$_{1.06}$O$_{0.35}$N$_{0.65}$ [15,16] prepared with Q(N$_2$) = 2 sccm and Q(O$_2$) = 0.4 sccm, crystallizes in a face-centered-cubic structure type NaCl with a lattice parameter estimated at 0.452 nm. Contrary to this film Fe$_{1.06}$O$_{0.35}$N$_{0.65}$, the chemical formula obtained by RBS is Fe$_{2.10}$O$_{2.34}$N$_{0.66}$ or Fe$_{2.80}$O$_{3.12}$N$_{0.88}$ for this film prepared with Q(N$_2$) = 8 sccm and Q(O$_2$) = 0.4 sccm. This reveals the difference between both samples (Q(N$_2$) = 2 sccm and Q(N$_2$) = 8 sccm) prepared with the same oxygen flow rate Q(O$_2$) = 0.4 sccm. In addition, the deposition rate for both series of films deposited with Q(N$_2$) = 2 sccm and Q(N$_2$) = 8 sccm, respectively, is shown in the insert in Figure 1. It is important to note that the initial deposition rate is around 2.1 μm.h^{-1} and 1.3 μm.h^{-1} for the iron nitrides ε-Fe$_{2.2}$N and γ''FeN, respectively. We also observe a progressive decrease in the deposition rate for samples prepared with Q(N$_2$) = 8 sccm. For the samples deposited with Q(N$_2$) = 2 sccm, increasing the oxygen content up to 0.4 sccm, the deposition rate remains constant and the value is close to that of the iron nitride ε-Fe$_{2.2}$N. On this basis, it is clear that the absence of oxynitride film type Fe(O, N) in the case of this work can be justified by the lower deposition rate of this process and by the absence of a rich phase on iron phase such as ε-Fe$_{2.2}$N. This result is in total agreement with the chemical composition and the X-ray diffraction results.

Finally, when Q(O$_2$) exceeds 0.6 sccm, the X-ray patterns confirm the oxide phase α-Fe$_2$O$_3$ formation with a progressive change on the preferential growth of the (104) peak [13,26]. Similar behavior [16] has been observed when preparing FeON films with Q(N$_2$) = 2 sccm. Regarding the evolution of the oxygen flow rate, there is an increase in the full width half maximum (FWHM) of the diffraction peaks. This is the evidence for the reduction in the grain size. In order to understand the similarities and the differences between these series (Q(N$_2$) = 8 sccm) and the series (Q(N$_2$) = 2 sccm) reported in the literature [15,16], a detailed analysis was carried out by comparing the samples of both series, with similar oxygen content in Figure 3. The diffractograms (Figure 3a) reveal that for the same oxygen flow rate Q(O$_2$) = 1 sccm, the sample deposited with high nitrogen flow rate is rather well-crystallized. To understand the Q(N$_2$) effect in the films structure and morphology, the coherent diffraction domain size was calculated considering the (104) plane (Figure 3a) and using Scherrer equation. From the equation, the crystalline domain size was found as approximately 5 nm and 18 nm for the films Q(N$_2$) = 2 sccm and Q(N$_2$) = 8 sccm, respectively.

These differences should result mainly from the presence of a higher quantity of nitrogen in the deposition chamber. It is thus clear that a high concentration of nitrogen plays a fundamental role in the crystallization of the hematite. Indeed, the crystallization of hematite is improved when the ratio (Q(O$_2$)/(Q(O$_2$) + Q(N$_2$))) decreases from 0.33 (when Q(N$_2$) = 2 sccm) to 0.11 ((when Q(N$_2$) = 8 sccm). This behavior could be the result of the catalytic effect of nitrogen on the crystallization of the hematite. In addition, such results have been observed by S. Venkataraj et al. [2] on the growth of zirconium oxynitride films prepared by reactive direct current magnetron sputtering. These authors show that the incorporation of nitrogen in the films improves the crystalline quality.

3.3. ^{57}Fe Mössbauer Spectrometry

Conversion Electron Mössbauer Spectrometry (CEMS) spectra for the films deposited with a nitrogen flow rate of 8 sccm are illustrated in Figure 4 at room temperature 300 K.

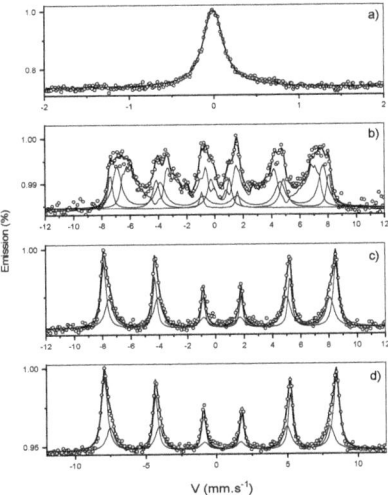

Figure 4. Mössbauer spectra of films prepared with a nitrogen flow rate of 8 sccm at room temperature 300 K. (**a**) $Q(O_2) = 0$ sccm, (**b**) $Q(O_2) = 0.4$ sccm, (**c**) $Q(O_2) = 0.8$ sccm, and (**d**) $Q(O_2) = 1$ sccm.

The film (Figure 4a) deposited without oxygen shows a paramagnetic spectrum at room temperature, which is almost single phase of ZnS-type nitride. The fitting is well-achieved by considering one singlet characterized by an isomer shift value of 0.09 mm/s. Based on the Mössbauer data reported in the literature by Schaaf et al. [27] or Hinomura et al. [28], this singlet can be attributed to γ''-FeN in the ZnS structure all having the nearest nitrogen neighbor sites occupied. Compared with the studies of different research groups [18,29], it is the first time that the monophased γ''-FeN has been evidenced.

When $Q(O_2)$ is fixed at 0.4 sccm, the spectrum (Figure 4b) obtained is broadened with a magnetic behavior at room temperature that cannot be satisfactorily fitted with the combination of magnetite (Fe_3O_4) and maghemite (γ-Fe_2O_3) [30]. This result confirms that this film is poorly crystallized and is consistent with the X-ray diffraction patterns. The best fit is achieved by using a paramagnetic doublet having an isomer shift (quadrupolar splitting) value of $0.40(2)$ mm s^{-1} ($1.19(2)$ mm s^{-1}), two sextets, and a distribution of hyperfine fields linearly correlated with a constant isomer shift. Thus, the doublet with isomer shift corresponds to ferric ion Fe^{3+}. Similar results with isomer shift of 0.46 mm s^{-1} and quadrupolar splitting of 1.19 mm s^{-1} have been obtained in Mössbauer spectra of γ-Fe_2O_3 nanoparticles. The values of the hyperfine parameters obtained for the sextets are as follows: IS = $0.35(2)$ mm/s ($B_{hf} = 47.7(2)$ T) and $0.39(2)$ mm/s ($45.0(2)$ T), respectively. The two sextets are unambiguous to Fe^{3+} sites characteristics of the γ-Fe_2O_3 phase. F. C. Voogt et al. [13] have produced magnetite-like oxynitride films with composition $Fe_{3+\delta}O_{4-y}N_y$ at low NO_2 fluxes ($0.08 < \delta < 0.16$ and $0.19 < y < 0.36$) by NO_2-assisted molecular beam epitaxy. In their paper, the Mössbauer spectra of the films, which are typical for the magnetite, were described with three components (Fe^{3+}, $Fe^{2.5+}$, and Fe-N) and the nitrogen atoms occupied substitutional sites of the oxygen anion sublattice by forming $Fe_{3+\delta}(O, N)_4$. Regarding the isomer shift, which provides valuable information of the oxidation state, one notes that the contribution of the $Fe^{2.5+}$ sites present in the magnetite phase with isomer shift of 0.66 mm s^{-1} is absent in the hyperfine parameters of this sample since there is clear indication of the absence of the magnetite phase or the intermediate compound $Fe_{3-\delta}O_4$ phase between Fe_3O_4 and γ-Fe_2O_3. Due to the presence of 12.9 at.% of nitrogen (Table 1), the last component corresponding to the distribution ($B_{hf} = 45(2) - 15(2)$ T) is attributed to iron having at least one nitrogen as first neighbor. No trace of the paramagnetic wüstite FeO with isomer shift of $0.93(2)$ mm s^{-1} and quadrupole splitting of $1.11(2)$ mm s^{-1} has been detected in the film. Indeed, the Mössbauer results indicate that the major contribution of this film is rather due to the maghemite-like oxynitride phase and that is consistent with the chemical formula obtained by RBS ($Fe_{2.10}O_{2.34}N_{0.66}$). This result confirms the disordered phase and indicates the correlation with the X-ray diffraction patterns and the RBS results.

Finally, for films prepared with higher oxygen flow rate, the Mössbauer spectra consist of a magnetic component with broadened and symmetrical lines that has to be described by at least two sextets. They correspond to two different Fe sites (Fe^{3+} and Fe-N). Analysis of the spectrum shown in Figure 4c,d yields B_{hf} = 50.6(2) T, IS = 0.37(2) mm s^{-1} and B_{hf} = 47.9(2) T, IS = 0.36(2) mm s^{-1} for the two sextets, respectively. The first subspectrum is interpreted as arising from Fe^{3+} ions and corresponds to the α-Fe_2O_3; in such a case, it is necessary to include the quadrupolar shift, which has a typical value of −0.15(2) mm s^{-1}. The second subspectrum is interpreted as arising from Fe ions that have at least one nitrogen nearest neighbor. Contrary to the previous paper [17], this result confirms the effective crystallization of the film prepared with Q(N_2) = 8 sccm and Q(O_2) = 1 sccm. Thus, as the sample prepared with 2 sccm of nitrogen, we can conclude that these films are hematite-like.

3.4. Electrical Properties

The effect of the oxygen flow rate on the electrical resistivity of the films is displayed in Figure 5.

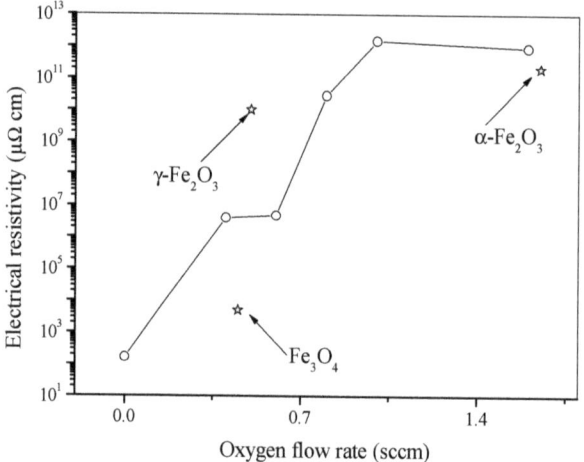

Figure 5. Influence of the oxygen flow rate on the electrical resistivity of Fe-O-N coatings deposited with 8 sccm. The star symbol is related to the electrical resistivity of amorphous iron oxide film deposited without nitrogen.

The nitride coating exhibits an electrical resistivity lower than 200 μΩ cm, which is consistent with the metallic character of iron nitride [31]. The behavior of the resistivity in nitrides is a very complex subject, depending not only on composition but also on parameter such as crystalline structure [32,33]. Thus, comparing this value (200 μΩ cm) to the value reported by I. Jouanny et al. [24] (270 ± 14 μΩ) in the γ″-FeN structure type, we note the correlation between the resistivity and the structure determined in this work.

Introduction of oxygen with flow rate of 0.4 sccm induces a strong increase in the films' electrical resistivity. From 0.4 sccm to 0.6 sccm, the electrical resistivity remains almost constant and the values are around 4.01 × 10^6 μΩ cm and 4.77 × 10^6 μΩ cm, respectively. The structure of these poorly crystallized films, correspond to maghemite or magnetite. It is well-known that the resistivity of the magnetite is 5 × 10^3 μΩ cm [34]. B. Mauvernay et al. [35] has shown that the resistivity versus the partial pressure of iron oxide (Fe_3O_4 + FeO) ranges from 2.8 × 10^5 to 12.3 × 10^5 μΩ cm. Note that the XRD results indicate the absence of the FeO in the samples (0.4 sccm and 0.6 sccm). Thus, this high value of electrical resistivity (4.01×10^6 µΩ cm or 4.77×10^6 µΩ cm) is an important indication for the absence of the magnetite in the sample. Based on the electrical resistivity results and the literature [34,35], we can say that the film deposited with 0.4 sccm and 0.6 sccm may be closed to maghemite.

With the further increase in $Q(O_2)$, the films crystallize in the α-Fe_2O_3 structure and the films exhibit an electrical resistivity higher than 10^{10} µΩ cm, which is consistent with the value measured on oxide films [35] deposited without nitrogen (2×10^{11} µΩ cm). One concludes that the Fe-O-N electrical properties are strongly correlated with the film's composition and structure. Similar results have been reported in literature; for instance, Chappé et al. [36] reported that the increase in oxygen content and the increase in ionic bonding character are responsible for the smooth increase in electrical resistivity in the transition between nitride and oxide sputtering regimes for the TiOxNy system.

In Figure 6, we display the electrical conductivity versus the inverse temperature of the disordered films prepared with these parameters:

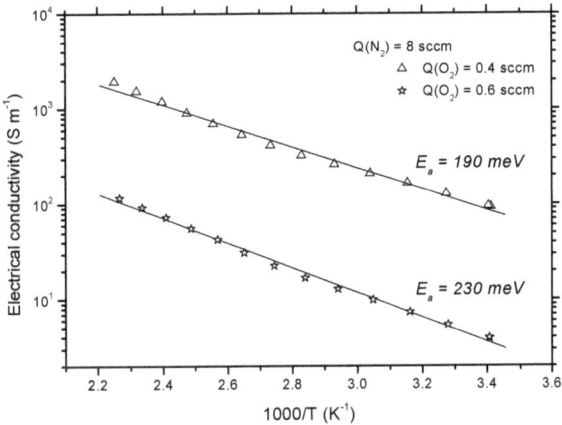

Figure 6. The electrical conductivity versus the inverse temperature of the oxynitride films deposited at $Q(N_2)$ = 8 sccm with $Q(O_2)$ = 0.4 and 0.6 sccm.

We can observe in Figure 6 an increase in the electrical conductivity when the temperature increases. This attributes a semiconductor behavior for these films. It is well-known that α-Fe_2O_3 (hematite) is a n-type semiconductor [37]. However, according to Morikawa et al. and Ogawa et al. [37,38], it can change to p-type conduction induced by N-doping in α-Fe_2O_3. Since we did not perform any hall effect measurements in this study, we can assume that our hematite-like films are of p-type.

In graph 6, the activation energies are calculated assuming a linear behavior in an Arrhenius plot. The activation energy of the disordered films is close and the values are important around 190 meV and 230 meV, respectively. Indeed, during our previous study [16] the measured activation energy of the well-crystallized film deposited with $Q(N_2)$ = 2 sccm and $Q(O_2)$ = 0.4 sccm was found to be 35 meV. In this present study, we increased the nitride flow rate ($Q(N_2)$ = 8 sccm), and the disordered film deposited with the same $Q(O_2)$ = 0.4 presented an activation energy of 190 meV. Therefore, we can assume that the high values of activation energy shown in Figure 6 are due to structural defects in the disordered films. Furthermore, the activation energy of conductivity increases when the oxygen flow rate or partial pressure increases. This may be due to the resistive nature of the O-rich films and to the effects of grain boundary scattering in the films. This evolution has been also observed by Miller et al. [39]. This resistive nature of O-rich films is also responsible for the decrease in the conductivity when the oxygen flow rate increases.

These results confirm the correlation between the structure of both disordered films and the catalytic effect of the nitrogen. Thus, coatings with $Fe_2(O, N)_3$ oxide-type structure exhibit very high electrical resistivity associated with a semiconducting behavior [39].

3.5. Optical Properties

Optical properties of Fe-O-N coatings have been investigated by UV–visible spectroscopy in the 300–1100 nm range. The transmittances of the coatings are plotted in Figure 7 as a function of the wavelength.

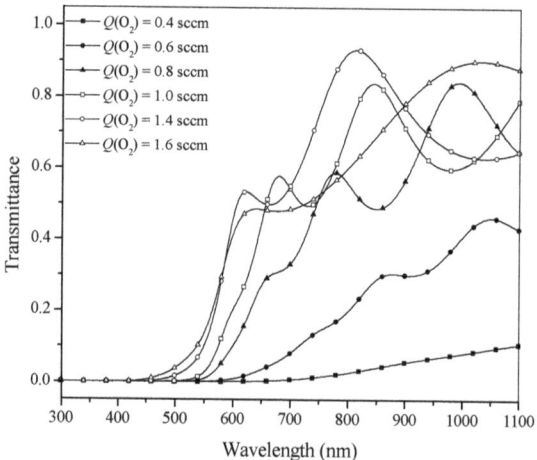

Figure 7. Influence of the oxygen flow rate on the transmittance of Fe-O-N films ($Q(N_2)$ = 8 sccm).

It is important to note the progressive change of the transmittance when the oxygen flow rate increases. Almost all the films are opaque in the UV range and in the visible range up to 450 nm. However, we note a small shift of the threshold of absorption toward smaller wavelengths, going with the increase in the oxygen flow rate. In addition, the films are slightly transparent for samples deposited with oxygen flow rate higher than 0.6 sccm, as previously reported [40]. In this plot, it is possible to observe that the last two samples present different optical behaviors, in total agreement with the composition and the crystalline results. Lei Zhang et al. [41] have shown that the absorption edge of the bulk γ-Fe_2O_3 reference is around 560 nm-wavelength. It is important to note that this absorption edge value is consistent with the value of this work (570 nm) for the disordered samples (0.4 sccm and 0.6 sccm) and confirms the maghemite phase of these films. One more time, the transmittance measurements confirm the difference between nitride, intermediate oxide-type (maghemite-like), and oxide-type (hematite-like) coatings.

From Figure 7, we have determined the energy gap between the highest filled valence band and the lowest total empty conduction band of the samples. Figure 8 shows the optical direct and indirect gap for representative samples as a function of the oxygen flow.

Let us remember that Yoko et al. [42] have reported in the literature the presence of two types of transitions in the α-Fe_2O_3. Thus, considering the direct transition, there is a progressive increase from 1.6 eV to 2.5 eV. Regarding the increase in the direct energy gap for the samples, the main reason should result from the decrease in the grain size versus the oxygen flow rates, as previously indicated. On the other hand, considering the indirect transition, it seems clear that the optical gap is unaffected by the oxygen flow higher than 0.6 sccm. Thus, the value of the indirect energy gap of the hematite-like films is around 1.7 eV. This result is similar to those of the literature [39,40].

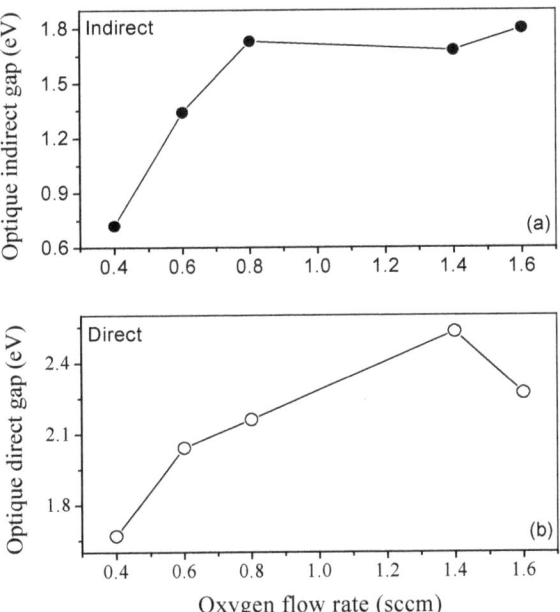

Figure 8. (a) Optical direct and (b) indirect gap for representative samples deposited with $Q(N_2) = 8$ sccm as a function of the oxygen flow.

A close look into Figure 8 shows that the reduction of oxygen partial pressure in the process leads to the decrease in the band gap. The energy gap of the magnetite film reported in the literature is 0.1 eV [11] or 0.3 eV [43], whereas it is higher for maghemite thin film. Optical properties of maghemite thin film were not studied in detail in the literature but the expected value of its direct optical band gap is larger than 2.46 eV [44]. Mirza et al. [45] obtained an optical bandgap of 2.3 eV for a direct transition. In this study, the direct optical bandgaps of the disordered films are 1.7 eV and 2 eV. Thus, based on the direct optical band gap, and since our direct optical bandgap values are close to those of Mirza et al. [45], we can assume that the disorder films could be maghemite-like films.

This optical band gap reduction observed between hematite-like film and maghemite-like film is correlated with the increase in the ratio N/O. It is important to note that this behavior has been also observed on transition metal oxynitrides coatings TiON [6], NbON [46], CrON [47], TiON [48].

The refractive index (n) and the extinction coefficient (k) have been deduced from spectroscopic ellipsometry analysis at an incidence angle of 70° in an energy range 0.75–4.5 eV (270–1700 nm), with a step of 0.02 eV. The results are displayed in Figure 9.

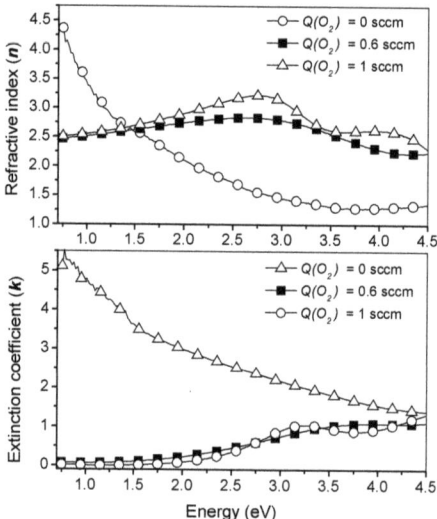

Figure 9. Effect of the oxygen flow rate on the refractive index and extinction coefficient of Fe-O-N films deposited with $Q(N_2) = 8$ sccm.

Indeed, for $Q(O_2) = 0.6$ sccm and 1 sccm, n and k of Fe-O-N coatings seem to be unaffected by the variation in $Q(O_2)$, since their evolution versus photon energy is very close. However, the nitride film deposited without oxygen shows the values of n and k different to those of the oxides. So, in accordance with XRD and RBS analysis, n and k evolution reveals the formation of nitride- and oxides-likes. Moreover, at low photon energy, k is very close to 0 for the films deposited with $Q(O_2) = 0.6$ sccm and 1 sccm. This indicates that these films exhibit dielectric properties observed in oxide films [49].

4. Conclusions

Fe-O-N films have been synthesized by reactive magnetron sputtering of an iron target at high nitrogen partial pressure of 1.11 Pa with different oxygen flow rates and compared with previous Fe-O-N films obtained at a low nitrogen partial pressure of 0.25 Pa. At low nitrogen partial pressure of 0.25 Pa, and before addition of oxygen flow rate, the films exhibited a ε-fe$_2$N structure with an iron to nitrogen atomic ratio close to 2.2 [16]. In this study, the sputtering process implemented with higher nitrogen partial pressure of 1.11 Pa excluded the presence of the ε-Fe$_2$N structure and promoted the emergence of a single phase of γ''-FeN with the ZnS-type structure. However, when the oxygen was introduced in the sputtering process with $Q(O_2) > 0.6$ sccm, the films exhibited a Fe/(O + N) chemical ratio close to 0.66 with a well-crystallized structure and their optical and electrical properties were consistent with those of an α-Fe$_2$O$_3$ structure. Comparatively to a previous study carried out at low nitrogen partial pressure (0.25 Pa), this behavior of films prepared at higher nitrogen partial pressure (1.11 Pa) could result from a catalytic effect of nitrogen on the crystallization of the hematite structure. At intermediate oxygen flow rates, 0.4 and 0.6 sccm, the deposited films were rather poorly crystallized, according to the XRD and Mössbauer results. The XRD results indicate that films deposited with 0.4 and 0.6 sccm can be attributed to a maghemite or magnetite phase or a mixture of these two phases. The results of XRD and composition are not sensible enough to conclusively determine the structural phase obtained. However, the electrical and optical results indicated that these films deposited with 0.4 and 0.6 sccm are very close to maghemite. The simulations of Mössbauer spectra do not reveal the presence of the intermediate ions denoted by the average state Fe$^{2.5+}$ in these samples. Therefore, it is unlikely that these films form the maghemite-like oxynitride phase and the chemical formula obtained by RBS is γ-

Fe2.10O2.34N0.66. The Mössbauer analysis of these samples, especially at low temperature, would be necessary to try to understand more precisely the structure of these two samples.

Due to the occurrence of high nitrogen partial pressure, the reactive sputtering processes can be accessed for the first time to the monophased γ''-FeN. Further, this process allows more easily, by a catalytic effect of nitrogen, the synthesis of hematite crystallized films with better optical and electrical properties.

Author Contributions: Conceptualization, C.R., J.F.P., M.G., and C.P.; methodology, C.R., J.F.P., M.G., and C.P.; validation, C.R., J.F.P., M.G., and C.P.; investigation, C.R., J.F.P., M.G., C.P., and K.B.J.-I.N.; resources, C.R., J.F.P., M.G., and C.P.; data curation, M.G. and C.R.; writing—original draft preparation, M.G. and C.P.; writing—review and editing, C.R., J.F.P., M.G., C.P., and K.B.J.-I.N.; visualization, C.R.; supervision, C.R., J.F.P., and M.G.; project administration, C.R. and J.F.P.; funding acquisition, C.R. and J.F.P. All authors have read and agreed to the published version of the manuscript.

Funding: This work has been financially supported by the European project "HARDECOAT": NMP3-CT-2003-505948, ITSFC, Pays Montbéliard Agglomération, Région Franche-Comté, DRIRE and FEDER.

Institutional Review Board Statement: Not applicable.

Informed Consent Statement: Not applicable.

Data Availability Statement: All data were presented in this manuscript.

Conflicts of Interest: The authors declare no conflict of interest.

References

1. Thobor, A.; Rousselot, C.; Clement, C.; Takadoum, J.; Martin, N.; Sanjines, R.; Levy, F. Enhancement of mechanical properties of TiN/AlN multilayers by modifying the number and the quality of interfaces. *Surf. Coat. Technol.* **2000**, *124*, 210–221. [CrossRef]
2. Venkataraj, S.; Kappertz, O.; Jayavel, R.; Wuttig, M.J. Growth and characterization of zirconium oxynitride films prepared by reactive direct current magnetron sputtering. *Appl. Phys.* **2002**, *92*, 2461–2466. [CrossRef]
3. Jong, C.A.; Chin, T.S. Optical characteristics of sputtered tantalum oxynitride Ta(N,O) films. *Mater. Chem. Phys.* **2002**, *74*, 201–209. [CrossRef]
4. Jurgons, R.; Seliger, C.; Hilpert, A.; Trahms, L.; Odenbach, S.; Alexiou, C.J. Drug loaded magnetic nanoparticles for cancer therapy. *Phys. Condens. Matter* **2006**, *18*, S2893. [CrossRef]
5. Nachtegaal, M.; Sparks, D.L.J. Effect of iron oxide coatings on zinc sorption mechanisms at the clay-mineral/water interface. *Colloid Interface Sci.* **2004**, *276*, 13–23. [CrossRef] [PubMed]
6. Martin, N.; Lintymer, J.; Gavoille, J.; Chappé, J.M.; Sthal, F.; Takadoum, J.; Vaz, F.; Rebouta, L. Reactive sputtering of TiO_xN_y coatings by the reactive gas pulsing process. Part I: Pattern and period of pulses. *Surf. Coat. Technol.* **2007**, *201*, 7720–7726. [CrossRef]
7. Mohamed, S.H.; Anders, A. Structural, optical, and electrical properties of WO_x (N_y) films deposited by reactive dual magnetron sputtering A. *Surf. Coat. Technol.* **2006**, *201*, 2977–2983. [CrossRef]
8. Yubero, F.; Ocana, M.; Caballero, A.; Gonzalez-Elipe, A.R. Structural modifications produced by the incorporation of Ar within the lattice of Fe_2O_3 thin films prepared by ion beam induced chemical vapour deposition. *Acta Mater.* **2000**, *48*, 4555–4561. [CrossRef]
9. Chebotkevich, L.A.; Vorob'ev, Y.D.; Pisarenko, I.V. Magnetic properties of iron nitride films obtained by reactive magnetron sputtering. *Phys. Solid State* **1998**, *40*, 650–651. [CrossRef]
10. Jiang, E.Y.; Sun, D.C.; Lin, C.; Tian, M.B.; Bai, H.L.; Liu, M.S. Facing targets sputtered Fe-N gradient films. *J. Appl. Phys.* **1995**, *78*, 2596–2600. [CrossRef]
11. Ogawa, Y.; Ando, D.; Sutou, Y.; Koike, J. The electrical and optical properties of Fe-O-N thin films deposited by RF magnetron sputtering. *Mater. Trans.* **2013**, *54*, 2055–2058. [CrossRef]
12. Sheftel, E.N.; Tedzhetov, V.A.; Harin, E.V.; Usmanova, G.S. Phase composition and magnetic structure in nanocrystalline ferromagnetic Fe–N–O films. *Curr. Appl. Phys.* **2020**, *20*, 1429–1434. [CrossRef]
13. Petitjean, C. Quantum Reversibility, Decoherence and Transport in Dynamical Systems. Ph.D. Thesis, University of Franche-Comté, Besançon, France, 2007.
14. Carretero, E.; Alonso, R.; Pelayo, C. Optical and electrical properties of stainless steel oxynitride thin films deposited in an in-line sputtering system. *Appl. Surf. Sci.* **2016**, *379*, 249–258. [CrossRef]
15. Grafoute, M.; Petitjean, C.; Rousselot, C.; Pierson, J.F.; Grenèche, J.M. Chemical environment of iron atoms in iron oxynitride films synthesized by reactive magnetron sputtering. *Scr. Mater.* **2007**, *56*, 153–156. [CrossRef]
16. Petitjean, C.; Grafouté, M.; Pierson, J.F.; Rousselot, C.; Banakh, O.J. Structural, optical and electrical properties of reactively sputtered iron oxynitride films. *Phys. D Appl. Phys.* **2006**, *39*, 1894. [CrossRef]

17. Grafouté, M.; Petitjean, C.; Rousselot, C.; Pierson, J.F.; Grenèche, J.M. Structural properties of iron oxynitride films obtained by reactive magnetron sputtering. *Phys. Condens. Matter* **2007**, *19*, 226207. [CrossRef]
18. Andrzejewska, E.; Gonzalez-Arrabal, R.; Borsa, D.; Boerma, D.O. Study of the phases of iron–nitride with a stoichiometry near to FeN. *Nucl. Instr. Meth. Phys. Res. B* **2006**, *249*, 838–842. [CrossRef]
19. Suzuki, K.; Yamaguchi, Y.; Kaneko, T.; Yoshida, H.; Obi, Y.; Fujimori, H.; Morita, H.J. Neutron diffraction studies of the compounds MnN and FeN. *Phys. Soc. Jpn.* **2001**, *70*, 1084. [CrossRef]
20. Voogt, F.C.; Fuji, T.; Smulders, P.J.M.; Nielsen, L.; James, M.A.; Hibma, T. NO_2-assisted molecular-beam epitaxy of Fe_3O_4, $Fe_{3-\delta}O_4$, and $\gamma-Fe_2O_3$ thin films on MgO(100). *Phys. Rev. B* **1999**, *60*, 11193. [CrossRef]
21. Paramês, M.L.; Mariano, J.; Rogalski, M.S.; Popovici, N.; Conde, O. UV pulsed laser deposition of magnetite thin films. *Mater. Sci. Eng. B* **2005**, *118*, 246. [CrossRef]
22. Easton, E.B.; Buhrmester, T.; Dahn, J.R. Preparation and characterization of sputtered $Fe_{1-x}N_x$ films. *Thin Solid Films* **2005**, *493*, 60–66. [CrossRef]
23. Wanga, X.; Zhenga, W.T.; Tiana, H.W.; Yua, S.S.; Xua, W.; Mengb, S.H.; Heb, X.D.; Hanb, J.C.; Sunc, C.Q.; Tay, B.K. Growth, structural, and magnetic properties of iron nitridethin films deposited by dc magnetron sputtering. *Appl. Surf. Sci.* **2003**, *220*, 30–39. [CrossRef]
24. Jouanny, I.; Weisbecker, P.; Demange, V.; Grafouté, M.; Peña, O.; Bauer-Grosse, E. Structural characterization of sputtered single-phase γ''' iron nitride coatings. *Thin Solid Films* **2010**, *518*, 1883–1891. [CrossRef]
25. Cornell, R.M.; Schwertmann, U. *The Iron Oxides: Structure, Properties, Reactions, Occurrences and Uses*, 2nd ed.; Wiley-VCH: Weinheim, Germany, 2003.
26. Sadykov, V.A.; Isupova, L.A.; Tsybulya, S.V.; Cherepanova, S.V.; Litvak, G.S.; Burgina, E.B.; Kustova, G.N.; Ko-lomiichuk, V.N.; Ivanov, V.P.; Paukshtis, E.A.; et al. Effect of mechanical activation on the real structure and reactivity of iron (III) oxide with corundum-type structure. *J. Solid State Chem.* **1996**, *123*, 191. [CrossRef]
27. Schaaf, P. Iron nitrides and laser nitriding of steel. *Hyperfine Interact.* **1998**, *111*, 113–119. [CrossRef]
28. Nakagawa, H.; Nasu, S.; Fuji, H.; Takahashi, M.; Kanamura, F. ^{57}Fe Mössbauer study of FeN_x ($x = 0.25 \approx 0.91$) alloys. *Hyperfine Interact.* **1991**, *69*, 455–458. [CrossRef]
29. Borsa, D.M. Nitride-Based Insulating and Magnetic Thin Films and Multilayers. Ph.D. Thesis, University of Groningen, Groningen, The Netherlands, 2004.
30. Daou, T.J. Synthèse et Fonctionnalisation de Nanoparticules D'oxydes de fer Magnétiques. Ph.D. Thesis, University of Louis Pasteur, Strasbourg, France, 2007.
31. Naganuma, H.; Nakatani, R.; Endo, Y.; Kawamura, Y.; Yamamoto, M. Magnetic and electrical properties of iron nitride films containing both amorphous matrices and nanocrystalline grains. *Sci. Technol. Adv. Mater.* **2004**, *5*, 101–106. [CrossRef]
32. Pilloud, D.; Dehlinger, A.S.; Pierson, J.F.; Roman, A.; Pichon, L. Reactively sputtered zirconium nitride coatings: Structural, mechanical, optical and electrical characteristics. *Surf. Coat. Technol.* **2003**, *174–175*, 338. [CrossRef]
33. Chuan-Pu, L.; Heng-Ghieh, Y. Systematic study of the evolution of texture and electrical properties of ZrN_x thin films by reactive DC magnetron sputtering. *Thin Solid Films* **2003**, *444*, 111.
34. Kingery, W.D.; Bowen, H.K.; Uhlmann, D.R. *Introduction to Ceramics*, 2nd ed.; Wiley & Sons: New York, NY, USA, 1976.
35. Mauvernay, B. Nanocomposites D'oxydes de fer en Couches Minces. Etudes de Leur élaboration et de Leurs Propriétés en Vue de Leur Utilisation Comme Matériaux Sensibles Pour la Détection Thermique. Ph.D. Thesis, University of Toulouse III, Paul Sabatier, France, 2007.
36. Chappe, J.-M.; Martin, N.; Pierson, J.F.; Terwagne, G.; Lintymer, J.; Gavoille, J.; Takadoum, J. Influence of substrate temperature on titanium oxynitride thin films prepared by reactive sputtering. *Appl. Surf. Sci.* **2004**, *225*, 29–38. [CrossRef]
37. Ogawa, Y.; Ando, D.; Sutou, Y.; Koike, J. Effects of O_2 and N_2 Flow Rate on the Electrical Properties of Fe-O-N Thin Films. *Mater. Trans.* **2014**, *55*, 1606–1610. [CrossRef]
38. Morikawa, T.; Kitazumi, K.; Takahashi, N.; Arai, T.; Kajino, T. p-type conduction induced by N-doping in $\alpha-Fe_2O_3$. *Appl. Phys. Lett.* **2011**, *98*, 242108. [CrossRef]
39. Miller, E.; Paluselli, D.; Marsen, B.; Rocheleau, R.E. Low-temperature reactively sputtered iron oxide for thin film devices. *Thin Solid Films* **2004**, *466*, 307–313. [CrossRef]
40. Dghoughi, L.; Elidrissi, B.; Bernède, C.; Addou, M.; Alaoui, M.L.; Regragui, M.; Erguig, H. Physico-chemical, optical and electrochemical properties of iron oxide thin films prepared by spray pyrolysis. *Appl. Surf. Sci.* **2006**, *253*, 1823–1829. [CrossRef]
41. Zhang, L.; Papaefthymiou, G.C.; Ying, J.Y. Size quantization and interfacial effects on a novel $\gamma-Fe2O3/SiO2\gamma-Fe2O3/SiO2$ magnetic nanocomposite via sol-gel matrix-mediated synthesis. *J. Appl. Phys.* **1997**, *81*, 6892. [CrossRef]
42. Yoko, T.; Kamiya, K.; Tanaka, K.; Sakka, S. Photoelectrochemical behavior of iron oxide thin film electrodes prepared by sol-gel method. *Bull. Inst. Chem. Res. Kyoto Univ.* **1989**, *67*, 5–6.
43. Chakrabarti, S.; Ganguli, D. Optical properties of $\gamma-Fe_2O_3$ nanoparticles dispersed on sol–gel silica spheres. *Phys. E* **2004**, *24*, 333. [CrossRef]
44. Balberg, I.; Pankove, J.I. Optical measurements on magnetite single crystals. *Phys. Rev. Lett.* **1971**, *27*, 596. [CrossRef]
45. Mirza, I.M.; Ali, K.; Sarfraz, A.K.; Ali, A.; Ul Haq, A. A study of dielectric, optical and magnetic characteristics of maghemite nanocrystallites. *Mater. Chem. Phys.* **2015**, *164*, 183–187. [CrossRef]

46. Fenker, M.; Kappl, H.; Petrikowski, K.; Bretzler, R. Pulsed power magnetron sputtering of a niobium target in reactive oxygen and/or nitrogen atmosphere. *Surf. Coat. Technol.* **2005**, *200*, 1356–1360. [CrossRef]
47. Wilhartitz, P.; Dreer, S.; Ramminger, P. Can oxygen stabilize chromium nitride? —Characterization of high temperature cycled chromium oxynitride. *Thin Solid Films* **2004**, *447–448*, 289–295. [CrossRef]
48. Vaz, F.; Cerqueira, P.; Rebouta, L.; Nascimento, S.M.C.; Alves, E.; Goudeau, P.; Riviere, J.P.; Pischow, K.; de Rijk, J. Structural, optical and mechanical properties of coloured TiN_xO_y thin films. *Thin Solid Films* **2004**, *447–448*, 449–454. [CrossRef]
49. Gordon, R. Chemical vapor deposition of coatings on glass. *J. Non-Cryst. Solids* **1997**, *218*, 81–91. [CrossRef]

Article

Structure, Microstructure, and Dielectric Response of Polycrystalline $Sr_{1-x}Zn_xTiO_3$ Thin Films

Olena Okhay [1,2,*], Paula M. Vilarinho [3,*] and Alexander Tkach [3,*]

1. TEMA–Centre for Mechanical Technology and Automation, Department of Mechanical Engineering, University of Aveiro, 3810-193 Aveiro, Portugal
2. LASI—Intelligent Systems Associate Laboratory, 4800-058 Guimaraes, Portugal
3. Department of Materials and Ceramic Engineering, CICECO–Aveiro Institute of Materials, University of Aveiro, 3810-193 Aveiro, Portugal
* Correspondence: olena@ua.pt (O.O.); paula.vilarinho@ua.pt (P.M.V.); atkach@ua.pt (A.T.)

Abstract: In a view of the research interest in the high-permittivity materials, continuous enhancement of the dielectric permittivity ε' with Zn content was reported for conventionally prepared $Sr_{1-x}Zn_xTiO_3$ ceramics with x up to 0.009, limited by the solubility of Zn on Sr site. Here, we use a sol-gel technique and a relatively low annealing temperature of 750 °C to prepare monophasic $Sr_{1-x}Zn_xTiO_3$ thin films with higher x of 0.01, 0.05, and 0.10 on $Pt/TiO_2/SiO_2/Si$ substrates. The incorporation of Zn on the Sr site is confirmed by the decrease of the lattice parameter, while the presence of Zn in the films is proven by energy dispersive spectroscopy. The film thickness is found to be ~330 nm by scanning electron microscopy, while the average grain size of 86–145 nm and roughness of 0.88–2.58 nm are defined using atomic force microscopy. ε' measured on the films down to 10 K shows a decreasing trend with Zn content in contrast to that for weakly doped $Sr_{1-x}Zn_xTiO_3$ ceramics. At the same time, the temperature dependence of the dissipation factor tanδ reveals a peak, which intensity and temperature increase with Zn content.

Keywords: perovskites; polar dielectrics; sol-gel thin films; doping; dielectric properties

Citation: Okhay, O.; Vilarinho, P.M.; Tkach, A. Structure, Microstructure, and Dielectric Response of Polycrystalline $Sr_{1-x}Zn_xTiO_3$ Thin Films. Coatings 2023, 13, 165. https://doi.org/10.3390/coatings13010165

Academic Editor: Xiaoding Qi

Received: 22 December 2022
Revised: 3 January 2023
Accepted: 10 January 2023
Published: 12 January 2023

Copyright: © 2023 by the authors. Licensee MDPI, Basel, Switzerland. This article is an open access article distributed under the terms and conditions of the Creative Commons Attribution (CC BY) license (https://creativecommons.org/licenses/by/4.0/).

1. Introduction

Dielectric thin films are widely used in modern technology and engineering aiming to decrease the size and weight of devices, although the physical phenomena in films are much more complicated to study and understand than the behaviour of bulk materials. Among them, $SrTiO_3$-based compounds have been attracting considerable interest both for a wide range of applications, particularly in energy conversion and storage as well as tunable electronic devices, and for break-through insights from a fundamental point of view [1,2]. Bulk undoped strontium titanate ($SrTiO_3$—ST) is known as incipient ferroelectric or quantum paraelectric due to paraelectric state stabilization by quantum fluctuations at low temperatures [3]. At the same time, whenever an electric field [4] or strain [5] is applied, or dopants [6–8] are introduced into the ST lattice, large changes are evident in the compound physical properties. In the case of ST thin films, these changes can be significantly enhanced, resulting e.g., in ferroelectricity induced at room temperature in epitaxial ST films by $DyScO_3$ substrate lattice misfit strain [9]. In addition to the strain [5,9,10], dopants can also induce a variety of phases, ranging from dipolar glass and relaxor to ferroelectric, at isovalent substitution for Sr^{2+} by Ba^{2+} [6,11], Pb^{2+} [6], Ca^{2+} [12,13], and Mn^{2+} ions [8,14] as well as at heterovalent substitution for Sr^{2+} by Bi^{3+} [7,15–17], Y^{3+} [18–20], Dy^{3+} [21] and Gd^{3+} ions [22,23]. We would wish to note here, that since the perovskite structure of $SrTiO_3$ is tolerant to substitutional dopants, whereas interstitial doping is not likely to occur in $SrTiO_3$ [24], doping in this work means ionic substitution.

Regarding Sr-site Zn-substituted ST, no ferroelectric phase transition but low-temperature dielectric permittivity increase with Zn content was reported in weakly (up to 0.9%) doped

$Sr_{1-x}Zn_xTiO_3$ ceramics sintered at 1400 °C by Guo et al. [25]. On the other hand, the room-temperature dielectric permittivity was reported to decrease with increasing $ZnTiO_3$ content in sol-gel derived $SrTiO_3/ZnTiO_3$ heterostructures annealed just at 750 °C by Li et al. [26] thus omitting the problem of high-temperature Zn volatility [27]. However, no low-temperature dielectric characterization was reported so far on moderately Sr-site Zn-substituted ST films. Therefore, in this work, we performed a structural, compositional, microstructural as well as variable temperature dielectric characterisation of sol-gel-derived $Sr_{1-x}Zn_xTiO_3$ thin films with x = 0.01, 0.05, and 0.10, deposited on $Pt/TiO_2/SiO_2/Si$ substrates and annealed at 750 °C.

2. Materials and Methods

For deposition of $Sr_{1-x}Zn_xTiO_3$, thin films with x = 0.01, 0.05, and 0.10, solutions with a concentration of about 0.2 M were prepared using strontium acetate $C_4H_6O_4Sr$ (98%, abcr GmbH, Karlsruhe, Germany), tetra-n-butyl orthotitanate $C_{16}H_{36}O_4Ti$ (98%, Merck KGaA, Darmstadt, Germany) and zinc acetate-2-hydrate $C_4H_6O_4Zn \times 2H_2O$ (99.5%, Riedel-de Haën, Seelze, Germany) as starting precursors. Acetic acid $C_2H_4O_2$ (99.8%, Merck KGaA, Darmstadt, Germany), 1,2-propanediol $C_3H_8O_2$ (99.5%, Riedel-de Haën, Seelze, Germany) and absolute ethanol C_2H_6O (99.8%, Merck KGaA, Darmstadt, Germany) were used as solvents. Strontium acetate was initially dissolved into heated acetic acid (T ~ 60 °C) followed by the addition of zinc acetate-2-hydrate under constant stirring to form a transparent solution. After cooling to room temperature, the former solution was diluted with 1,2-propanediol and then titanium isopropoxide was added. The resultant solution was continuously stirred in a closed flask for 12 h, at the end of which ethanol was added as a final step. Using these transparent and homogeneous solutions, layers of Sr-site Zn-substituted $SrTiO_3$ were deposited on $Pt/TiO_2/SiO_2/Si$ substrates (Inostek Inc., Seoul, Republic of Korea) by spin-coating at 4000 rpm for 30 s, using spin-coater KW-4A (Chemat Technology, Los Angeles, CA, USA). Before the deposition, the substrates were cleaned in boiling ethanol and dried on a hot plate. After the deposition of each wet layer on the substrate, they were heated on a hot plate at 350 °C for ~1 min to ensure the complete removal of the volatile species between the layers. After the complete deposition of 10 layers, they were annealed in air at 750 °C for 60 min with a heating/cooling rate of 5 °C/min.

The thin film crystal phase was analysed at room temperature using a Rigaku D/Max-B X-ray diffractometer (Rigaku, Tokyo, Japan), using Cu Kα radiation. The X-ray diffraction (XRD) data were recorded in 0.02° step mode with a scanning rate of 1°/min from 20° to 80° using Cu Kα radiation. The lattice parameters were calculated from the XRD peak positions in the range of 30–60° using Bragg's law. Compositional analysis of the films was conducted using an energy dispersive spectroscopy (EDS) system (QUANTAX 75/80, Bruker, Ettlingen, Germany) in the top-view geometry under an acceleration voltage of 10 kV of a scanning electron microscope (SEM, Hitachi TM4000Plus, Tokyo, Japan) to reduce the substrate contribution. The thickness of the thin films was determined and their cross-sectional morphology was also observed using SEM (Hitachi S4100, Tokyo, Japan) but under the acceleration voltage of 25 kV. For the film roughness and average grain size determination, a modified commercial atomic force microscope (AFM, Multimode Nanoscope IIIa, Veeco, Santa Barbara, CA, USA) with conductive hard Si tip cantilevers was employed. The topography images were processed using WSxMbeta6_0 software. Dielectric spectroscopy measurements of Sr-site Zn-substituted ST films were performed using Au, sputtered through a mask onto the films, as top electrodes, and the substrate Pt layer as the bottom one. Complex dielectric permittivity, consisting of real part ε' and imaginary part ε'', as well as the dissipation factor $\tan\delta = \varepsilon''/\varepsilon'$, were measured under an oscillation voltage of 50 mV at a frequency of 10 kHz, using a precision LCR-meter (HP 4284A, Hewlett Packard, Palo Alto, CA, USA). A He closed-cycle cryogenic system (Displex APD-Cryostat HC-2, Allentown, PA, USA) equipped with silicon diode temperature sensors and a digital temperature controller, Scientific Instruments Model 9650, was used for temperature variation in the range of 10–300 K. Part of preparation and characterization procedure was performed according to the methodology in Ref. [28].

3. Results and Discussion

XRD profiles of 1%, 5%, and 10% Sr-site Zn-substituted ST thin films on Pt/TiO$_2$/SiO$_2$/Si substrates reveal the cubic perovskite-related peaks corresponding to Pm-3m structure of the films (PDF#35-0734) and those from the substrate, particularly Pt layer, as presented in Figure 1a. With increasing Zn content, positions of the peaks related to the perovskite structure slightly shift toward higher 2θ values as displayed in Figure 1b for the (200) peak. Accordingly, the lattice parameter decreases with Zn content, as shown in the inset of Figure 1b. Such a decrease proves that at least the major part of Zn is incorporated onto the Sr site of the SrTiO$_3$ lattice, taking into account that Sr^{2+} ionic size is larger than that of Zn^{2+}, while Zn^{2+} ionic size is larger than that of Ti^{4+} [29]. Therefore, the formation of Sr$_{1-x}$Zn$_x$TiO$_3$ solid solution is reasonable to be supposed from the lattice parameter decrease, since for SrTi$_{1-x}$Zn$_x$O$_{3-\delta}$ solid solution the lattice parameter should increase with Zn content.

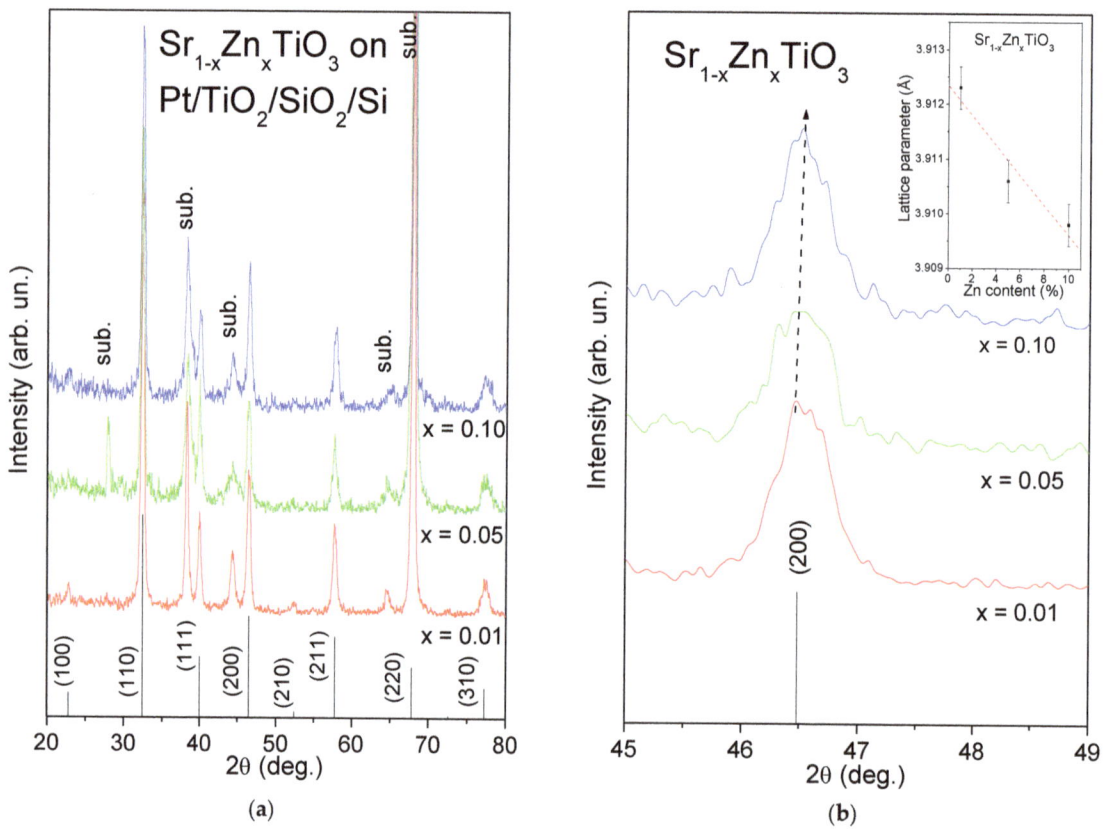

Figure 1. XRD profiles (**a**) and magnified view at (200) reflections (**b**) for Sr$_{1-x}$Zn$_x$TiO$_3$ thin films with x = 0.01, 0.05, and 0.10 deposited on Pt/TiO$_2$/SiO$_2$/Si substrates as well as reflections related to perovskite structure of SrTiO$_3$ card PDF#35-0734 marked by corresponding indexes. Inset in (**b**) shows the lattice parameter variation with Zn content.

EDS analysis of Sr$_{1-x}$Zn$_x$TiO$_3$ thin films, presented in Figure 2a, clearly displays the Zn peak, which intensity increases with the x-value. According to the spectra semi-quantitative analysis, Zn concentrations in Sr-site Zn-substituted ST thin films increase together with nominal ones, while overall estimated elemental contents indicate the proximity of all the film compositions to the nominal ones. Just in the case of x = 0.10 the real substitution content looks to be slightly lower than the nominal one, in agreement with a lower decrement of the lattice parameter compared to that for x = 0.05. Figure 2b shows the SEM cross-sectional microstructure of $Sr_{1-x}Zn_xTiO_3$ thin films with x = 0.01, 0.05, and 0.10 grown on platinized silicon substrates. The film thickness for Sr-site Zn-substituted ST films is about 330 nm on average as seen in Table 1. Moreover, the morphology of all the films reveals several rounded and closely packed grains across the film thickness.

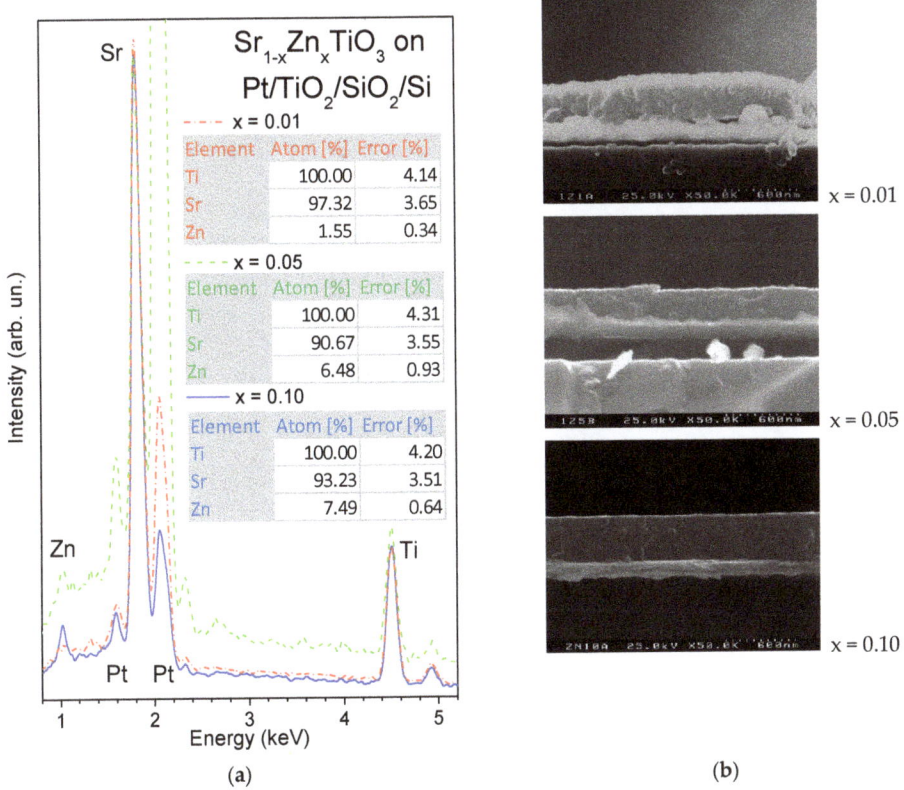

Figure 2. Energy-dispersive spectra (**a**) and SEM cross-section micrographs (**b**) for $Sr_{1-x}Zn_xTiO_3$ thin films with x = 0.01, 0.05, and 0.10 deposited on $Pt/TiO_2/SiO_2/Si$ substrates. The spectra quantification results are also presented in (**a**).

Table 1. Average thickness, grain size, root mean square (RMS) roughness, peak dielectric permittivity value, peak dissipation factor value, and temperature at 10 kHz for $Sr_{1-x}Zn_xTiO_3$ thin films deposited on platinized Si substrates.

Zn Content, x	Average Film Thickness (nm)	Average Grain Size (nm)	RMS Roughness (nm)	Peak ε'	Peak tan δ	tan δ Peak Temperature (K)
0.01	345 ± 30	141 ± 38	2.58 ± 0.08	412	0.046	85
0.05	260 ± 10	145 ± 41	0.88 ± 0.23	277	0.078	101
0.10	375 ± 5	86 ± 22	1.26 ± 0.36	229	0.097	105

From the AFM topological images, shown in Figure 3, the root mean square (RMS) roughness of the films is estimated to be below 2.7 nm, implying the applicability of the films for reliable macroscopic dielectric characterisation. The films reveal also a dense and crack-free microstructure with an average grain size of about 141–145 nm for x = 0.01 and 0.05 decreasing to 86 nm for Zn content of 10%, as also shown in Table 1 together with the standard deviation values.

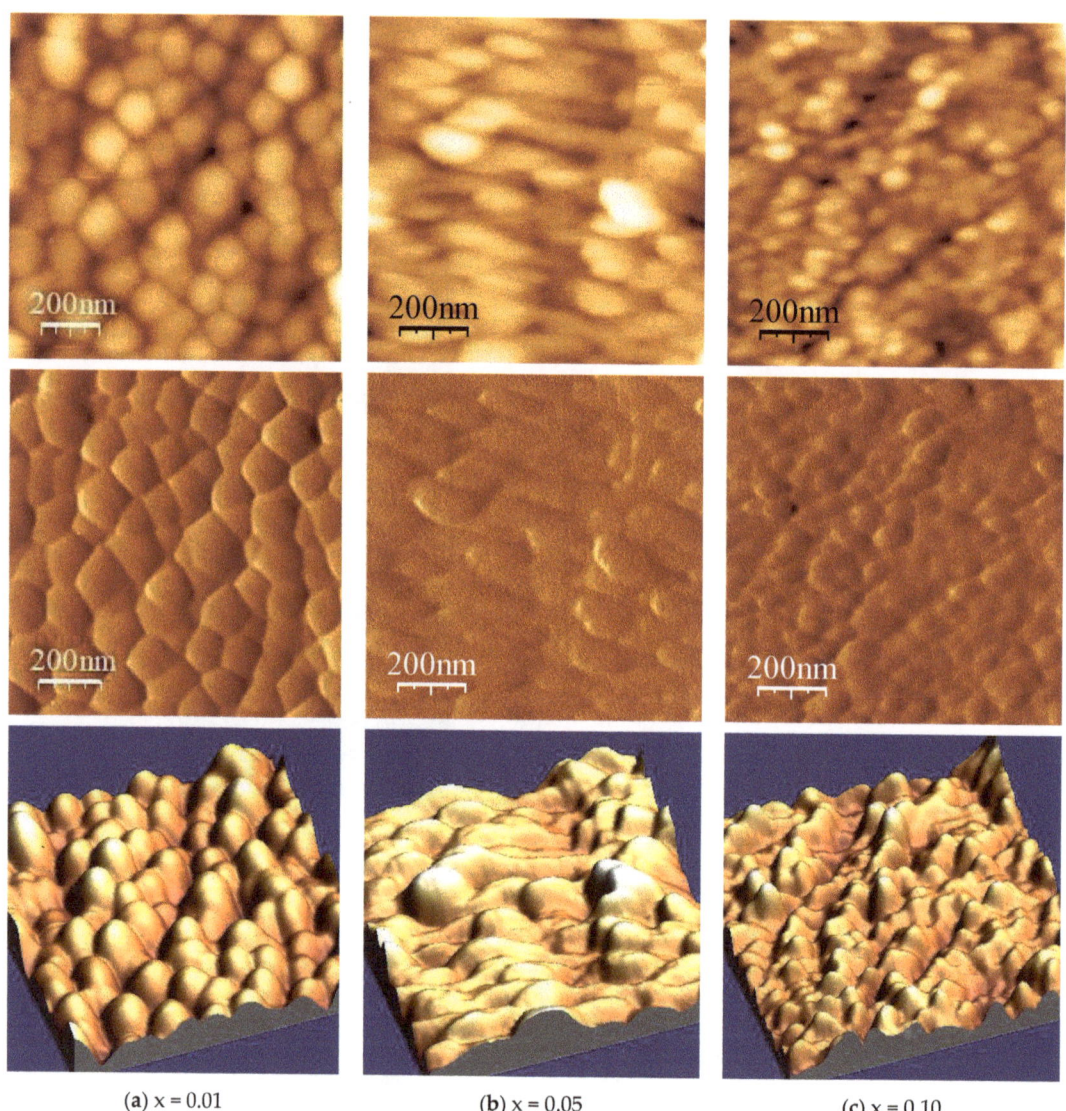

(a) x = 0.01   (b) x = 0.05   (c) x = 0.10

Figure 3. AFM amplitude (*top panel*), phase (*middle panel*), and 3D (*bottom panel*) topography of $Sr_{1-x}Zn_xTiO_3$ thin films with x = 0.01 (**a**), 0.05 (**b**), and 0.10 (**c**).

The $\varepsilon'(T)$ dependence of $Sr_{1-x}Zn_xTiO_3$ thin films is shown in Figure 4 for a frequency of 10 kHz revealing an increase upon cooling until a diffuse peak at about 50 K. The permittivity value, increasing to 412 in peak for x = 0.01 compared with that of 267 in peak for identically prepared undoped ST film [1,16], decreases monotonously with further increasing Zn content. The temperature dependence of dissipation factor tanδ of Sr-site Zn-substituted ST films shown in Figure 4 presents up to three peaks at ~30 K, ~95 K, and ~185 K. While edge tanδ peaks diminish, the middle peak increases with Zn content, revealing values from about 0.5% to about 1.0%, while its position shifts from 85 to 105 K at 10 kHz.

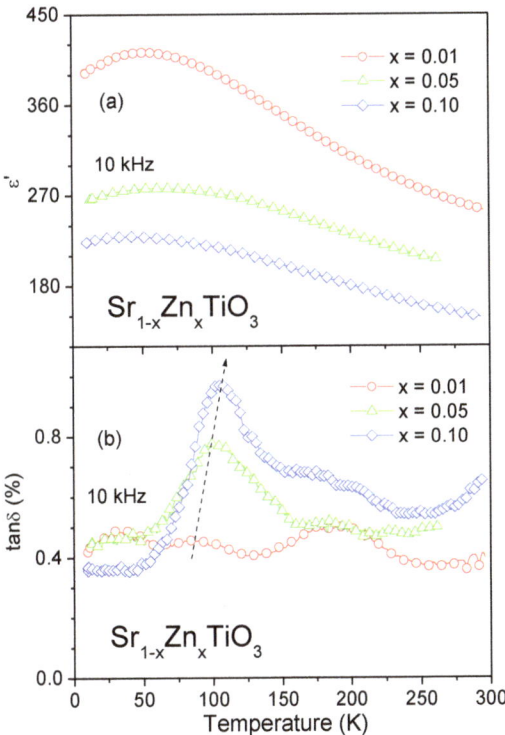

Figure 4. Temperature dependence of the real part of dielectric permittivity ε' (**a**) and dissipation factor tanδ (**b**) for $Sr_{1-x}Zn_xTiO_3$ thin films with x = 0.01, 0.05, and 0.10 at 10 kHz.

The key microstructural and dielectric response parameters of the studied Sr-site Zn-substituted ST thin films are listed in Table 1. They show that increasing Zn content in $Sr_{1-x}Zn_xTiO_3$ films from 1% to 5% and further to 10%, leads to lower dielectric permittivity and higher peak losses. Therefore, the permittivity increase reported by Guo et al. for the $Sr_{1-x}Zn_xTiO_3$ system [25] should be valid for low x-values only, whereas for moderately Sr-site Zn-substituted ST system the permittivity decreases towards that reported by Li et al. for $SrTiO_3/ZnTiO_3$ heterostructures [26].

A similar variation of the permittivity with dopant content originated from the off-centrality of smaller ions on large Sr sites was reported for Ca-, Bi-, Gd-, Dy-, and Y-substituted ST systems [12,15,18,19,21–23]. At low content of the off-central ions, they induce independent polar dipoles, which can enhance the dielectric permittivity of the material. However, when the off-central ion concentration increases, the polar dipoles start to interact with each other that inhibits their ability to be re-oriented by an external electric field and hence makes the dielectric permittivity lower. In the case of $Sr_{1-x}Ca_xTiO_3$, the permittivity was reported to be the highest for Ca content x = 0.0107, further dropping with x increase toward 0.12 [12]. For $Sr_{1-1.5x}Bi_xTiO_3$ ceramics, the highest permittivity was observed at Bi content x = 0.0067 [15], while for $Sr_{1-1.5x}M_xTiO_3$ ceramics with M = Gd, Dy, or Y, the highest permittivity was also observed for x as low as 0.01 [19,21,23]. Therefore, the dielectric behaviour of $Sr_{1-x}Zn_xTiO_3$ films of this study is in the line with the permittivity variations of other ST-based systems.

4. Conclusions

Sr-site Zn substitution was successfully performed in about 330 nm thick and below 2.7 nm rough sol-gel derived $Sr_{1-x}Zn_xTiO_3$ films with x = 0.01, 0.05, and 0.10, deposited on $Pt/TiO_2/SiO_2/Si$ substrates, and found to have a significant effect on the dielectric response. The observation of the Zn peak by EDS and the lattice parameter decrease with Zn content confirmed the substitution for Sr^{2+} by Zn^{2+} ions. Due to such substitution, the relative permittivity was found to decrease in the whole range from 10 K to room temperature for Zn content exceeding 1%. Additionally, a peak, which intensity increases and position shifts to a higher temperature with Zn content, is observed in the temperature dependence of the dissipation factor. Thus, this study helps to fill the gap in the literature regarding low-temperature dielectric characterization on moderately Sr-site Zn-substituted ST, showing that this system behaves similarly to other ST-based systems with Sr-site substitution by smaller ions.

Author Contributions: Conceptualization, P.M.V.; methodology, O.O.; validation, O.O.; formal analysis, O.O., A.T. and P.M.V.; investigation, O.O.; resources, P.M.V.; data curation, O.O.; writing—original draft preparation, A.T. and O.O.; writing—review and editing, A.T., O.O. and P.M.V.; visualization, A.T. and O.O.; funding acquisition, P.M.V. All authors have read and agreed to the published version of the manuscript.

Funding: This work was supported by national funds, through FCT (Fundação para a Ciência e a Tecnologia) in the scope of the framework contract foreseen in numbers 4, 5, and 6 of article 23 of the Decree Law 57/2016, of 29 August, UIDB/00481/2020 and UIDP/00481/2020; and CENTRO-01-0145- FEDER-022083—Centro Portugal Regional Operational Programme (Centro2020), under the PORTUGAL 2020 Partnership Agreement, through the European Regional Development Fund and developed within the scope of the project CICECO-Aveiro Institute of Materials, UIDB/50011/2020, UIDP/50011/2020 & LA/P/0006/2020, financed by national funds through the FCT/MEC (PIDDAC) as well as within FCT independent researcher grant 2021.02284.CEECIND and FLEXIDEVICE project PTDC/CTMCTM/29671/2017.

Institutional Review Board Statement: Not applicable.

Informed Consent Statement: Not applicable.

Data Availability Statement: The data presented in this study are available on request from the corresponding author.

Conflicts of Interest: The authors declare no conflict of interest.

References

1. Tkach, A.; Vilarinho, P. (Eds.) *Strontium Titanate: Synthesis, Properties and Uses*; Nova Science Publishers: New York, NY, USA, 2019.
2. Kleemann, W.; Dec, J.; Tkach, A.; Vilarinho, P.M. $SrTiO_3$—Glimpses of an inexhaustible source of novel solid state phenomena. *Condens. Matter* **2020**, *5*, 58. [CrossRef]
3. Muller, K.A.; Burkard, H. $SrTiO_3$: An intrinsic quantum paraelectric below 4 K. *Phys Rev. B* **1979**, *19*, 3593–3602. [CrossRef]
4. Worlock, J.M.; Fleury, P.A. Electric field dependence of optical-phonon frequencies. *Phys. Rev. Lett.* **1967**, *19*, 1176–1179. [CrossRef]
5. Uwe, H.; Sakudo, T. Stress-induced ferroelectricity and soft phonon modes in $SrTiO_3$. *Phys. Rev. B* **1976**, *13*, 271–286. [CrossRef]
6. Lemanov, V.V. Phase transitions in $SrTiO_3$ quantum paraelectric with impurities. *Ferroelectrics* **1999**, *226*, 133–146. [CrossRef]
7. Porokhonskyy, V.; Pashkin, A.; Bovtun, V.; Petzelt, J.; Savinov, M.; Samoukhina, P.; Ostapchuk, T.; Pokorny, J.; Avdeev, M.; Kholkin, A.; et al. Broad-band dielectric spectroscopy of $SrTiO_3$: Bi ceramics. *Phys. Rev. B* **2004**, *69*, 144104. [CrossRef]
8. Tkach, A.; Vilarinho, P.M.; Nuzhnyy, D.; Petzelt, J. Sr- and Ti-site substitution, lattice dynamics, and octahedral tilt transition relationship in $SrTiO_3$:Mn ceramics. *Acta Mater.* **2010**, *58*, 577–582. [CrossRef]
9. Haeni, J.H.; Irvin, P.; Chang, W.; Uecker, R.; Reiche, P.; Li, Y.L.; Choudhury, S.; Tian, W.; Hawley, M.E.; Craigo, B.; et al. Room-temperature ferroelectricity in strained $SrTiO_3$. *Nature* **2004**, *430*, 758–761. [CrossRef]
10. Tkach, A.; Okhay, O.; Reaney, I.; Vilarinho, P.M. Mechanical strain engineering of dielectric tunability in polycrystalline $SrTiO_3$ thin films. *J. Mater. Chem. C* **2018**, *6*, 2467–2475. [CrossRef]
11. Lemanov, V.V.; Smirnova, E.P.; Syrnikov, P.P.; Tarakanov, E.A. Phase transitions and glasslike behavior in $Sr_{1-x}Ba_xTiO_3$. *Phys. Rev. B* **1996**, *54*, 3151–3157. [CrossRef]
12. Bednorz, J.G.; Müller, K.A. $Sr_{1-x}Ca_xTiO_3$: An XY quantum ferroelectric with transition to randomness. *Phys. Rev. Lett.* **1984**, *52*, 2289–2293. [CrossRef]

13. Kleemann, W.; Schäfer, F.J.; Müller, K.A.; Bednorz, J.G. Domain state properties of the random-field xy-model system $Sr_{1-x}Ca_xTiO_3$. *Ferroelectrics* **1988**, *80*, 297–300. [CrossRef]
14. Tkach, A.; Vilarinho, P.M.; Kholkin, A.L. Polar behavior in Mn-doped $SrTiO_3$ ceramics. *Appl. Phys. Lett.* **2005**, *86*, 172902. [CrossRef]
15. Ang, C.; Yu, Z.; Vilarinho, P.M.; Baptista, J.L. $Bi:SrTiO_3$: A quantum ferroelectric and a relaxor. *Phys. Rev. B* **1998**, *57*, 7403–7406. [CrossRef]
16. Okhay, O.; Wu, A.; Vilarinho, P.M.; Tkach, A. Dielectric relaxation of $Sr_{1-1.5x}Bi_xTiO_3$ sol-gel thin films. *J. Appl. Phys.* **2011**, *109*, 064103. [CrossRef]
17. Tkach, A.; Okhay, O.; Nuzhnyy, D.; Petzelt, J.; Vilarinho, P.M. Polar phonon behaviour in polycrystalline Bi-doped strontium titanate thin films. *Materials* **2021**, *14*, 6414. [CrossRef]
18. Burn, I.; Neirman, S. Dielectric properties of donor-doped polycrystalline $SrTiO_3$. *J. Mater. Sci.* **1982**, *17*, 3510–3524. [CrossRef]
19. Tkach, A.; Vilarinho, P.M.; Almeida, A. Low-temperature dielectric relaxations in Y-doped strontium titanate ceramics. *J. Phys. D Appl. Phys.* **2015**, *48*, 085302. [CrossRef]
20. Tkach, A.; Okhay, O.; Almeida, A.; Vilarinho, P.M. Giant dielectric permittivity and high tunability in Y-doped $SrTiO_3$ ceramics tailored by sintering atmosphere. *Acta Mater.* **2017**, *130*, 249–260. [CrossRef]
21. Tkach, A.; Amaral, J.S.; Zlotnik, S.; Amaral, V.S.; Vilarinho, P.M. Enhancement of the dielectric permittivity and magnetic properties of Dy substituted strontium titanate ceramics. *J. Eur. Ceram. Soc.* **2018**, *38*, 605–611. [CrossRef]
22. Fang, L.; Dong, W.; Zheng, F.; Shen, M. Effects of Gd substitution on microstructures and low temperature dielectric relaxation behaviours of $SrTiO_3$ ceramics. *J. Appl. Phys.* **2012**, *112*, 034114. [CrossRef]
23. Tkach, A.; Amaral, J.S.; Amaral, V.S.; Vilarinho, P.M. Dielectric spectroscopy and magnetometry investigation of Gd-doped strontium titanate ceramics. *J. Eur. Ceram. Soc.* **2017**, *37*, 2391–2397. [CrossRef]
24. Tkach, A. Antiferrodistortive phase transition in doped strontium titanate ceramics: The role of the perovskite lattice vacancies. In *Perovskite Ceramics*; Clabel, J., Rivera, V., Eds.; Elsevier: Amsterdam, The Netherlands, 2022.
25. Guo, Y.Y.; Guo, Y.J.; Liu, J.-M. Zn doping-induced enhanced dielectric response of quantum paraelectric $SrTiO_3$. *J. Appl. Phys.* **2012**, *111*, 074108. [CrossRef]
26. Li, Y.; Xu, B.; Xia, S.; Shi, P. Microwave dielectric properties and optical transmittance of $SrTiO_3/ZnTiO_3$ heterolayer thin films fabricated by sol–gel processing. *J. Adv. Dielect.* **2020**, *10*, 2050027. [CrossRef]
27. Tkach, A.; Okhay, O. Comment on "Giant dielectric response in (Nb + Zn) co-doped strontium titanate ceramics tailored by atmosphere". *Scr. Mater.* **2020**, *185*, 19–20. [CrossRef]
28. Okhay, O.; Vilarinho, P.M.; Tkach, A. Low-temperature dielectric response of strontium titanate thin films manipulated by Zn doping. *Materials* **2022**, *15*, 859. [CrossRef]
29. Shannon, R.D. Revised effective ionic radii and systematic studies of interatomic distances in halides and chalcogenides. *Acta Crystallogr. A* **1976**, *32*, 751–767. [CrossRef]

Disclaimer/Publisher's Note: The statements, opinions and data contained in all publications are solely those of the individual author(s) and contributor(s) and not of MDPI and/or the editor(s). MDPI and/or the editor(s) disclaim responsibility for any injury to people or property resulting from any ideas, methods, instructions or products referred to in the content.

Article

Role of the Alkylation Patterning in the Performance of OTFTs: The Case of Thiophene-Functionalized Triindoles

Roger Bujaldón [1,2,†], Alba Cuadrado [1,2,†], Dmytro Volyniuk [3], Juozas V. Grazulevicius [3], Joaquim Puigdollers [4] and Dolores Velasco [1,2,*]

1. Grup de Materials Orgànics, Departament de Química Inorgànica i Orgànica, Secció de Química Orgànica, Universitat de Barcelona, Martí i Franquès, 1, E-08028 Barcelona, Spain; rr.bujaldon@ub.edu (R.B.); a.cuadradosantolaria@gmail.com (A.C.)
2. Institut de Nanociència i Nanotecnologia (IN2UB), Universitat de Barcelona, E-08028 Barcelona, Spain
3. Department of Polymer Chemistry and Technology, Kaunas University of Technology, Radvilenu pl. 19, LT-50254 Kaunas, Lithuania; juozas.grazulevicius@ktu.lt (J.V.G.)
4. Departament d'Enginyeria Electrònica, Universitat Politècnica de Catalunya, Jordi Girona, 1–3, E-08034 Barcelona, Spain; joaquim.puigdollers@upc.edu
* Correspondence: dvelasco@ub.edu
† These authors contributed equally to this work.

Abstract: Organic semiconductors have emerged as potential alternatives to conventional inorganic materials due to their numerous assets and applications. In this context, the star-shaped triindole core stands as a promising system to design new organic materials with enticing charge-transporting properties. Herein, we present the synthesis of three thiophene-containing triindole derivatives that feature *N*-alkyl chains of different lengths, from methyl to decyl. The impact of the alkylation patterning on the crystallinity of the thin films and their resultant performance as semiconductor have been analyzed. All derivatives displayed p-type semiconductor properties, as demonstrated via both TOF measurements and integration in organic thin-film transistor (OTFT) devices. The attachment of longer alkyl chains and the functionalization of the silicon substrate with octadecyltrichlorosilane (OTS) prompted better OTFT characteristics, with a hole mobility value up to 5×10^{-4} cm^2 V^{-1} s^{-1}. As elucidated from the single crystal, this core is arranged in a convenient cofacial packing that maximizes the π-overlapping. The analysis of the thin films also corroborates that derivatives possessing longer *N*-alkyl chains confer a higher degree of order and a more adequate morphology.

Keywords: alkylation patterning; carbazole; organic chemistry; organic semiconductors; OTFTs; triindole; thin-film morphology

Citation: Bujaldón, R.; Cuadrado, A.; Volyniuk, D.; Grazulevicius, J.V.; Puigdollers, J.; Velasco, D. Role of the Alkylation Patterning in the Performance of OTFTs: The Case of Thiophene-Functionalized Triindoles. *Coatings* **2023**, *13*, 896. https://doi.org/10.3390/coatings13050896

Academic Editors: Zohra Benzarti and Ali Khalfallah

Received: 31 March 2023
Revised: 3 May 2023
Accepted: 5 May 2023
Published: 9 May 2023

Copyright: © 2023 by the authors. Licensee MDPI, Basel, Switzerland. This article is an open access article distributed under the terms and conditions of the Creative Commons Attribution (CC BY) license (https://creativecommons.org/licenses/by/4.0/).

1. Introduction

The current upsurge in organic electronics acknowledges the potential of organic semiconductors as a realistic alternative to conventional inorganic materials [1,2]. Even though the state-of-the-art technology in this area still cannot compete with the performance of crystalline silicon in terms of charge mobility, several materials have already surpassed the milestone of 10 cm^2 V^{-1} s^{-1} [3,4]. Furthermore, the aim of organic materials is not substituting inorganic ones in high-performing applications but rather granting desired features in next-generation devices, such as flexibility or transparency. Their many advantages also encompass uses from the fabrication of large-area displays with a lower production cost to the modulation of their properties via facile synthetic methods [2,5–7].

The π-conjugated backbone of an organic compound rules fundamental properties, such as the electronic profile, the arrangement in the solid state, the air stability, and the optical characteristics [8–10]. Since the structural design represents the cornerstone of an organic semiconductor, the investigation of this topic has led to countless molecular structures developed from a wide array of building blocks [3,6,11–15]. Apart from the key role of the main aromatic nucleus, however, additional structural features can be equally decisive in governing the semiconductor characteristics of a material. A prime example of this is the inclusion of flexible alkyl chains, which not only improve the solubility but also influence the intermolecular interactions in the solid state [16–20]. Therefore, the alkylation patterning in a particular core should be carefully addressed to extract its full potential in terms of charge carrier mobility.

In this way, 9H-carbazole emerges as a highly promising nucleus that can supply several characteristics needed in novel materials, e.g., hole-transporting properties with stability against oxidative doping by atmospheric oxygen [13,21] and high fluorescence [22,23]. Additionally, it embeds nitrogen as a heteroatom that provides a suitable point to insert alkyl chains. The case of the diindolo[3,2-a:3′,2′-c]carbazole core, also known as triindole, represents a particularly enticing system constructed from carbazole that features a star-shaped π-extension [24,25]. The successful integration of triindole-based structures in assorted devices, such as OSCs [26–28], OLEDs [29–31], and, more recently, OTFTs [18,32–34], clearly reinforces their potential.

Herein, we report the synthesis of a set of new triindole derivatives functionalized with thiophene moieties and their integration in OTFT devices. The inclusion of sulfur-containing heterocycles, such as thiophene, represents a well-known strategy in the search for novel materials within organic electronics [35–40]. The advantages of attaching sulfurated moieties, such as 5-methylthien-2-yl and benzothien-2-yl, to the triindole core, instead of aromatic hydrocarbons, have been already demonstrated in our research group [33]. In this particular study, the 3,8,13-tri(thiophen-2-yl)-10,15-dihydro-5H-diindolo[3,2-a:3′,2′-c] carbazole nucleus has been characterized as a p-type semiconductor and structurally ameliorated considering both the molecular and the device design. Synthetically, this core has been N-alkylated with chains of different lengths, from methyl to decyl, as shown in Figure 1. Regarding the architecture of the device, the focus has been placed on optimizing the interface between these semiconductors and the Si/SiO$_2$ substrate in a bottom-gate top-contact OTFT. In fact, the presence of a passivation coating or anchored groups in the interface can greatly influence the growth and features of a vacuum-deposited thin film [41–45]. Specifically, the SiO$_2$ surface has been either functionalized with octadecyltrichlorosilane (OTS) as an aliphatic self-assembled monolayer (SAM) or coated with polystyrene (PS) as an aromatic polymeric layer. The effects of both the N-alkylation patterning and the passivation layer over the performance of this core have been correlated with the degree of order and morphology of the thin films, confirming the relevance of an adequate structural design.

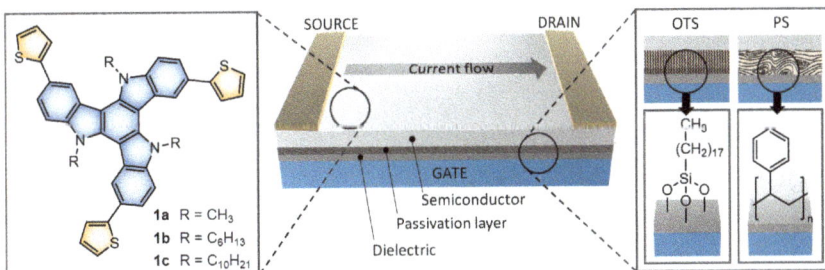

Figure 1. Molecular design of the studied thiophene-substituted triindoles **1a–c** and the architecture of the OTFT devices, constructed over Si/SiO$_2$ substrates coated with either OTS or PS.

2. Materials and Methods

2.1. Synthesis and Characterization

The commercially available chemicals were employed as received. Chemicals and reagents were as follows: 5-bromoisatin (Alfa Aesar, 91.2%), sodium hydride (Aldrich, 60% dispersion in mineral oil), methyl iodide (Acros Organics, 99%), 1-bromohexane (Aldrich, 98%), 1-bromodecane (Aldrich, 98%), hydrazine hydrate (Sigma-Aldrich, reagent grade,

55% N$_2$H$_4$), phosphorous(V) oxychloride (Acros Organics, 99%), 2-thienylboronic acid (TCI, 98.7%), and Pd(PPh$_3$)$_4$ (Acros Organics, 99%). Anhydrous DMF (Thermo Scientific) was kept under nitrogen atmosphere over a molecular sieve. Dichloromethane (VWR) was distilled from CaH$_2$. Flash chromatography was carried out over commercial silica gel (VWR, 40–63 µm). All synthetic procedures were carried out in open-air atmosphere unless otherwise stated.

2.2. Instrumentation and Methods

Here, ^{1}H NMR (400 MHz) and ^{13}C NMR (100 MHz) spectra were collected in a Varian Mercury spectrophotometer (Varian Inc., Palo Alto, CA, USA). In the case of compound **1a**, the ^{13}C NMR (100 MHz) spectrum was recorded in a Bruker 400 MHz Avance III. The analysis of the NMR spectra was achieved using MestRec Nova software (version 14.2.0). The solvent signal was used to reference all the spectra. Absorption spectra were registered on a Varian Cary UV–Vis–NIR 500E spectrophotometer (Palo Alto, CA, USA) and emission spectra were registered using a PTI fluorimeter (Birmingham, AL, USA). 1,4-Bis(5-phenyl-2-oxazolyl)benzene (POPOP) was selected as the standard (λ_{Ex} = 300 nm, Φ = 0.93 in cyclohexane) for the analysis of the fluorescence quantum yields, as suggested in the literature protocol [46]. Fluorescence quantum yields of thin films were determined by means of an integrating sphere. Cyclic voltammograms were registered in a cylindrical three-electrode cell using the following electrodes: an Ag/Ag$^+$ electrode (1 mM AgNO$_3$ in acetonitrile) as the reference electrode, a glassy-carbon electrode as the working electrode, and a platinum wire as the counter electrode. All voltammetric curves were recorded with a microcomputer-controlled potentiostat/galvanostat Autolab with PGSTAT30 equipment (Metrohm Autolab BV, Utrecht, The Netherlands) and GPES software (version 4.9) under quiescent conditions, at a scan rate of 100 mV s^{-1} and under an argon atmosphere. The solutions were prepared in distilled dichloromethane (1 mM) with tetrabutylammonium hexafluorophosphate (TBAP) as the supporting electrolyte (0.1 M). The potentials were referred to the Fc$^+$/Fc redox couple. The ionization potentials (IP) were calculated from the onset of the first oxidation peak as IP = $^{ox}E_{onset}$ + 5.39, where 5.39 eV stands as the formal potential of the Fc$^+$/Fc couple in the Fermi scale [47]. The electron affinities (EA) were estimated as EA = IP − E_{gap}. The optical gap energies (E_{gap}) were estimated from the absorption spectra (λ_{onset}). Ionization potentials (IP) in the solid state were measured by the photoelectron emission method in air. Thin films for IP measurements were prepared by vacuum thermal evaporation (10^{-6} mbar) of the organic compounds on glass slides coated with fluorine-doped tin oxide. A negative voltage of 300 V was applied to the sample substrate. A deep UV deuterium light source (ASBN-D130-CM) and a CM110 1/8m monochromator (Spectral Products, Putnam, CT, USA) were used for illumination of the samples with monochromatic light. A 6517B Keithley electrometer (Keithley, Solon, OH, USA) was connected to the counter electrode for the photocurrent measurement, which was flowing in the circuit under illumination. An energy scan of the incident photons was performed while increasing the photon energy. Thermogravimetric analyses (TGA) were recorded at a heating rate of 20 °C min^{-1} under nitrogen atmosphere using a TA Instruments Q50 (New Castle, DE, USA). The extraction of the charge drift mobility was performed by means of the TOF technique. The organic compounds were vacuum-deposited (10^{-6} mbar) on pre-cleaned indium tin oxide (ITO)-coated glass substrates, and then 80 nm of an aluminum layer was also deposited via thermal vacuum evaporation using a mask (area = 0.06 cm^2). Photo generation of charge carriers was performed by a light pulse through the ITO. A Keithley 6517B electrometer was used to apply external voltages with a pulsed third-harmonic Nd:YAG laser EKSPLA NL300 (pulse duration of 3–6 ns, λ = 355 nm). A Tektronix TDS 3032C digital storage oscilloscope was used to record the TOF transients. The transit times (t_t) were calculated using the kink on the curve of the transient in the log–log scale. The drift mobilities were calculated as $\mu = d^2/Ut_t$, where d is the layer thickness and U is the surface potential at the moment of illumination. Zero-field mobilities (μ_0) and field dependence parameters (α) were extracted from $\mu = \mu_0 e^{\alpha E^{1/2}}$. The single crystal was analyzed using a D8 Venture System (Bruker AXS, Karlsruhe, Germany) equipped with a multilayer monochromator and a Mo microfocus ($\lambda = 0.71073$ Å). The frames were integrated with the Bruker SAINT software package (version SAINT V8.38A) via a narrow-frame algorithm. The structure was elucidated and refined using the Bruker SHELXTL software package. Out-of-plane GIXRD measurements were performed on vacuum-deposited thin films (semiconductor thickness = 75 nm) with a PANalytical X'Pert PRO MRD diffractometer (Almelo, the Netherlands) possessing a PIXcel detector, a parabolic Göbel mirror at the incident beam, and a parallel plate collimator at the diffracted beam (Cu Kα radiation ($\lambda = 1.5418$ Å), with a work power of 45 kV × 40 mA). The optimized angle of incidence used was 0.20° (**1a**) or 0.18° (**1b,c**). The morphology of the layers, analyzed by means of atomic force microscopy (AFM), was profiled using an AFM Dimension 3100 system connected to a Nanoscope IVa electronics unit (Bruker, Billerica, MA, USA).

2.3. OTFT Fabrication and Characterization

OTFTs were constructed in a bottom-gate top-contact geometry using thermally-oxidized crystalline-silicon wafers with a SiO$_2$ layer as the gate dielectric. The gate side of the substrates was partially unprotected with ammonium fluoride. Then, the wafers were cleansed by subsequent ultrasonic treatments in acetone, isopropyl alcohol, and water, dried by a nitrogen blow, and heated at 100 °C for 5 min. The SiO$_2$ surface was then either functionalized with octadecyltrichlorosilane (OTS) or coated with polystyrene (PS). The functionalization with OTS SAMs [48,49] was achieved by immersing the substrates in a solution of OTS in toluene (2 mM) for 24 h at room temperature. Then, the substrates were cleaned by subsequent ultrasonic treatments in toluene, acetone, and isopropyl alcohol, and finally dried by a nitrogen blow and heated at 100 °C for 5 min. The coating with PS was carried out with a solution of PS in toluene (4 mg mL^{-1}), which was spin-coated onto the wafer. The substrate was spun at 1500 rpm for 5 s and 2500 rpm for 33 s with a P6700 spin-coater (Specialty Coating System, Indianapolis, IN, USA). The organic semiconductor was deposited by thermal evaporation in a vacuum system with a base pressure below 10^{-6} mbar. The temperature was manually controlled to ensure a stable deposition rate of 0.3 Å s^{-1} until a thickness of ca. 75 nm was obtained. Then, the substrates were transferred to a different vacuum chamber to deposit the metallic contacts. Gold was chosen as the metal for the drain and source electrodes, which were defined with a metallic mask possessing a channel length (L) of 80 μm and width (W) of 2 mm. The OTFTs were electrically characterized in the dark under ambient conditions using a Keithley 2636A source meter (Solon, OH, USA). The charge carrier mobility was extracted in the saturation regime (μ_{sat}) from Equation (1), as follows:

$$I_D = \frac{W\, C_{ox}\, \mu}{2\, L}(V_G - V_{th})^2 \qquad (1)$$

where W and L are the channel width and length, respectively, and C_{ox} is the unit dimensional dielectric capacitance of the gate insulator.

3. Results and Discussion

3.1. Synthesis and Characterization

The synthesis towards the final compounds **1a–c**, which started with the commercially available 5-bromoisatin, is presented in Scheme 1. The alkylation of the starting material was conducted under standard conditions, using NaH and CH$_3$I to obtain **2a** or K$_2$CO$_3$ and the corresponding 1-bromoalkane in the case of **2b** and **2c**. Then, the reduction via the Wolff–Kishner reaction led to the 5-bromooxindole derivatives **3a–c**. The cyclocondensation towards the brominated triindole systems, accomplished with POCl$_3$ under reflux, prompted a higher yield for the derivatives featuring shorter alkyl chains. Finally, the attachment of the thiophene moieties was achieved through the Suzuki–Miyaura cross-coupling reaction [50]. It should be mentioned that the severe insolubility of the methylated precursor **4a** required the use of microwave irradiation to provide the desired compound **1a** in a comparable yield.

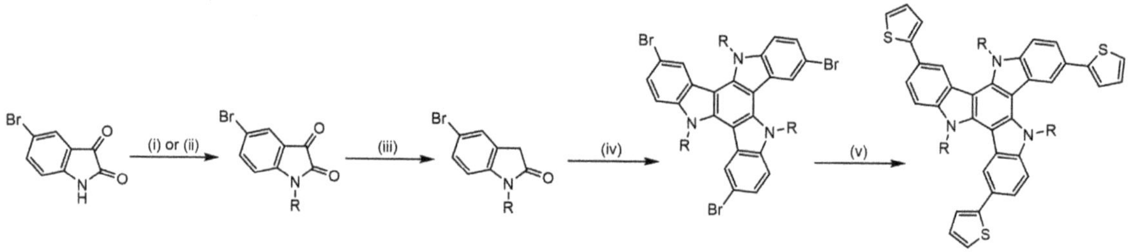

2a R = CH₃ (98%)	3a R = CH₃ (65%)	4a R = CH₃ (58%)	1a R = CH₃ (50%)	
2b R = C₆H₁₃ (92%)	3b R = C₆H₁₃ (90%)	4b R = C₆H₁₃ (39%)	1b R = C₆H₁₃ (53%)	
2c R = C₁₀H₂₁ (99%)	3c R = C₁₀H₂₁ (73%)	4c R = C₁₀H₂₁ (21%)	1c R = C₁₀H₂₁ (55%)	

Scheme 1. Synthetic route followed to furnish the organic semiconductors **1a–c**. Reagents and conditions: (i) NaH and CH$_3$I in DMF at RT for the synthesis of **2a** or (ii) K$_2$CO$_3$ and R-Br in DMF at RT for the synthesis of **2b,c**; (iii) NH$_2$NH$_2$·H$_2$O, reflux; (iv) POCl$_3$, reflux; (v) 2-thienylboronic acid, Pd(PPh$_3$)$_4$ and K$_2$CO$_3$ in THF:H$_2$O, μW (**1a**) or reflux (**1b,c**).

3.2. Physical Characterization

The thermal and optical properties in solution and in the solid state of compounds **1a–c** are compiled in Table 1. Remarkably, all compounds are suitable for the vacuum-evaporation process, with decomposition temperatures (T_d) that surpass 425 °C (TGA scans are depicted in the Supporting Information).

Table 1. Thermal and optical properties in solution and in the solid state of derivatives **1a–c**.

Compound	T_d (°C) [1]	Solution [2]			Solid State [3]		
		$\lambda_{abs,max}$ (nm)	$\lambda_{em,max}$ (nm)	Φ_f [4]	$\lambda_{abs,max}$ (nm)	$\lambda_{em,max}$ (nm)	Φ_f [4]
1a	476	317, 344	395, 414	0.14	315, 353	432	0.02
1b	433	318, 344	397, 415	0.15	316, 350	424	0.03
1c	427	318, 344	397, 415	0.15	313, 352	424	0.03

[1] Onset decomposition temperature obtained from TGA. [2] Optical properties determined from a 10 μM solution in CH$_2$Cl$_2$. [3] Optical properties determined from a vacuum-evaporated thin film of the compound deposited over quartz. [4] Fluorescence quantum yield (Φ_f) obtained with an integration sphere (λ_{ex} = 330 nm).

In terms of the optical properties, derivatives **1a–c** display similar characteristics regardless of the length of the N-alkyl chains. All of them emit into the UV–blue region, peaking at 415 nm in CH$_2$Cl$_2$ and displaying a considerable fluorescence quantum yield of ca. 0.15. In the solid state, the emission spectra are slightly red-shifted and broadened, with a decay of the fluorescence quantum yield to ca. 0.03, which is associated to the aggregation of the triindole systems. The absorption and emission spectra of derivatives **1a–c** are represented in the Supporting Information.

The electrochemical properties of **1a–c**, analyzed by means of cyclic voltammetry in CH$_2$Cl$_2$, are listed in Table 2. All compounds exhibited a first quasi-reversible oxidation process and a second irreversible one, whereas no reduction process could be observed. Consequently, the ionization potentials could be estimated. Their values are slightly conditioned by the length of the N-alkyl chains, ranging from 5.58 to 5.66 eV in derivatives **1a** and **1c**, respectively. These values are translated into quite low-lying HOMO energy levels, conferring stability against atmospheric oxygen to the structure while also being suitable for hole injection. The energy of the optical band gaps (E_{gap}) was estimated to be 3.22 eV in all three cases, which is smaller than that of the bare triindole system (3.53 eV [20]) because of the more extended π-conjugation of the system. The ionization potentials were also estimated in the solid state though the photoelectron emission technique as a closer approximation of the device conditions. The featured values, which go from 5.11 eV (**1b,c**) to 5.18 eV (**1a**), should ensure an optimal injection process due to the closeness to the gold work function (5.1 eV). The resulting energy levels are represented in Figure 2. The cyclic voltammograms and the photoelectron emission spectra are compiled in the Supporting Information.

Table 2. Electrochemical characterization of derivatives **1a–c**.

Compound	E_{gap} (eV) [1]	$^{ox}E_{onset}$ (V) [2]	IP (eV) [3]	EA (eV) [4]	IP (eV) [5]
1a	3.22	0.18	5.58	2.36	5.18
1b	3.22	0.21	5.60	2.38	5.11
1c	3.22	0.27	5.66	2.44	5.11

[1] Optical energy gap (E_{gap}) estimated from the absorption spectrum ($\lambda_{abs,onset}$) in CH_2Cl_2. [2] Onset oxidation potential ($^{ox}E_{onset}$) vs. Fc^+/Fc determined from CV in dichloromethane (1 mM). [3] Ionization potential (IP) estimated as IP = $^{ox}E_{onset\ vs.\ Fc+/Fc}$ + 5.39. [4] Electron affinity (EA) estimated as EA = IP − E_{gap}. [5] Ionization potential (IP) in the solid state, determined via the photoelectron emission technique.

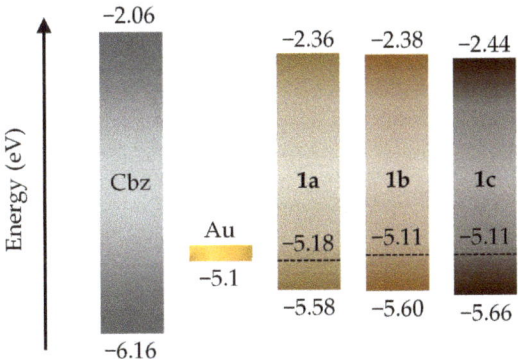

Figure 2. Energy levels of 9H-carbazole (Cbz) [34] and compounds **1a–c** with respect to the gold work function (the dashed lines correspond to the energy levels calculated from the photoelectron emission technique in the solid state).

The charge-transporting properties of the final compounds **1a–c** were analyzed via the time-of-flight (TOF) technique as a first approximation (Table 3). The graphics containing the TOF transients and the dependence of the hole drift mobilities are collected in the Supporting Information. All three compounds showed a non-dispersive behavior and anticipated promising p-type semiconductor properties. As observed, the extracted mobility values are highly conditioned by the N-alkylation patterning. Derivative **1a**, featuring the shortest N-alkyl chain, displays a maximum μ_h of 7×10^{-5} cm^2 V^{-1} s^{-1} that is clearly outdone by its analogs **1b,c** (1×10^{-3} cm^2 V^{-1} s^{-1}). This enhancement, which goes up to an order of magnitude, manifests the preference of longer N-alkyl chains in this particular core. Nevertheless, a further elongation from hexyl to decyl did not show a substantial modification using this technique.

Table 3. Charge carrier mobility values of derivatives **1a–c** obtained from TOF measurements.

Compound	μ_h (cm^2 V^{-1} s^{-1}) [E (V cm^{-1})] [1]	μ_0 (cm^2 V^{-1} s^{-1}) [2]	α ((cm V^{-1})$^{1/2}$) [3]
1a	7×10^{-5} [7×10^5]	4×10^{-6}	0.0034
1b	1×10^{-3} [2×10^5]	6×10^{-4}	0.0017
1c	1×10^{-3} [2×10^5]	2×10^{-4}	0.0053

[1] Hole mobility (μ_h) at the specified electric field. [2] Zero-field mobility (μ_0). [3] Field dependence (α). The measurements were performed under ambient conditions at room temperature. The layers were prepared under vacuum evaporation, with thicknesses ranging from 0.56 to 1.4 μm.

3.3. Organic Thin-Film Transistors

Considering the appropriate hole-transporting properties in conjunction with the adequate energy levels of compounds **1a–c**, all of them were tested as p-type semiconductors in OTFT. The characteristics of the OTFT devices fabricated from **1a–c** over OTS- and PS-treated Si/SiO$_2$ substrates are compiled in Table 4.

Table 4. OTFT characteristics of devices integrating compounds **1a–c** over Si/SiO$_2$ substrates passivated with either OTS or PS.

Compound	SiO$_2$/OTS		SiO$_2$/PS	
	μ_h (cm^2 V^{-1} s^{-1})	I_{on}/I_{off}	μ_h (cm^2 V^{-1} s^{-1})	I_{on}/I_{off}
1a	7×10^{-5}	10^3	6×10^{-5}	10^2
1b	2×10^{-4}	10^4	1×10^{-4}	10^3
1c	5×10^{-4}	10^4	2×10^{-4}	10^4

As observed, the length of the N-alkyl chains appears as the main factor to modulate the hole mobility. In fact, the substitution of the N-methyl by the N-decyl implies an increase in the μ_h up to an order of magnitude over OTS-treated substrates, from 7×10^{-5} to 5×10^{-4} cm^2 V^{-1} s^{-1}. The trend displayed by OTFT devices is, therefore, consistent with the results collected via TOF measurements. The effect of the N-alkylation on the performance of triindole also surpasses that of the nature of the aromatic moieties. Specifically, the attachment of alternative sulfurated scaffolds (i.e., 5-methylthien-2-yl and benzothien-2-yl) or aromatic hydrocarbons (i.e., phenyl and naphtyl) could only modulate the μ_h from 2×10^{-4} to 4×10^{-4} cm^2 V^{-1} s^{-1} [33]. The I_{on}/I_{off} ratios follow the same tendency, with higher and more suitable values of ca. 10^4 in the case of the N-decylated **1c**. Considering the effect of the passivation layer, OTS-containing devices slightly outperform their PS counterparts. This effect is also more significant with derivatives featuring longer N-alkyl chains. Another point to highlight is that devices fabricated with the N-methylated derivative **1a** display very linear saturation characteristics and low threshold voltage. Contrarily, derivatives **1b,c** exhibit a kink, so the charge mobility values were extracted from the region at higher V_G, as suggested in the literature [51,52]. The OTFT characteristics of devices fabricated from compounds **1a–c** over OTS-treated substrates are illustrated in Figure 3, whereas those of their PS counterparts can be found in the Supporting Information. The evolution of the hole mobility through time was also monitored to evaluate the air stability of this core, a feature that is highly coveted in organic electronics. As shown in Figure 3d, compounds **1a–c** exhibit minimal fluctuation of the hole mobility throughout a month. Their notorious air-stability was also corroborated by a shelf lifetime surpassing 100 days (Figure S7).

3.4. Solid-State Characterization
3.4.1. Crystallographic Data

The crystal structure of derivative **1a** could be elucidated by means of single-crystal X-ray diffraction. The ORTEP projection and the molecular packing detailing the main intermolecular interactions are depicted in Figure 4. Compound **1a** crystallized in space group $P–1$ of the triclinic system, with dimensions a = 7.2546 (7) Å, b = 14.2460 (14) Å, c = 14.4915 (15) Å, α = 95.662 (5)°, β = 100.412 (4)°, and γ = 98.766 (4)°, and a total volume of 1443.5(3) Å^3. The intermolecular packing could be classified as a β-type or sheet [53,54], featuring slightly displaced cofacial interactions. This type of arrangement is often considered as optimal for charge transport due to the strong π-overlap between molecules [54,55]. Specifically, the π–π interactions show conveniently short distances as close as 3.34 Å (Figure 4c). The methyl chains also assist and reinforce the packing with CH$\cdots$π interactions. It should be noted that the triindole backbone is not entirely planar, since the peripheral benzene rings display a deviation of ca. 9° with respect to the central one. The thiophene moieties also show torsion angles with respect to the triindole nucleus that go from 13.4 to 40.8° (Figure 4a). In spite of this, the arrangement of derivative **1a** is prone to facilitating the charge transport throughout the material.

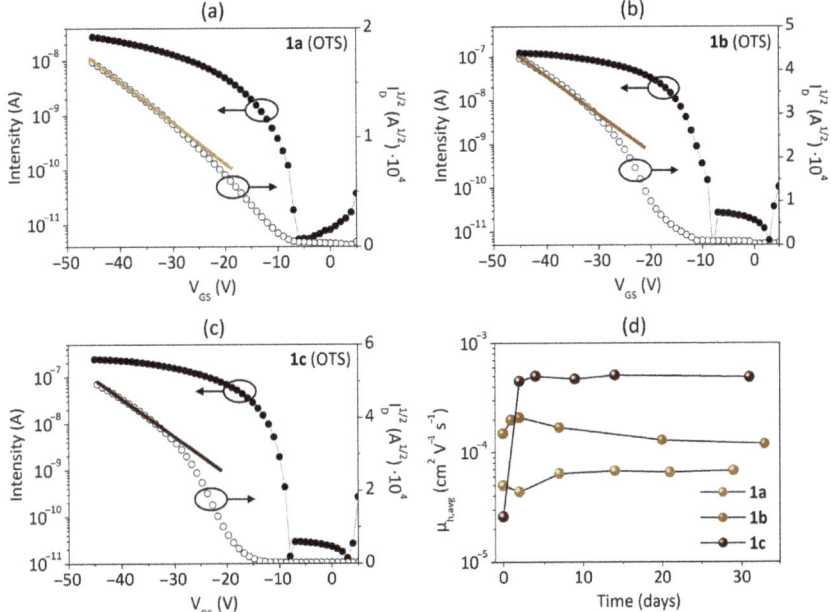

Figure 3. OTFT characteristics of OTS-functionalized devices based on compounds: (**a**) **1a**; (**b**) **1b**; (**c**) **1c**. The evolution of μ_h of representative devices throughout a month is represented in (**d**).

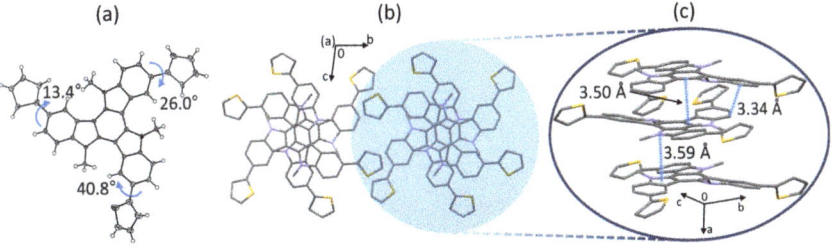

Figure 4. Crystal structure of derivative **1a**: (**a**) ORTEP projection, indicating the torsion angles between the triindole nucleus and the thiophene scaffolds; (**b**) intermolecular packing observed from the plane (100); (**c**) amplified region indicating the π-stacking distances.

3.4.2. Order and Morphology of the Thin Films

In order to correlate the characteristics of the OTFT devices based on derivatives **1a–c** with the analyzed structural variations, the crystallinity and disposition within the thin films was evaluated by means of grazing incidence X-ray diffraction (GIXRD). The GIXRD patterns, illustrated in Figure 5a,b, are consistent with the tendency exhibited by the OTFT devices. The longer the N-alkyl chains, the sharper and more intense the diffraction peaks, which indicates a more prominent degree of order and crystallinity in the thin films. Comparing the nature of the passivation layer, the outperforming OTS-treated devices also provide more crystalline films than their PS analogs in derivatives **1b,c**.

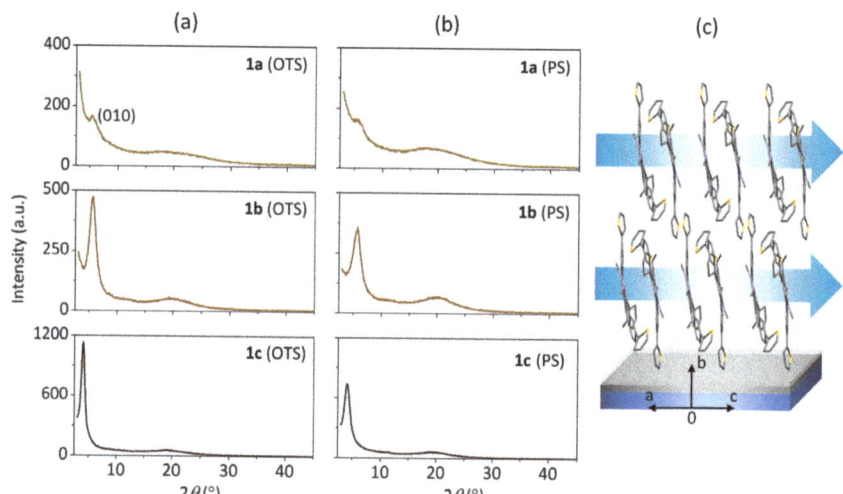

Figure 5. GIXRD patterns of thin films of **1a–c** deposited over Si/SiO$_2$ substrates passivated with OTS (**a**) and PS (**b**), and the proposed arrangement of **1a** over the substrate and the π-stacking direction (**c**), based on the GIXRD patterns and the elucidated single-crystal structure.

The *N*-methylated **1a** provides a rather amorphous arrangement regardless of the passivation layer, agreeing with the lower OTFT performance and the small difference between OTS and PS. Nevertheless, the availability of the single-crystal structure of **1a** permitted a closer insight into the arrangement of this material. From the simulated powder diffractogram, the diffraction signal peaking at $2\theta = 5.52°$ could be assigned as the plane (010), which stands parallel to the substrate. As represented in Figure 5c, the π-system of **1a** adopts a perpendicular disposition with respect to the substrate. Consequently, the π-stacking direction lays parallel to the substrate, ensuring a proper pathway for charge transport. The morphology of the vacuum-deposited thin films of **1a–c** over OTS- and PS-treated substrates was analyzed by means of atomic force microscopy (AFM). The AFM images, depicted in Figure 6, reveal quite homogeneous layers with small protrusions or hillocks.

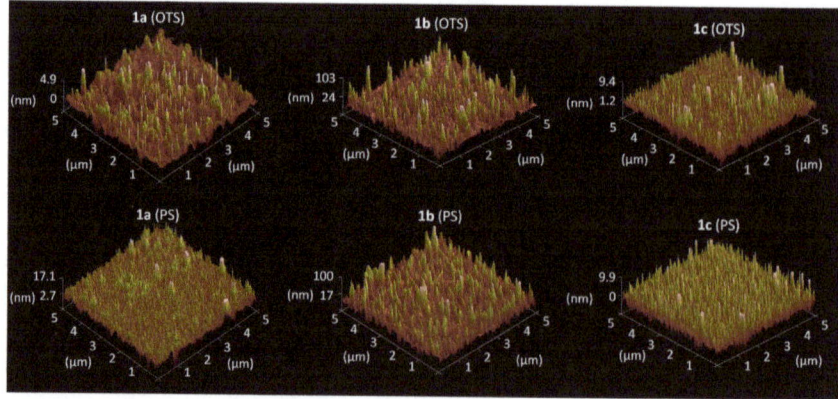

Figure 6. Morphology of the thin films based on compounds **1a–c** deposited over OTS (**above**) and PS (**below**).

The structural differences in derivatives **1a–c** are mainly translated into distinct grain sizes. The methylated derivative **1a** presents a quite uniform surface with undefined domains of small grain sizes from 0.03 to 0.06 μm over PS, coinciding with the notoriously low degree of order detected via GIXRD. The deposition over OTS constitutes a more irregular layer but with slightly larger grains (0.1 μm). On the other hand, derivatives **1b,c** exhibit surface profiles with more defined regions and larger grain sizes from 0.15 to 0.3 μm over both OTS and PS, agreeing with the higher degree of order of the films. In addition, the minor roughness and morphology of **1c** films are again accordant to the superior hole mobilities extracted from the OTFTs.

4. Conclusions

Three thiophene-functionalized triindoles possessing different *N*-alkylation patterning (**1a–c**) were successfully synthesized in a facile four-step route. The incorporation of the thiophene moieties placed the ionization potential of the triindole nucleus at 5.11 eV in the solid state, making it perfectly suitable for hole injection while still preserving stability against oxidation. Particularly, the *N*-decylated derivative **1c** reached a hole mobility up to 1×10^{-3} and 5×10^{-4} cm^2 V^{-1} s^{-1} based on TOF and OTFT measurements, respectively. In fact, the presence of longer *N*-alkyl chains not only endowed the thin films with a higher degree of order and crystallinity, but also granted a more adequate morphology. The deposition of the thin film and its characteristics were also ameliorated by functionalizing the Si/SiO$_2$ surface with OTS in comparison with the PS coating. All these factors directly contributed to promoting the hole mobility in the respective OTFTs. Structurally, the strong π–π interactions and the favorable β packing found in the single crystal correlate with the good OTFT performance. Finally, the potential of these materials in organic electronics is supported by their outstanding air stability, which implies a minimal decrease in the OTFT performance and a shelf lifetime surpassing 100 days.

Supplementary Materials: The following Supporting Information can be downloaded at: https://www.mdpi.com/article/10.3390/coatings13050896/s1, Synthetic procedures and characterization; Figure S1: TGA scans of compounds **1a–c**. The decomposition temperatures (T_d) were estimated from the onset; Figure S2: Absorption and emission spectra of compounds **1a–c** in: (a) dichloromethane (10 μM) and (b) vacuum-evaporated thin films; Figure S3: Electrochemical data of compounds **1a–c**: (a) cyclic voltammogram of **1b** as representative (the inset shows the first oxidation step) and (b) photoemission spectra of compounds **1a–c**; Figure S4: Estimation of the hole mobility of compounds **1a–c** via the TOF technique: (a) TOF transients (the inset shows one of the transient curves in the linear plot) and (b) electric field dependence of the hole mobility; Figure S5: Output characteristics (V_G from 0 to −40 V) of OTS-treated devices based on compounds: (a) **1a**; (b) **1a**; and (c) **1c**; Figure S6: OTFT characteristics of PS-treated devices incorporating derivatives **1a–c**: (a) transfer and saturation (V_{DS} = −40 V) and (b) output characteristics; Figure S7: Transfer and saturation (V_{DS} = −40 V) characteristics of devices fabricated with compound **1b** after 132 days (above) and **1c** after 122 days (below) over: (a) OTS- and (b) PS-treated substrates.

Author Contributions: Conceptualization, R.B., A.C. and D.V. (Dolores Velasco); validation, J.V.G., J.P. and D.V. (Dolores Velasco); formal analysis, R.B., A.C. and D.V. (Dmytro Volyniuk), investigation, R.B. and A.C.; resources, J.V.G., J.P. and D.V. (Dolores Velasco); data curation, R.B., A.C. and D.V. (Dmytro Volyniuk); writing—original draft preparation, R.B. and A.C.; writing—review and editing, R.B., J.V.G., J.P. and D.V. (Dolores Velasco); supervision, J.V.G., J.P. and D.V. (Dolores Velasco); project administration, D.V. (Dolores Velasco); funding acquisition, J.V.G., J.P. and D.V. (Dolores Velasco). All authors have read and agreed to the published version of the manuscript.

Funding: This research was funded by the Ministerio de Economía, Industria y Competitividad, grant number FUNMAT-PGC2018-095477-B-I00, and the Ministerio de Ciencia e Innovación, grant number PID2020-116719RB-C41.

Data Availability Statement: Not applicable.

Acknowledgments: The authors want to thank Mercè Font-Bardía and Josep M. Bassas of the DRX unit of the CCiTUB for their assistance with the crystallographic elucidation and GIXRD measurements, respectively. Furthermore, thank you to Jordi Díaz of the CCiTUB for the assistance in AFM measures. R.B. and A.C. are thankful for the grant FI AGAUR from Generalitat de Catalunya.

Conflicts of Interest: The authors declare no conflict of interest.

References

1. Wang, C.; Dong, H.; Jiang, L.; Huanli, D. Organic semiconductor crystals. *Chem. Soc. Rev.* **2017**, *47*, 422–500. [CrossRef] [PubMed]
2. Wu, F.; Liu, Y.; Zhang, J.; Duan, S.; Ji, D.; Yang, H. Recent advances in high-mobility and high-stretchability organic field-effect transistors: From materials, devices to applications. *Small Methods* **2021**, *5*, 2100676. [CrossRef]
3. Peng, H.; He, X.; Jiang, H. Greater than 10 cm^2 V^{-1} s^{-1}: A breakthrough of organic semiconductors for field-effect transistors. *Infomat* **2021**, *3*, 613–630. [CrossRef]
4. Paterson, A.F.; Singh, S.; Fallon, K.J.; Hodsden, T.; Han, Y.; Schroeder, B.C.; Bronstein, H.; Heeney, M.; McCulloch, I.; Anthopoulos, T.D. Recent progress in high-mobility organic transistors: A reality check. *Adv. Mater.* **2018**, *30*, 1801079. [CrossRef]
5. Scaccabarozzi, A.D.; Basu, A.; Aniés, F.; Liu, J.; Zapata-Arteaga, O.; Warren, R.; Firdaus, Y.; Nugraha, M.I.; Lin, Y.; Campoy-Quiles, M.; et al. Doping approaches for organic semiconductors. *Chem. Rev.* **2021**, *122*, 4420–4492. [CrossRef]
6. Jiang, H.; Zhu, S.; Cui, Z.; Li, Z.; Liang, Y.; Zhu, J.; Hu, P.; Zhang, H.-L.; Hu, W. High-performance five-ring-fused organic semiconductors for field-effect transistors. *Chem. Soc. Rev.* **2022**, *51*, 3071–3122. [CrossRef] [PubMed]
7. Tahir, M.; Din, I.U.; Zeb, M.; Aziz, F.; Wahab, F.; Gul, Z.; Alamgeer; Sarker, M.R.; Ali, S.; Ali, S.H.M.; et al. Thin Films Characterization and Study of N749-Black Dye for Photovoltaic Applications. *Coatings* **2022**, *12*, 1163. [CrossRef]
8. Bronstein, H.; Nielsen, C.B.; Schroeder, B.C.; McCulloch, I. The role of chemical design in the performance of organic semiconductors. *Nat. Rev. Chem.* **2020**, *4*, 66–77. [CrossRef] [PubMed]
9. Ueberricke, L.; Mastalerz, M. Triptycene end-capping as strategy in materials chemistry to control crystal packing and increase solubility. *Chem. Rec.* **2021**, *21*, 558–573. [CrossRef] [PubMed]
10. Danac, R.; Leontie, L.; Carlescu, A.; Shova, S.; Tiron, V.; Rusu, G.G.; Iacomi, F.; Gurlui, S.; Șușu, O.; Rusu, G.I. Electric conduction mechanism of some heterocyclic compounds, 4,4′-bipyridine and indolizine derivatives in thin films. *Thin Solid Film.* **2016**, *612*, 216–222. [CrossRef]
11. Zhao, X.; Chaudhry, S.T.; Mei, J. Chapter Five—Heterocyclic Building Blocks for Organic Semiconductors. In *Heterocyclic Chemistry in the 21st Century: A Tribute to Alan Katritzky*; Advances in Heterocyclic, Chemistry; Scriven, E.F.V., Ramsden, C.A., Eds.; Academic Press: Cambridge, MA, USA, 2017; Volume 121, pp. 133–171. [CrossRef]
12. Takimiya, K.; Osaka, I.; Nakano, M. π-Building Blocks for Organic Electronics: Revaluation of "Inductive" and "Resonance" Effects of π-Electron Deficient Units. *Chem. Mater.* **2014**, *26*, 587–593. [CrossRef]
13. Salman, S.; Sallenave, X.; Bucinskas, A.; Volyniuk, D.; Bezvikonnyi, O.; Andruleviciene, V.; Grazulevicius, J.V.; Sini, G. Effect of methoxy-substitutions on the hole transport properties of carbazole-based compounds: Pros and cons. *J. Mater. Chem. C* **2021**, *9*, 9941–9951. [CrossRef]
14. Vegiraju, S.; Luo, X.-L.; Li, L.-H.; Afraj, S.N.; Lee, C.; Zheng, D.; Hsieh, H.-C.; Lin, C.-C.; Hong, S.-H.; Tsai, H.-C.; et al. Solution processable pseudo n-thienoacenes via intramolecular S···S lock for high performance organic field effect transistors. *Chem. Mater.* **2020**, *32*, 1422–1429. [CrossRef]
15. Kim, Y.; Wang, B.; Suo, J.; Jatautiene, E.; Simokaitiene, J.; Durgaryan, R.; Volyniuk, D.; Hagfeldt, A.; Sini, G.; Grazulevicius, J.V. Additives-free indolo[3,2-b]carbazolebased hole-transporting materials for perovskite solar cells with three yeses: Stability, efficiency, simplicity. *Nano Energy* **2022**, *101*, 107618. [CrossRef]
16. Lei, T.; Wang, J.-Y.; Pei, J. Roles of Flexible Chains in Organic Semiconducting Materials. *Chem. Mater.* **2014**, *26*, 594–603. [CrossRef]
17. Mišicák, R.; Novota, M.; Weis, M.; Cigáň, M.; Šiffalovič, P.; Nádaždy, P.; Kožíšek, J.; Kožíšková, J.; Pavúk, M.; Putala, M. Effect of alkyl side chains on properties and organic transistor performance of 2,6-bis(2,2′-bithiophen-5-yl)naphthalene. *Synth. Met.* **2017**, *233*, 1–14. [CrossRef]
18. Reig, M.; Bagdziunas, G.; Ramanavicius, A.; Puigdollers, J.; Velasco, D. Interface engineering and solid-state organization for triindole-based p-type organic thin-film transistors. *Phys. Chem. Chem. Phys.* **2018**, *20*, 17889–17898. [CrossRef]
19. Shaik, B.; Park, J.H.; An, T.K.; Noh, Y.R.; Yoon, S.B.; Park, C.E.; Yoon, Y.J.; Kim, Y.-H.; Lee, S.-G. Small asymmetric anthracene-thiophene compounds as organic thin-film transistors. *Tetrahedron* **2013**, *69*, 8191–8198. [CrossRef]
20. Shaik, B.; Han, J.-H.; Song, D.J.; Kang, H.-M.; Kim, Y.B.; Park, C.E.; Lee, S.-G. Synthesis of donor–acceptor copolymer using benzoselenadiazole as acceptor for OTFT. *RSC Adv.* **2016**, *6*, 4070–4076. [CrossRef]
21. Ong, B.S.; Wu, Y.W.; Li, Y. Organic Semiconductors Based on Polythiophene and Indolo[3,2-b]carbazole. In *Organic Electronics: Materials, Manufacturing and Applications*; Klauk, H., Ed.; WILEY-VCH: Weinheim, Germany, 2006; pp. 75–107. [CrossRef]
22. Wex, B.; Kaafarani, B.R. Perspective on carbazole-based organic compounds as emitters and hosts in TADF applications. *J. Mater. Chem. C* **2017**, *5*, 8622–8653. [CrossRef]
23. Grybauskaite-Kaminskiene, G.; Volyniuk, D.; Mimaite, V.; Bezvikonnyi, O.; Bucinskas, A.; Bagdziunas, G.; Grazulevicius, J.V. Aggregation-enhanced emission and thermally activated delayed fluorescence of derivatives of 9-phenyl-9H-carbazole: Effects of methoxy and tert- butyl substituents. *Chem. Eur. J.* **2018**, *24*, 9581–9591. [CrossRef]
24. Górski, K.; Mech-Piskorz, J.; Pietraszkiewicz, M. From truxenes to heterotruxenes: Playing with heteroatoms and the symmetry of molecules. *New J. Chem.* **2022**, *46*, 8939–8966. [CrossRef]
25. Li, X.-C.; Wang, C.-Y.; Lai, W.-Y.; Huangab, W. Triazatruxene-based materials for organic electronics and optoelectronics. *J. Mater. Chem. C* **2016**, *4*, 10574–10587. [CrossRef]

26. Qian, X.; Zhu, Y.-Z.; Song, J.; Gao, X.-P.; Zheng, J.-Y. New donor-π-acceptor type triazatruxene derivatives for highly efficient dye-sensitized solar cells. *Org. Lett.* **2013**, *15*, 6034–6037. [CrossRef] [PubMed]
27. Bulut, I.; Chávez, P.; Mirloup, A.; Huaulmé, Q.; Hébraud, A.; Heinrich, B.; Fall, S.; Méry, S.; Ziessel, R.; Heiser, T.; et al. Thiazole-based scaffolding for high performance solar cells. *J. Mater. Chem. C* **2016**, *4*, 4296–4303. [CrossRef]
28. Bura, T.; Leclerc, N.; Bechara, R.; Lévêque, P.; Heiser, T.; Ziessel, R. Triazatruxene-Diketopyrrolopyrrole Dumbbell-Shaped Molecules as Photoactive Electron Donor for High-Efficiency Solution Processed Organic Solar Cells. *Adv. Energy Mater.* **2013**, *3*, 1118–1124. [CrossRef]
29. Lai, W.-Y.; He, Q.-Y.; Zhu, R.; Chen, Q.-Q.; Huang, W. Kinked star-shaped fluorene/triazatruxene co-oligomer hybrids with enhanced functional properties for high performance, solution-processed, blue organic light-emitting diodes. *Adv. Funct. Mater.* **2008**, *18*, 265–276. [CrossRef]
30. Chen, Y.; Wang, S.; Wu, X.; Xu, Y.; Li, H.; Liu, Y.; Tong, H.; Wang, L. Triazatruxene-based small molecules with thermally activated delayed fluorescence, aggregation-induced emission and mechanochromic luminescence properties for solution-processable nondoped OLEDs. *J. Mater. Chem. C* **2018**, *6*, 12503–12508. [CrossRef]
31. Hu, Y.-C.; Lin, Z.-L.; Huang, T.-C.; Lee, J.-W.; Wei, W.-C.; Ko, T.-Y.; Lo, C.-Y.; Chen, D.-G.; Chou, P.-T.; Hung, W.-Y.; et al. New exciplex systems composed of triazatruxene donors and N-heteroarene-cored acceptors. *Mater. Chem. Front.* **2020**, *4*, 2029–2039. [CrossRef]
32. Ruiz, C.; Arrechea-Marcos, I.; Benito-Hernández, A.; Gutierrez-Puebla, E.; Monge, M.A.; Navarrete, J.L.; Ruiz Delgado, M.C.; Ponce Ortiz, R.; Gómez-Lor, B. Solution-processed N-trialkylated triindoles for organic field effect transistors. *J. Mater. Chem. C* **2018**, *6*, 50–56. [CrossRef]
33. Cuadrado, A.; Cuesta, J.; Puigdollers, J.; Velasco, D. Air stable organic semiconductors based on diindolo[3,2-a:3′,2′-c]carbazole. *Org. Electron.* **2018**, *62*, 35–42. [CrossRef]
34. Reig, M.; Puigdollers, J.; Velasco, D. Molecular order of air-stable p-type organic thin-film transistors by tuning the extension of the π-conjugated core: The cases of indolo[3,2-b]carbazole and triindole semiconductors. *J. Mater. Chem. C* **2015**, *3*, 506–513. [CrossRef]
35. Takimiya, K.; Shinamura, S.; Osaka, I.; Miyazaki, E. Thienoacene-Based Organic Semiconductors. *Adv. Mater.* **2011**, *23*, 4347–4370. [CrossRef] [PubMed]
36. Ryu, S.; Yun, C.; Ryu, S.; Ahn, J.; Kim, C.; Seo, S. Characterization of [1]Benzothieno[3,2-b]benzothiophene (BTBT) Derivatives with End-Capping Groups as Solution-Processable Organic Semiconductors for Organic Field-Effect Transistors. *Coatings* **2023**, *13*, 181. [CrossRef]
37. Bujaldón, R.; Puigdollers, J.; Velasco, D. Towards the Bisbenzothienocarbazole Core: A Route of Sulfurated Carbazole Derivatives with Assorted Optoelectronic Properties and Applications. *Materials* **2021**, *14*, 3487. [CrossRef] [PubMed]
38. Borchert, J.W.; Peng, B.; Letzkus, F.; Burghartz, J.N.; Chan, P.K.L.; Zojer, K.; Ludwigs, S.; Klauk, H. Small contact resistance and high-frequency operation of flexible low-voltage inverted coplanar organic transistors. *Nat. Commun.* **2019**, *10*, 1119. [CrossRef]
39. Choi, E.; Jang, Y.; Ho, D.; Chae, W.; Earmme, T.; Kim, C.; Seo, S. Development of Dithieno[3,2-b:2′,3′-d]thiophene (DTT) Derivatives as Solution-Processable Small Molecular Semiconductors for Organic Thin Film Transistors. *Coatings* **2021**, *11*, 1222. [CrossRef]
40. Sugiyama, M.; Jancke, J.; Uemura, T.; Kondo, M.; Inoue, Y.; Namba, N.; Araki, T.; Fukushima, T.; Sekitani, T. Mobility enhancement of DNTT and BTBT derivative organic thin-film transistors by triptycene molecule modification. *Org. Electron.* **2021**, *96*, 106219. [CrossRef]
41. Singh, M.; Kaur, N.; Comini, E. The role of self-assembled monolayers in electronic devices. *J. Mater. Chem. C* **2020**, *8*, 3938–3955. [CrossRef]
42. Hasan, M.M.; Islam, M.M.; Li, X.; He, M.; Manley, R.; Chang, J.; Zhelev, N.; Mehrotra, K.; Jang, J. Interface Engineering with Polystyrene for High-Performance, Low-Voltage Driven Organic Thin Film Transistor. *IEEE Trans. Electron Devices* **2020**, *67*, 1751–1756. [CrossRef]
43. Kim, D.; Kim, C.A. Ladder Type Organosilicate Copolymer Gate Dielectric Materials for Organic Thin-Film Transistors. *Coatings* **2018**, *8*, 236. [CrossRef]
44. Feriancová, L.; Kmentová, I.; Micjan, M.; Pavúk, M.; Weis, M.; Putala, M. Synthesis and Effect of the Structure of Bithienyl-Terminated Surfactants for Dielectric Layer Modification in Organic Transistor. *Materials* **2021**, *14*, 6345. [CrossRef] [PubMed]
45. Sun, H.; Liao, J.; Hou, S. Single-Molecule Field-Effect Transistors with Graphene Electrodes and Covalent Pyrazine Linkers. *Acta Phys. Chim. Sin.* **2021**, *37*, 1906027. [CrossRef]
46. Demas, J.N.; Crosby, G.A. Measurement of photoluminescence quantum yields. Review. *J. Phys. Chem.* **1971**, *75*, 991–1024. [CrossRef]
47. Cardona, C.M.; Li, W.; Kaifer, A.E.; Stockdale, D.; Bazan, G.C. Electrochemical considerations for determining absolute frontier orbital energy levels of conjugated polymers for solar cell applications. *Adv. Mater.* **2011**, *23*, 2367–2371. [CrossRef]
48. Ghalgaoui, A.; Shimizu, R.; Hosseinpour, S.; Álvarez-Asencio, R.; McKee, C.; Johnson, C.M.; Rutland, M.W. Monolayer Study by VSFS: In Situ Response to Compression and Shear in a Contact. *Langmuir* **2014**, *30*, 3075–3085. [CrossRef]
49. Song, D.; Wang, H.; Zhu, F.; Yang, J.; Tian, H.; Geng, Y.; Yan, D. Phthalocyanato Tin(IV) Dichloride: An Air-Stable, High-Performance, n-Type Organic Semiconductor with a High Field-Effect Electron Mobility. *Adv. Mater.* **2008**, *20*, 2142–2144. [CrossRef]

50. Miyaura, N.; Suzuki, A. Palladium-Catalyzed Cross-Coupling Reactions of Organoboron Compounds. *Chem. Rev.* **1995**, *95*, 2457–2483. [CrossRef]
51. McCulloch, I.; Salleo, A.; Chabinyc, M. Avoid the kinks when measuring mobility. *Science* **2016**, *352*, 1521–1522. [CrossRef] [PubMed]
52. Choi, H.H.; Cho, K.; Frisbie, C.D.; Sirringhaus, H.; Podzorov, V. Critical assessment of charge mobility extraction in FETs. *Nat. Mater.* **2018**, *17*, 2–7. [CrossRef] [PubMed]
53. Desiraju, G.R.; Gavezzotti, A. Crystal structures of polynuclear aromatic hydrocarbons. Classification, rationalization and prediction from molecular structure. *Acta Cryst.* **1989**, *45*, 473–482. [CrossRef]
54. Campbell, J.E.; Yang, J.; Day, G.M. Predicted energy–structure–function maps for the evaluation of small molecule organic semiconductors. *J. Mater. Chem. C* **2017**, *5*, 7574–7584. [CrossRef]
55. Mas-Torrent, M.; Rovira, C. Role of Molecular Order and Solid-State Structure in Organic Field-Effect Transistors. *Chem. Rev.* **2011**, *111*, 4833–4856. [CrossRef] [PubMed]

Disclaimer/Publisher's Note: The statements, opinions and data contained in all publications are solely those of the individual author(s) and contributor(s) and not of MDPI and/or the editor(s). MDPI and/or the editor(s) disclaim responsibility for any injury to people or property resulting from any ideas, methods, instructions or products referred to in the content.

Article

Mechanical Properties and Creep Behavior of Undoped and Mg-Doped GaN Thin Films Grown by Metal–Organic Chemical Vapor Deposition

Ali Khalfallah [1,2,3,*] and Zohra Benzarti [1,4,5]

[1] CEMMPRE, Department of Mechanical Engineering, University of Coimbra, Rua Luís Reis Santos, 3030-788 Coimbra, Portugal; zohra.benzarti@ua.pt
[2] Laboratoire de Genie Mécanique, Ecole Nationale d'Ingénieurs de Monastir, Université de Monastir, Av. Ibn El-Jazzar, Monastir 5019, Tunisia
[3] DGM, Institut Supérieur des Sciences Appliquées et de Technologie de Sousse, Université de Sousse, Cité Ibn Khaldoun, Sousse 4003, Tunisia
[4] Laboratory of Multifunctional Materials and Applications (LaMMA), Department of Physics, Faculty of Sciences of Sfax, University of Sfax, Soukra Road km 3.5, B.P. 1171, Sfax 3000, Tunisia
[5] Department of Electronics, Telecommunications and Informatics (DETI), University of Aveiro, Campus Universitário de Santiago, 3810-193 Aveiro, Portugal
* Correspondence: ali.khalfallah@dem.uc.pt; Tel.: +351-931-491-650

Abstract: This paper investigates the mechanical properties and creep behavior of undoped and Mg-doped GaN thin films grown on sapphire substrates using metal–organic chemical vapor deposition (MOCVD) with trimethylgallium (TMG) and bis(cyclopentadienyl)magnesium (Cp2Mg) as the precursors for Ga and Mg, respectively. The Mg-doped GaN layer, with a [Mg]/[TMG] ratio of 0.33, was systematically analyzed to compare its mechanical properties and creep behavior to those of the undoped GaN thin film, marking the first investigation into the creep behavior of both GaN and Mg-doped GaN thin films. The results show that the incorporated [Mg]/[TMG] ratio was sufficient for the transition from n-type to p-type conductivity with higher hole concentration around 4.6×10^{17} cm^{-3}. Additionally, it was observed that Mg doping impacted the hardness and Young's modulus, leading to an approximately 20% increase in these mechanical properties. The creep exponent is also affected due to the introduction of Mg atoms. This, in turn, contributes to an increase in pre-existing dislocation density from 2×10^8 cm^{-2} for undoped GaN to 5×10^9 cm^{-2} for the Mg-doped GaN layer. The assessment of the creep behavior of GaN and Mg-doped GaN thin films reveals an inherent creep mechanism governed by dislocation glides and climbs, highlighting the significance of Mg doping concentration in GaN thin films and its potential impact on various technological applications.

Keywords: MOCVD; Mg-doped GaN layers; point defects; nanoindentation; creep behavior

Citation: Khalfallah, A.; Benzarti, Z. Mechanical Properties and Creep Behavior of Undoped and Mg-Doped GaN Thin Films Grown by Metal–Organic Chemical Vapor Deposition. *Coatings* **2023**, *13*, 1111. https://doi.org/10.3390/coatings13061111

Academic Editor: Andrey V. Osipov

Received: 18 May 2023
Revised: 12 June 2023
Accepted: 14 June 2023
Published: 16 June 2023

Copyright: © 2023 by the authors. Licensee MDPI, Basel, Switzerland. This article is an open access article distributed under the terms and conditions of the Creative Commons Attribution (CC BY) license (https://creativecommons.org/licenses/by/4.0/).

1. Introduction

Group III-nitrides, such as GaN, are wide-bandgap semiconductor materials (3.4 eV) that gained significant attention in optoelectronic devices due to their superior optical properties. The Mg-doped GaN is particularly important as it can enable reproducible p-type conductivity, which is essential for the development of high-efficiency light emitting diode LEDs and laser diodes [1–3]. Successful preparation of p-type Mg-doped GaN layers is one of the vital challenges for realizing blue, green LEDs, and high-power devices [4,5]. The electrical and optical characteristics of GaN and Mg-doped GaN were extensively explored to better understand their behavior and optimize their quality [6–8]. In addition to probing the latter features, the mechanical properties of GaN layers are also important for functional components in optoelectronic devices. Therefore, investigating the impact of Mg doping on the mechanical properties of GaN layers is crucial. Specifically, when GaN layers are subjected to high gradients of temperature and electric and/or stress fields, their performance may deteriorate and they may experience changes in their electric, optical properties, and structural damage. These changes potentially lead to device failure. The analysis of the mechanical behavior of GaN and Mg-doped GaN layers contributes to optimizing film quality and realizing the full potential of these devices.

However, the available literature indicates that there is limited research addressing the determination of the mechanical properties of Young's modulus and hardness of GaN thin films [9], and more specifically, Mg-doped GaN thin films [10]. Furthermore, there is a paucity of literature on plasticity-dependent time tests such as creep nanoindentation for GaN thin films [11]. Despite the significance of investigating their creep behavior, which is critical for designing and optimizing sensors, actuators, and RF devices, there are limited studies on it. This lack impedes our understanding of the creep behavior and its impact on the long-term stability and reliability of power electronics, MEMS, and optoelectronics devices [12–14]. Therefore, the justification for conducting further investigation into the creep behavior of material thin films, especially for semiconductors, is evident.

This study attempts to fill this gap in the literature by investigating the creep behavior of undoped GaN and Mg-doped GaN thin films. The study demonstrates that Mg doping in GaN layers leads, on one hand, to a transition from n-type (GaN layer) to p-type conductivity (Mg-doped GaN layer), resulting in high hole concentration favorable for high blue luminescence, and on the other hand, to an increase in defects and the disappearance of pop-ins observed in nanoindentation curves of Mg-doped GaN layers. Indeed, the presence of controlled defects in GaN layers can be advantageous in achieving a balance between optimal physical properties and mechanical characteristics for the development of functional and high-strength semiconductor materials. These materials showed potential for use in optoelectronic devices, as was demonstrated for Al-doped GaN films [15]. In addition, it is shown that the creep behavior of the GaN layer was influenced by the introduction of defects, leading to a decrease in creep stress exponent and creep depth, which suggests a creep mechanism governed by dislocation glides and climbs. These findings can provide valuable insights into the design and optimization for a wide range of GaN-based technological applications.

2. Materials and Methods

Mg-doped GaN layers were grown on c-(0001) plane sapphire substrates using a vertical MOCVD reactor. TMG and ammonia (NH_3) were used as the precursors for Ga and N, respectively, while Cp_2Mg was used as the Mg source. The flow rate of Cp_2Mg was set to 20 µmol/min, while the flow rate of TMG was maintained at 7 µmol/min, resulting in a [Mg]/[TMG] ratio of 0.33. Prior to the growth of GaN layers, a 30 nm-thick low-temperature buffer layer was grown at 600 °C for 10 min for both undoped and Mg-doped GaN layers to enhance the quality of the substrate–layer interface and thus to reduce the mismatch. Subsequently, the temperature was increased from 600 °C to 1100 °C to initiate the deposition of the undoped GaN or the Mg-doped GaN layers. Both samples were grown under a mixture of N_2 and H_2 carrier gas. The carrier gas flux of N_2 and H_2 was 3.5 L/min and 0.4 L/min, respectively. Mg acceptors were activated by annealing the Mg-doped GaN layer for 30 min at 900 °C in the presence of N_2 gas. The thickness of each layer, determined through cross-sectional SEM imaging, was found to be approximately 2 µm.

Physical properties and nanomechanical characteristics of both undoped GaN and Mg-doped GaN layers were analyzed by various techniques, including high-resolution X-ray diffraction (HRXRD), Raman spectroscopy, scanning electron microscopy (SEM), atomic force microscopy (AFM), and nanoindentation technique. Mg concentration was measured using a secondary ion mass spectrometry (SIMS) technique by means of EVANS EAST using an oxygen ion bombardment with an impact energy of 5.5 KeV. HRXRD was conducted on a Bruker D8 Advance diffractometer using CuKα radiation ($\lambda = 0.154$ nm). It is used to investigate the crystalline quality and structural properties of layers, where rocking curves of symmetric and asymmetric reflections were utilized to determine the dislocation densities of edge and screw varieties. Raman spectroscopy was carried out using a micro-Raman spectrometer LABRAMHRT 4600 HR 800 with a laser excitation line of 518 nm in the frequency range of 100–1000 cm^{-1}. Measurements were performed at room temperature to assess the residual stress within both samples. Zeiss Merlin Gemini II equipment high-resolution field emission scanning electron microscopy (FE-SEM) and Nanoman AFM controlled by a Nanoscope V electronics from Bruker Instruments (Bruker, Billerica, MA, USA) were used to investigate the surface morphology and roughness of the studied layers. Both samples were subjected to van der Pauw (Hall) measurement to determine their carrier concentration and resistivity. The NanoTest NT1, NanoMaterials, Ltd. instrument with a diamond Berkovich-type indenter was used to conduct nanoindentation tests. The maximum load reached was 10 mN. To obtain an average and assess the uncertainty of the measured properties, at least twenty nanoindentation tests were performed for each experimental point. The indentation cycle consisted of a loading part of 0.5 mN/s, followed by a dwell period of 30 s at the maximum load, and then an unloading phase of 0.5 mN/s until the load reached 10% of the maximum. A second hold period of 30 s was conducted to correct depth drift. In the study, the thermal drift was maintained below 0.03 nm/s for all nanoindentations. Additionally, during the first hold stage at the maximum load of 10 mN, creep behavior for both samples was examined for 30 s. To prevent possible interaction that could affect the measurement results, the indentation marks were separated by 30 μm. SEM was utilized to investigate the indentation imprints for both undoped GaN and Mg-doped GaN layers.

3. Analytical Background

The technique of nanoindentation is widely used for determining the mechanical properties of nanomaterials and thin films due to its high sensitivity and precision in measuring hardness, elastic modulus, and elastic/plastic deformation characteristics. The process involves pressing a diamond indenter tip into the sample until it reaches a specified load or depth and then recording the load as a function of indenter displacement during both loading and unloading stages. The Oliver and Pharr method [16] is typically used to analyze the nanoindentation test results and determine the hardness and Young's modulus. The analysis of the test results is performed based on this method as follows:

The hardness is defined as the ratio between the maximum load P_{max} and the projected area of the indentation imprint A_c.

$$H = \frac{P_{max}}{A_c} \quad (1)$$

The ideal Berkovich indenter is only a theoretical concept that is not used in practice and results in an underestimation of the actual contact area. In reality, the imperfections in the indenter tip are considered when calculating the projected contact area using the Oliver and Pharr empirical relationship. Further details could be found in [16].

The reduced Young's modulus E_r of the thin film is expressed as follows:

$$\frac{1}{E_r} = \frac{1-v^2}{E} + \frac{1-v_i^2}{E_i} \quad (2)$$

where v and v_i are the Poisson coefficients of the thin film and the indenter, respectively. E and E_i are the Young's moduli of the thin film and the indenter, respectively. For the diamond indenter, Young's modulus and Poisson coefficient are E_i = 1141 GPa and v_i = 0.07, respectively [16].

Nanoindentation is also a useful tool to study the creep behavior of a wide range of materials at the nanoscale; it can provide insights into their underlying deformation mechanism during the creep. To perform a nanoindentation creep test, the indenter is pressed into the material surface with a constant load during a period of time. The indentation depth is measured as a function of time, allowing for the determination of creep displacement curves. Moreover, the creep rate along with the stress exponent can also be derived. It is widely accepted that uniaxial tensile tests are commonly employed to characterize the creep behavior of materials. During the steady-state creep stage, a commonly used empirical law that is applicable to a wide range of materials is as follows:

$$\dot{\varepsilon} = k\sigma^n \tag{3}$$

where $\dot{\varepsilon}$ is the strain rate, σ is the applied stress, n is the creep stress exponent, and k is a fitting material constant. The slope of the log $\dot{\varepsilon}$-log σ curve is identified as the creep stress exponent. This parameter is very useful for giving insight into the creep mechanism of the material.

$$\dot{\varepsilon} \approx \frac{1}{h}\left(\frac{dh}{dt}\right) \text{ and } \sigma \approx \frac{F}{h^2} \tag{4}$$

The computation of the displacement rate uses the derivation with respect to time of the following empirical equation given as follows [17]:

$$h(t) = h_i + a(t - t_i)^m + bt \tag{5}$$

where h_i and t_i are the initial displacement and instant of the creep; a and b are material constants of fitting.

4. Results and Discussion

4.1. Morphological, Electrical, Structural and, Vibrational Exams

Figure 1 illustrates the surface morphology of undoped and Mg-doped GaN layers. The root mean square (RMS) value of the surface roughness increases from 0.5 nm for GaN sample to 5 nm for the Mg-doped GaN sample. Previous studies reported that highly Mg-doped GaN films grown by the MOCVD process exhibit increased surface roughness and decreased free carrier concentration. This is attributed to the accumulation of Mg beyond a critical flow rate [18]. However, it is important to note that a balance between incorporating enough Mg to create p-type material and maintaining an acceptable surface morphology when manufacturing light-emitting diodes is of crucial interest [2]. The room temperature Hall measurements are used to determine the electrical properties of both samples. It was found that the hole concentration in the p-type Mg-doped GaN layer was around 4.6×10^{17} cm^{-3}. The obtained result shows a slight improvement over the value reported in the literature [19]. Additionally, the electron concentration in the n-type undoped GaN layer is equal to 4×10^{17} cm^{-3}. The following values of the mobility, 280 cm^2/V.s, and 9 cm^2/V.s are determined for undoped GaN and the Mg-doped GaN layer, respectively. The measured resistivity highly increases from 0.1 Ω·cm for undoped GaN layer to 2.4 Ω·cm for the Mg-doped GaN layer.

The threading dislocations for each sample were computed based on X-ray ω-scans rocking curve on the (0002) and (10$\bar{1}$2) diffraction peaks, which are shown in Figure 2. From these curves, the full width at half maximum of the reflections (FWHM) was assessed and used to determine the screw and edge dislocations. The threading dislocations were found equal to 2×10^8 cm^{-2} and 5×10^9 cm^{-2} for undoped GaN and Mg doped GaN layers, respectively. The observed increase of the threading dislocation density in Mg doped GaN layer is attributed to the presence of defects introduced during the incorporation of Mg atoms. Moreover, the use of N$_2$ in the mixture of the carrier gas during the growth process of Mg doped GaN layer contributes to the increase of the threading dislocation density [20]. It has been earlier reported by Svensk et al. [21] that the introduction of Mg atoms contributes to the incorporation of point defects as long as N$_2$ carrier gas is used and then the crystalline quality is spoiled. Moreover, Smorchkova et al. [22] reported that high incorporation of Mg atoms in the GaN layer leads to the formation of point defects and defect complexes, resulting in self-compensation, which is responsible for the observed low mobility and high resistivity in the p-type GaN layers, as similarly shown in this study. Kaufmann et al. [23] suggested that the nitrogen vacancy (V$_N$) is a native defect with a significant concentration in p-type GaN layers grown by MOCVD. It was found that the concentration of V_N is practically the same as that of the incorporated Mg, which is introduced through a substitutional mechanism with Ga atoms (Mg_{Ga}). As a result, the Mg_{Ga}-V_N defect complex is formed through a self-compensation mechanism. Consequently, the observed decrease of electron mobility and the increase of resistivity of Mg doped GaN compared to undoped GaN layer corroborates the fact of introducing defects along with Mg atoms' incorporation.

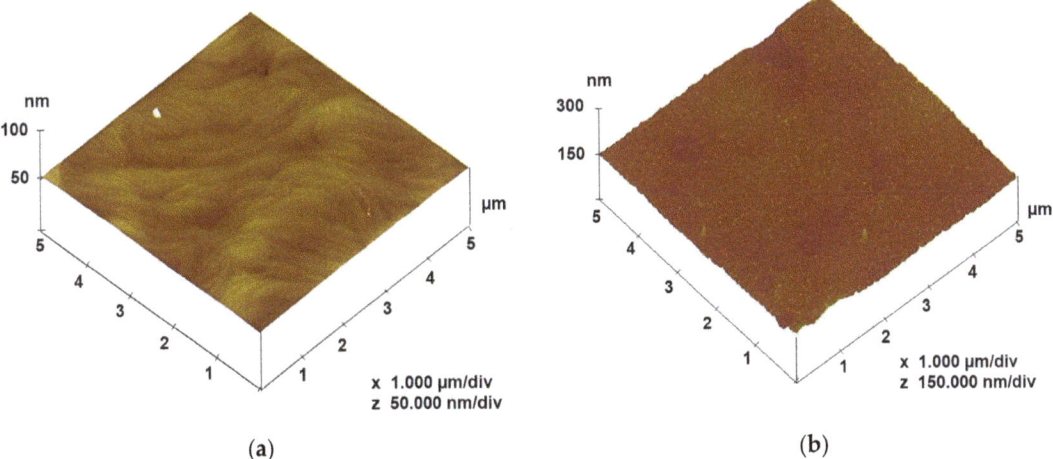

Figure 1. AFM images displaying the surface morphology of the grown layers: (**a**) undoped GaN; (**b**) Mg-doped GaN layer.

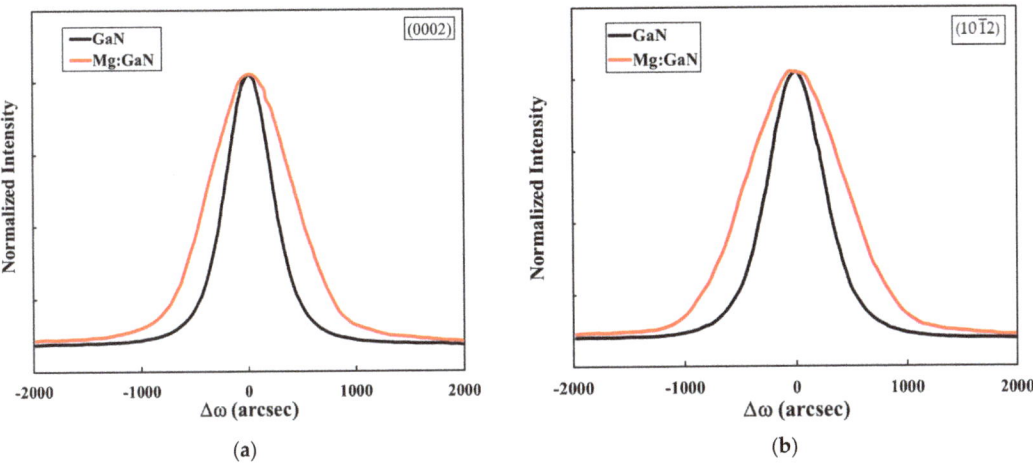

Figure 2. X–ray ω scans (rocking curve) of: (**a**) (0002) reflection; (**b**) ($10\bar{1}2$) reflection of undoped GaN and Mg-doped GaN layers.

Figure 3 shows the SIMS depth profiles for the Mg-doped GaN layer. It reveals that the concentration of Mg atoms is about 2×10^{19} atom/cm^3. In addition, the Mg concentration was found homogeneously distributed across the thickness of the sample.

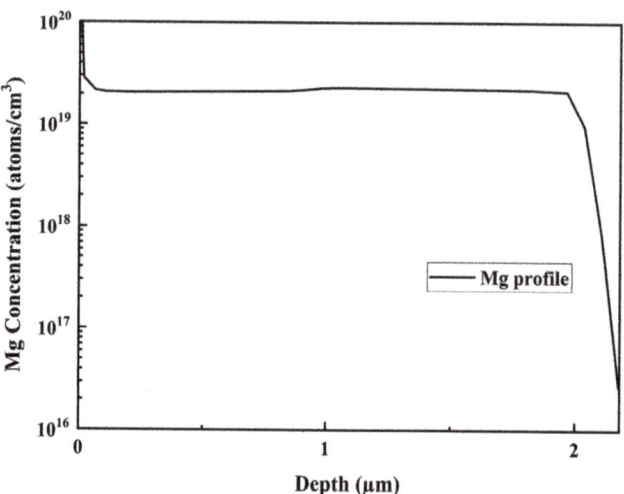

Figure 3. SIMS depth profiles of Mg atoms for the Mg-doped GaN layer.

Figure 4 exhibits the Raman spectra of undoped GaN and Mg-doped GaN layers. The analysis of these spectra allows for the determination the stress states in both samples using the intense E_2 (high) phonon frequency mode. This enables a direct comparison of the stress states relative to a reference peak [20]. The strain-free frequency of the E_2 (high) phonon mode for the GaN bulk layer is located at 566.34 cm^{-1} [21] and it is represented by the dashed line. Both layers are under compressive stresses as they are located on the right side of the bulk GaN position. The spectrum of the Mg-doped GaN layer showed weakening and broadening of the E_2 (high) peak at 568.4 cm^{-1}, indicating lower crystalline quality compared to the E_2 (high) peak at 566.8 cm^{-1} for the undoped GaN layer. Moreover, the introduction of Mg atoms induced higher compressive stress in the Mg-doped GaN layer compared to undoped GaN, leading to a shift of the E_2 (high) peak towards higher wave numbers. To quantitatively assess the residual stresses inherent in both layers, the E_2 (high) peaks are used to calculate these stresses using the method as reported in [22]. It was found that $\sigma_{\text{undoped GaN}} = -0.10$ GPa and $\sigma_{Mg \text{ doped GaN}} = -0.47$ GPa. The minus sign (−) indicates that the residual stresses are of a compressive nature.

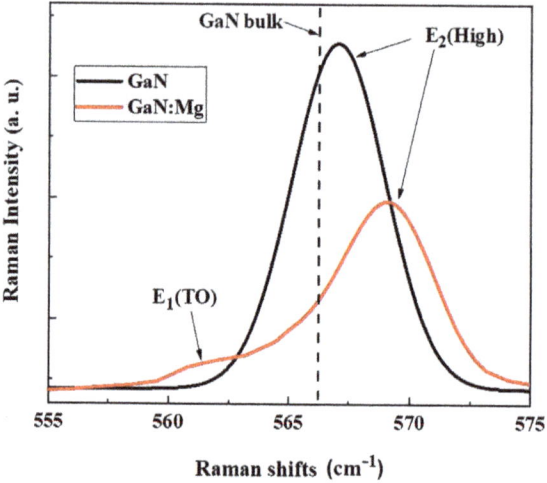

Figure 4. Raman spectra of undoped and Mg-doped GaN layers measured at room temperature.

This behavior was explained by the difference in size between the dopant atom radius (r_{Mg} = 0.136 nm) and the replaced host atom radius (r_{Ga} = 0.126 nm). Additionally, it was observed that the incorporated Mg atoms contributed to disclose the E_1 (TO) phonon mode at 561 cm^{-1}, indicating disorder in the Mg-doped GaN layer accredited to the presence of high concentrations of defects [22], when compared to the undoped GaN layer. Subsequently, the analysis of Raman spectra provides valuable insight into the effects of Mg doping on the lattice vibrational properties of GaN layers, including the emergence of new phonon modes.

4.2. Nanoindentation Tests Analysis

Figure 5 shows the load–unload nanoindentation curves obtained for both undoped and Mg-doped GaN layers carried out at the maximal load P_{max} = 10 mN. It clearly reveals the different features between both grown layers. The load–unload curve for the undoped GaN layer depicts the first pop-in events (sudden burst during the loading curve) corresponding to the critical load P_c around 4 mN and the length of the plateau is about 20 nm, for which the incipient plasticity of the GaN layer was initiated under the effect of yielding shear stress τ_{max}. The dissipated energy for the formation of the pop-ins is approximately 8×10^{-12} J. However, for the Mg-doped GaN layer, the elastic–plastic transition occurred without the appearance of the pop-in events.

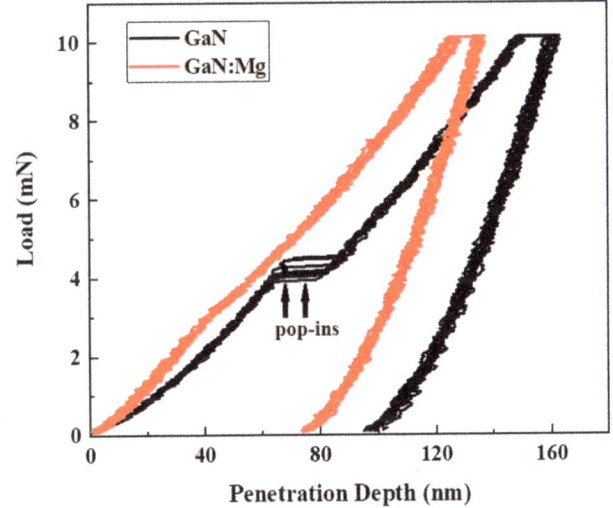

Figure 5. Load–unload nanoindentation curves recorded for undoped GaN and Mg-doped GaN layers. The indentation curve for the undoped GaN layer clearly discloses pop-in events that indicate the onset of the elastic–plastic transition. For the Mg-doped GaN curve, the elastic–plastic transition occurred without the appearance of pop-in events.

The yielding shear stress τ_{max}, for which the plastic deformation initiated locally underneath the indenter tip, is calculated using Johnson's relationship [23] and is expressed as follows:

$$\tau_{max} = 0.31 \left(\frac{6 P_c E^2}{\pi^3 R^2} \right)^{1/3} \qquad (6)$$

where P_c, E, and R are the critical load, the Young's modulus and the indenter tip radius, respectively. The critical load corresponds to the load that initiates the elastic–plastic transition. For the load–unload curves that present pop-in events, the critical load is directly observed. However, for the load–unload curves for which no pop-in events are revealed, the load for the transition between the elastic and plastic is obtained using the divergence point between loading curve and the curve representing the Hertz equation

(Hertz contact theory) [23]. Moreover, it is noticed that the maximal depth penetration for the Mg-doped GaN layer is lower than that for the undoped GaN layer. This latter assumes that the resistance to the penetration of the indenter tip to further plastically deform the Mg-doped GaN layer is higher for the same applied load than that for the undoped GaN layer.

Thus, it is expected that the hardness of the Mg-doped GaN layer is higher than that for the GaN layer. In addition, it is worth noting that at the maximal load P_{max} and for holding time equal to 30 s, the depth penetration obtained for the Mg-doped GaN layer is lower than that revealed for the GaN layer. In this stage, the evolution of the plastic deformation for a constant load and during a period of time is attributed to the creep phenomenon, which is thoroughly investigated in the next section.

In order to corroborate these qualitative observations, mechanical properties were determined for both samples; namely the hardness H, the Young's modulus E, the maximal shear stress τ_{max}, the plastic work Wp, and the H/E ratio. Table 1 lists these mechanical properties along with findings documented in literature for comparison purposes.

Table 1. Mechanical properties of undoped and Mg-doped GaN layers obtained in this study compared to those reported in the literature.

Sample	Hardness H (GPa)	Young's Modulus E (GPa)	Plastic Work ($\times 10^{-12}$ J) Wp	Maximal Shear Stress τ_{max} (GPa)	H/E
Undoped GaN #	17.7 ± 0.9	340.3 ± 10.7	472 ± 28	23.1 ± 0.7	0.052
[24]	20.0	323.8	-	18.9 ± 1.3	-
[10]	19 ± 1	286 ± 25	0.7	6.3	-
Mg-doped GaN #	21.3 ± 1.4	410.2 ± 11.3	437 ± 21	20.7 ± 0.5	0.052
[10]	22.3 ± 1.2	333.2 ± 8.2	4.1	7.5 ± 0.4	-

This study.

The analysis of the results presented in this table shows that the hardness and Young's modulus of the Mg-doped GaN layer are higher than those measured for the undoped GaN layer. In other words, Mg-doped GaN thin film is harder and stiffer due to the strengthening effect of the Mg atoms in the crystal lattice, but it may also be less ductile due to the pinning of dislocations by the Mg atoms that can act to inhibit dislocations' motion [25]. As a result, the plastic work of the Mg-doped GaN layer is lower than that of the undoped GaN layer. Additionally, it is important to note that the maximal shear stress calculated for the undoped GaN layer is slightly higher than that obtained for the Mg-doped GaN layer. The maximal shear stress is the stress at which the transition from the elastic to the plastic regime is initiated. The observed difference in the maximal shear stress is attributed to the appearance of pop-in events in the undoped GaN layer, which are absent in the Mg-doped GaN layer due to the dissimilar defect densities; and thus due to the different dislocation mechanisms that are undergone in each sample during nanoindentation tests. Moreover, it is worth noting that the values of hardness and Young's modulus in Table 1 show some degree of scattering compared to the literature. Several examples are documented in reference [9]. This scattering can be attributed to various factors, such as different growth conditions and processes of GaN [26], the applied maximal forces [27], nanoindentation instrumentation techniques (e.g., continuous contact stiffness measurements (CSM)) [28], and different indenter types [29], etc. However, our findings are included in the range of the mechanical properties documented in the literature. Furthermore, it is worth mentioning that there is limited literature available regarding the mechanical properties of Mg-doped GaN layers [10].

Pop-in events are sudden and large plastic deformations can occur during nanoindentation, usually for materials featured by low dislocation density. These events are often associated with the homogeneous nucleation and emission of dislocations is activated [30]. In the case of undoped GaN layers, the presence of low dislocation density as low as

2×10^8 cm^{-2} can lead to the formation of pop-in events during nanoindentation and can lead to a higher degree of plastic deformation. These dislocations are nucleated and propagated through the material in response to the shear stress applied by the indenter, leading to the observed pop-in events. However, in Mg-doped GaN layers, the absence of these pop-ins suggests a different mechanism of plastic deformation as what could be associated to the undoped GaN layer. The pinning of dislocations by Mg atoms can inhibit the homogeneous nucleation of dislocations, preventing the formation of pop-in events. Instead, plastic deformation in Mg-doped GaN may occur through a collective activation of mobile dislocations due to its higher dislocation density compared to that of the undoped GaN layer. In this mechanism, dislocations that are already present in the material are activated by the applied stress and move collectively to accommodate the accumulated strains. To support these findings regarding the observed pop-in events, it is worth noting that previous studies demonstrated a significant influence of the pre-existing density of dislocations on the occurrence of pop-in events [9,31]. Lorenz et al. [32] indicated that pop-in events will not occur when the dislocation density is greater than 10^9 cm^{-2} for a high indenter tip radius. Here, it is worth noting that the radius R of the Berkovich indenter tip used in this study was evaluated to be 500 ± 10 nm using the Hertz contact theory.

The calculated H/E ratio is found similar for both thin films. This ratio is a measure of a material's resistance to plastic deformation, with higher values indicating greater resistance. Furthermore, this ratio can be also used to compute the fracture toughness of materials K_{IC}, as it is inversely proportional to fracture toughness [33]. The fact that the H/E ratio is the same for both types of thin films indicates that they likely have similar resistance to crack propagation despite the observed differences in their mechanical strength and stiffness. Nevertheless, it is important to note that in the current study, our aim is not measuring the fracture toughness K_{IC} of both samples, which would require indentations with an appropriate load to induce cracks.

In summary, the presence or absence of pop-in events during the nanoindentation of undoped and Mg-doped GaN layers is indicative of different mechanisms of plastic deformation, which can be attributed to the presence or absence of dislocation pinning by magnesium atoms. The mechanism of plastic deformation can affect the maximal shear stress of the material, with undoped GaN having a higher maximal shear stress due to its ability to deform plastically through homogeneous dislocation nucleation. Though, no effect was observed on the H/E ratio of both types of thin films. Thus, the findings could have implications for the design and development of GaN thin films for various applications, including microelectromechanical devices, sensors, and optoelectronics.

Figure 6 shows imprints of the Berkovich indenter tip on both the undoped and Mg-doped GaN layers, which were subjected to a maximal loading charge of 500 mN. No cracks were detected during this high loading, which was likely due to the good fracture toughness of both samples. Such behavior could be interpreted by the lower H/E ratio calculated for both samples. The undoped GaN shows a slightly ductile behavior, where it is possible to undergo more plastic strains by applying local severe shear stresses, when compared to the Mg-doped GaN layer.

4.3. Creep Behavior Analysis

Figure 7 exhibits a comparison of the creep displacement versus time during a 30 s dwell period for both undoped GaN and Mg-doped GaN layers. For both curves, the fitting curve shows a very good adjustment for the experimental points, with correlation coefficients above 0.98. The results indicate that both types of layers undergo an initial rapid increase in creep displacement, followed by a subsequent reduction in the rate of creep displacement. Notably, the displacement curve for the undoped GaN layer is positioned higher than that for the Mg-doped GaN layer, indicating that at a constant load of 10 mN, undoped GaN is more susceptible to plastic deformation during creep than Mg-doped GaN. The maximum indentation depth observed during the 30 s dwell period is higher for the undoped GaN than for the Mg-doped GaN. This variation is likely due to different dislocation mechanisms that are undergone at the initiation of the creep stage and to the dislocation density formed beneath the indenter in both types of GaN layers, which subsequently affect the creep behavior.

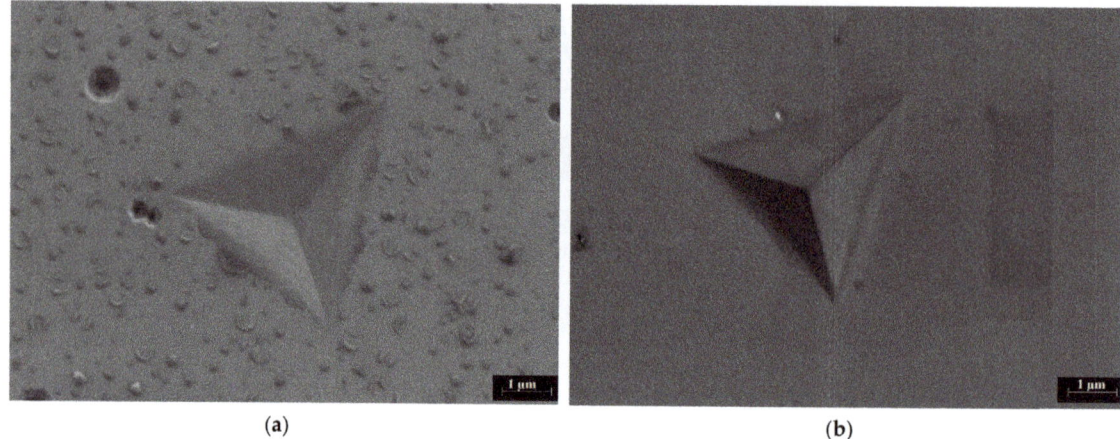

Figure 6. SEM images of imprinted zones using a 500 mN loading charge with Berkovich indenter showing no microcracks: (**a**) undoped GaN layer; (**b**) Mg-doped GaN layer.

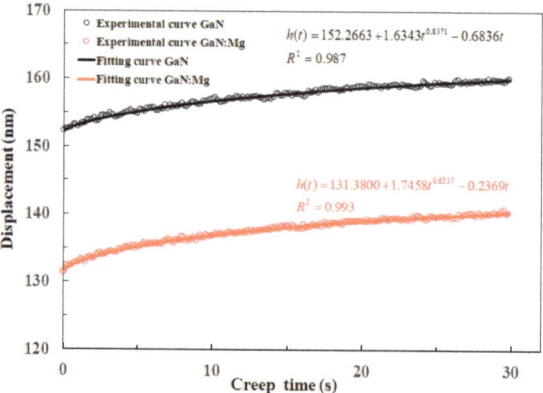

Figure 7. Creep displacement vs. the creep time for both undoped GaN and Mg-doped GaN (GaN:Mg) layers using a maximum indentation force of 10 mN.

Furthermore, we plotted in Figure 8 the creep strain rate $\dot{\varepsilon}$ as a function of creep time (during the dwell time) for both types of samples and calculated it using Equation (4). During the initial stage of the creep process, the strain rate was very high, then it abruptly decreased to reach a strain rate of around 2×10^{-3} s^{-1} in less than 10 s. This initial stage of creep is called transient creep. Thereafter, the strain rate is gradually decreased to converge progressively towards a steady-state creep regime, which is called secondary creep. In the steady-state regime, the creep strain rate is so low that its effect on the modulus result is negligible.

Figure 9 shows the evolution of $\ln \dot{\varepsilon}$ vs. $\ln \sigma$ to derive the creep stress exponent during the steady-state regime, revealing the real nanoindentation creep properties of both samples. The stress exponent $n = (\partial \ln \dot{\varepsilon} / \partial \ln \sigma)$ is instantaneous property, which varies along with the applied stress underneath the indenter tip. To gain a deeper understanding of the observed phenomenon and to question the time-dependent plastic deformation mechanism developed during the creep process, it is crucial to probe the creep stress exponent in the steady-state creep regime. It is evident that the creep stress exponent is higher for the undoped GaN layer. Furthermore, it is observed that higher hardness and Young's modulus are associated with a lower stress exponent. Since the maximum applied load is similar for both types of samples, the correlation between hardness and the related stress exponent is likely due to the severity encountered to further plastically deform the volume localized underneath the indenter. It was previously established that the stress exponent is proportional to the sheared volume [34]. As the indent size increases (i.e., for lower hardness materials), the plastically deformed volume also increases, leading to a decrease in the stress exponent. Additionally, the creep stress exponent is typically considered as a useful indicator for identifying the predominant creep mechanism. For instance, for n = 1, the observed dominant mechanism is the diffusion creep mechanism involving vacancy flow through the lattice, known as the Nabarro–Herring creep. However, for a creep stress exponent greater than 3, the creep mechanism is governed by dislocation glides and climbs (power-law creep equation) [35].

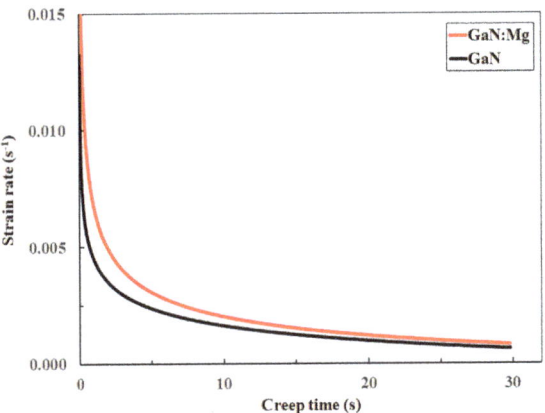

Figure 8. Creep strain rate as a function of the creep time for undoped GaN and Mg-doped GaN (GaN:Mg) layers using a maximum indentation force of 10 mN.

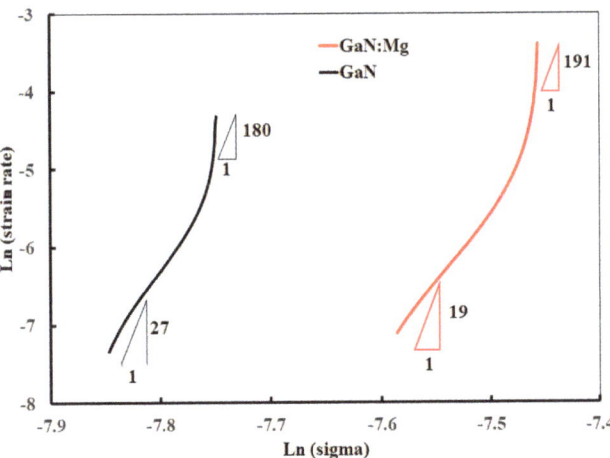

Figure 9. Logarithmic creep strain rate versus logarithmic stress curves for undoped GaN and Mg-doped GaN (GaN:Mg) samples under a load of 10 mN during the creep time of 30 s.

The tested samples exhibited a high creep stress exponent. In the Mg-doped GaN layer, the stress exponent was found to be 19, compared to 27 for the undoped GaN layer (i.e., greater than three), indicating that the creep behavior observed in both samples is governed by the mechanism of dislocation glides and climbs.

In addition, it should be noted that the nanoindentation creep is sensitive to experimental conditions, including the initial loading and unloading phases of the cycle. During the nanoindentation test, the hardness and Young's modulus were measured and found to increase in the case of Mg-doped GaN compared with the undoped GaN layer. However, the steady-state regime of the nanoindentation creep is influenced by significant creep effects, which can affect the measured Young's modulus. The Oliver–Pharr method [16] was employed to measure the Young's modulus, which assumes a purely elastic contact during the unloading stage. Nevertheless, this assumption is not always valid due to thermal drift effects, and creep effects can cause an overestimation of the contact stiffness, thereby influencing the measured Young's modulus. To obtain accurate measurements of the hardness and Young's modulus, a holding time of 30 s was considered sufficient to mitigate any potential creep effects on the measured values, as suggested by Chudoba et al. [36]. The drift rate during this time was estimated to be 0.01 nm/s, which is typically low.

5. Conclusions

This paper investigates the physical and nanomechanical properties of undoped and Mg-doped GaN layers deposited on sapphire substrates using the MOCVD technique. Various probing techniques were employed to analyze the effect of Mg incorporation compared to the undoped GaN layer, on the morphological, electrical, structural, lattice vibrational properties, nanomechanical properties, and creep behavior. The incorporation of Mg resulted in a transition from n-type to p-type conductivity, which led to an increase in defects and the disappearance of pop-ins that were observed in the loading–unloading nanoindentation curve of the undoped GaN layer. Furthermore, a rise in the nanomechanical properties of the Mg-doped GaN layer was revealed, as seen in the increased hardness and Young's modulus. This is attributed to the different dislocation mechanisms that occurred in both GaN samples during the nanoindentation tests. However, the H/E ratio was found to be alike for both layers, assuming a likely similar fracture resistance feature. The introduction of controlled defects in Mg-doped GaN layers can offer a favorable balance between improved physical properties and high-strength semiconductor materials. This suggests the potential application of Mg-doped GaN layers in optoelectronic devices.

From a mechanical standpoint, the hardening effect was attributed to an increase in the pre-existing dislocation density and the introduction of point defects underneath the indenter tip during the incorporation of Mg doping into the GaN layers. This led to a decrease in the creep stress exponent and creep depth, which are related to the inherent creep mechanism governed by dislocation glides and climbs. Therefore, this study emphasizes the importance of understanding the correlation between the physical properties and mechanical characteristics of undoped and Mg-doped GaN thin films, which may have significant implications for a wide range of technological applications.

Author Contributions: Conceptualization, A.K. and Z.B.; methodology, A.K. and Z.B.; investigation, A.K. and Z.B.; validation, A.K. and Z.B.; formal analysis, A.K. and Z.B.; data curation, A.K. and Z.B.; writing—original draft preparation, A.K.; writing—review and editing, A.K. and Z.B. All authors have read and agreed to the published version of the manuscript.

Funding: This research received no external funding.

Institutional Review Board Statement: Not applicable.

Informed Consent Statement: Not applicable.

Data Availability Statement: Not applicable.

Acknowledgments: Zohra Benzarti gratefully acknowledge her support by FEDER funds through the program COMPETE and by the project AM4SP (code POCI-01-0247-FEDER-070521), co-funded by FCT, Portuguese Foundation for Science and Technology and POCI 2020 through the program

Portugal 2020. The authors thank the support to this investigation by M. Evaristo and by A. Cavaleiro from CEMMPRE @ University of Coimbra in Portugal.

Conflicts of Interest: The authors declare no conflict of interest.

References

1. Yoshida, S. Growth of Cubic III-Nitride Semiconductors for Electronics and Optoelectronics Application. *Phys. E Low-Dimens. Syst. Nanostructures* **2000**, *7*, 907–914. [CrossRef]
2. Benzarti, Z.; Sekrafi, T.; Bougrioua, Z.; Khalfallah, A.; El Jani, B. Effect of SiN Treatment on Optical Properties of InxGa1−xN/GaN MQW Blue LEDs. *J. Electron. Mater.* **2017**, *46*, 4312–4320. [CrossRef]
3. Zheng, Y.; Sun, C.; Xiong, B.; Wang, L.; Hao, Z.; Wang, J.; Han, Y.; Li, H.; Yu, J.; Luo, Y. Integrated Gallium Nitride Nonlinear Photonics. *Laser Photonics Rev.* **2022**, *16*, 2100071. [CrossRef]
4. Benzarti, Z.; Halidou, I.; Bougrioua, Z.; Boufaden, T.; El Jani, B. Magnesium Diffusion Profile in GaN Grown by MOVPE. *J. Cryst. Growth* **2008**, *310*, 3274–3277. [CrossRef]
5. Pengelly, R.S.; Wood, S.M.; Milligan, J.W.; Sheppard, S.T.; Pribble, W.L. A Review of GaN on SiC High Electron-Mobility Power Transistors and MMICs. *IEEE Trans. Microw. Theory Tech.* **2012**, *60*, 1764–1783. [CrossRef]
6. Abdul Amir, H.A.A.; Fakhri, M.A.; Abdulkhaleq Alwahib, A. Review of GaN Optical Device Characteristics, Applications, and Optical Analysis Technology. *Mater. Today Proc.* **2021**, *42*, 2815–2821. [CrossRef]
7. Azimah, E.; Zainal, N.; Hassan, Z.; Shuhaimi, A.; Bahrin, A. Electrical and Optical Characterization of Mg Doping in GaN. *Adv. Mater. Res.* **2013**, *620*, 453–457. [CrossRef]
8. Sun, L.; Weng, G.E.; Liang, M.M.; Ying, L.Y.; Lv, X.Q.; Zhang, J.Y.; Zhang, B.P. Influence of P-GaN Annealing on the Optical and Electrical Properties of InGaN/GaN MQW LEDs. *Phys. E Low-Dimens. Syst. Nanostructures* **2014**, *60*, 166–169. [CrossRef]
9. Boughrara, N.; Benzarti, Z.; Khalfallah, A.; Evaristo, M.; Cavaleiro, A. Comparative Study on the Nanomechanical Behavior and Physical Properties Influenced by the Epitaxial Growth Mechanisms of GaN Thin Films. *Appl. Surf. Sci.* **2022**, *579*, 152188. [CrossRef]
10. Jian, S.R.; Ke, W.C.; Juang, J.Y. Mechanical Characteristics of Mg-Doped Gan Thin Films by Nanoindentation. *Nanosci. Nanotechnol. Lett.* **2012**, *4*, 598–603. [CrossRef]
11. Li-Jun, D.; Yuzhen, L.; Dapeng, C.; Wang, X. Green-Blue Luminescence from the Silicon-Rich Silicon Nitride Thin Films. *Chin. J. Lumin.* **2005**, *26*, 380–384.
12. Prado, E.O.; Bolsi, P.C.; Sartori, H.C.; Pinheiro, J.R. An Overview about Si, Superjunction, SiC and GaN Power MOSFET Technologies in Power Electronics Applications. *Energies* **2022**, *15*, 5244. [CrossRef]
13. Wang, S.; Hung, N.T.; Tian, H.; Islam, M.S.; Saito, R. Switching Behavior of a Heterostructure Based on Periodically Doped Graphene Nanoribbon. *Phys. Rev. Appl.* **2021**, *16*, 024030. [CrossRef]
14. Lin, J.T.; Wang, P.; Shuvra, P.; McNamara, S.; McCurdy, M.; Davidson, J.; Walsh, K.; Alles, M.; Alphenaar, B. Impact of X-Ray Radiation on GaN/AlN MEMS Structure and GaN HEMT Gauge Factor Response. In Proceedings of the IEEE 33rd International Conference on Micro Electro Mechanical Systems (MEMS), Vancouver, BC, Canada, 18–22 January 2020; pp. 968–971. [CrossRef]
15. Boughrara, N.; Benzarti, Z.; Khalfallah, A.; Oliveira, J.C.; Evaristo, M.; Cavaleiro, A. Thickness-Dependent Physical and Nanomechanical Properties of AlxGa1−xN Thin Films. *Mater. Sci. Semicond. Process.* **2022**, *151*, 107023. [CrossRef]
16. Oliver, W.C.; Pharr, G.M. An Improved Technique for Determining Hardness and Elastic Modulus Using Load and Displacement Sensing Indentation Experiments. *J. Mater. Res.* **1992**, *7*, 1564–1583. [CrossRef]
17. Li, H.; Ngan, A.H.W. Size Effects of Nanoindentation Creep. *J. Mater. Res.* **2004**, *19*, 513–522. [CrossRef]
18. Bethoux, J.M.; Vennéguès, P.; Natali, F.; Feltin, E.; Tottereau, O.; Nataf, G.; De Mierry, P.; Semond, F. Growth of High Quality Crack-Free AlGaN Films on GaN Templates Using Plastic Relaxation through Buried Cracks. *J. Appl. Phys.* **2003**, *94*, 6499–6507. [CrossRef]
19. Liu, S.T.; Yang, J.; Zhao, D.G.; Jiang, D.S.; Liang, F.; Chen, P.; Zhu, J.J.; Liu, Z.S.; Liu, W.; Xing, Y.; et al. Influence of Carrier Gas H$_2$ Flow Rate on Quality of P-Type GaN Epilayer Grown and Annealed at Lower Temperatures. *Chin. Phys. B* **2018**, *27*, 127803. [CrossRef]
20. Kozawa, T.; Kachi, T.; Kano, H.; Nagase, H.; Koide, N.; Manabe, K. Thermal Stress in GaN Epitaxial Layers Grown on Sapphire Substrates. *J. Appl. Phys.* **1995**, *77*, 4389–4392. [CrossRef]
21. Cho, Y.S.; Hardtdegen, H.; Kaluza, N.; Thillosen, N.; Steins, R.; Sofer, Z.; Lüth, H. Effect of Carrier Gas on GaN Epilayer Characteristics. *Phys. Status Solidi* **2006**, *3*, 1408–1411. [CrossRef]
22. Kunert, H.W.; Brink, D.J.; Auret, F.D.; Maremane, M.; Prinsloo, L.C.; Barnas, J.; Beaumont, B.; Gibart, P. Photoluminescence and Raman Spectroscopy of Mg-Doped GaN.; As Grown, Hydrogen Implanted and Annealed. *Mater. Sci. Eng. B Solid State Mater. Adv. Technol.* **2003**, *102*, 293–297. [CrossRef]
23. Johnson, K.L. *Contact Mechanics*; Cambridge University Press: Cambridge, UK, 1985; ISBN 9780521255769.
24. Fujikane, M.; Inoue, A.; Yokogawa, T.; Nagao, S.; Nowak, R. Mechanical Properties Characterization of c-Plane (0001) and m-Plane (10-10) GaN by Nanoindentation Examination. *Phys. Status Solidi* **2010**, *7*, 1798–1800. [CrossRef]
25. Shuldiner, A.V.; Zakrevskii, V.A. The Mechanism of Interaction of Dislocations with Point Defects in Ionic Crystals. *J. Phys. Condens. Matter* **2002**, *14*, 9555–9562. [CrossRef]

26. García Hernández, S.A.; Compeán García, V.D.; López Luna, E.; Vidal, M.A. Elastic Modulus and Hardness of Cubic GaN Grown by Molecular Beam Epitaxy Obtained by Nanoindentation. *Thin Solid Films* **2020**, *699*, 137915. [CrossRef]
27. Jian, S.R.; Fang, T.H.; Chuu, D.S. Analysis of Physical Properties of III-Nitride Thin Films by Nanoindentation. *J. Electron. Mater.* **2003**, *32*, 496–500. [CrossRef]
28. Yang, P.F.; Huang, C.C.; Chen, D.L.; Lin, T.C.; Tsai, Y.H.; Jian, S.R. Nanomechanical Properties and Fracture Behaviors of GaN and Al0.08Ga0.92N Thin Films Deposited on c-Plane Sapphire by MOCVD. In Proceedings of the 2016 11th International Microsystems, Packaging, Assembly and Circuits Technology Conference (IMPACT), Taipei, Taiwan, 26–28 October 2016; Technical Paper. pp. 349–352. [CrossRef]
29. Fujikane, M.; Leszczyński, M.; Nagao, S.; Nakayama, T.; Yamanaka, S.; Niihara, K.; Nowak, R. Elastic-Plastic Transition during Nanoindentation in Bulk GaN Crystal. *J. Alloys Compd.* **2008**, *450*, 405–411. [CrossRef]
30. Benzarti, Z.; Sekrafi, T.; Khalfallah, A.; Bougrioua, Z.; Vignaud, D.; Evaristo, M.; Cavaleiro, A. Growth Temperature Effect on Physical and Mechanical Properties of Nitrogen Rich InN Epilayers. *J. Alloys Compd.* **2021**, *885*, 160951. [CrossRef]
31. Lodes, M.A.; Hartmaier, A.; Göken, M.; Durst, K. Influence of Dislocation Density on the Pop-in Behavior and Indentation Size Effect in CaF2 Single Crystals: Experiments and Molecular Dynamics Simulations. *Acta Mater.* **2011**, *59*, 4264–4273. [CrossRef]
32. Lorenz, D.; Zeckzer, A.; Hilpert, U.; Grau, P.; Johansen, H.; Leipner, H.S. Pop-in Effect as Homogeneous Nucleation of Dislocations during Nanoindentation. *Phys. Rev. B Condens. Matter Mater. Phys.* **2003**, *67*, 172101. [CrossRef]
33. Lawn, B.R.; Evans, A.G.; Marshall, D.B. Elastic/Plastic Indentation Damage in Ceramics: The Median/Radial Crack System. *J. Am. Ceram. Soc.* **1980**, *63*, 574–581. [CrossRef]
34. Argon, A. Plastic Deformation in Metallic Glasses. *Acta Metall.* **1979**, *27*, 47–58. [CrossRef]
35. Choi, I.C.; Yoo, B.G.; Kim, Y.J.; Jang, J. Il Indentation Creep Revisited. *J. Mater. Res.* **2012**, *27*, 3–11. [CrossRef]
36. Chudoba, T.; Richter, F. Investigation of Creep Behaviour under Load during Indentation Experiments and Its Influence on Hardness and Modulus Results. *Surf. Coat. Technol.* **2001**, *148*, 191–198. [CrossRef]

Disclaimer/Publisher's Note: The statements, opinions and data contained in all publications are solely those of the individual author(s) and contributor(s) and not of MDPI and/or the editor(s). MDPI and/or the editor(s) disclaim responsibility for any injury to people or property resulting from any ideas, methods, instructions or products referred to in the content.

Article

Impact Resistance of CVD Multi-Coatings with Designed Layers

Jiedong Deng [1], Feng Jiang [1,*], Xuming Zha [2], Tao Zhang [1,3], Hongfei Yao [3,*], Dongwei Zhu [3], Hongmei Zhu [3], Hong Xie [4], Fuzeng Wang [1], Xian Wu [5] and Lan Yan [5]

1. Institute of Manufacturing Engineering, Huaqiao University, Xiamen 361021, China
2. College of Marine Equipment and Mechanical Engineering, Jimei University, Xiamen 361021, China
3. Zhejiang Xinxing Tools Co., Ltd., Jiaxing 314300, China
4. Dongfang Turbine Co., Ltd., Deyang 618000, China
5. College of Mechanical Engineering and Automation, Huaqiao University, Xiamen 361021, China
* Correspondence: jiangfeng@hqu.edu.cn (F.J.); yhf@ch-tools.com (H.Y.)

Abstract: Coated cutting tools are widely used in the manufacturing industry due to their excellent properties of high heat resistance, high hardness, and low friction. However, the milling process is a dynamic process, so the coatings of milling tools suffer severe cyclic impact loads. Impact resistance is important for the life of milling tools. Multi-coatings with different layer thickness may influence their impact resistance, but few studies focus on this topic. In this study, CVD coating with a structure of TiN layer, Al_2O_3 layer, and TiCN layer was selected as the research objective. Four different CVD coatings with different layer thicknesses were designed and prepared. The impact resistance test method was proposed to simulate the impact due to cut-in during down the milling process. We obtained the load by setting an impact depth of 25/30/35 μm, recording the impact force during the impact process, and calculating the contact stress; it was found that, at the impact depth of 25/30/35 μm, the download loads were around 9/11/13 N, while the contact stresses were all around 1 GPa. The failure morphology of the coating surface was investigated after the impact process. By comparing the contact stress and the surface morphology of the designed coatings, the impact resistance of four kinds of designed CVD coatings were evaluated. Experiments have shown that an increase in coating thickness and total coating thickness reduces impact resistance by about 10%. The impact resistance of coating samples without a TiN surface layer also decreased by about 10%. When the surface layer of TiN was thinner than 1 μm, the surface layer of TiN was prone to chipping and peeling off. Decreasing the thickness of the middle layer of Al_2O_3 and increasing the thickness of the inner layer of TiCN obviously lowered the impact resistance.

Keywords: CVD multi-coatings; milling tool; impact test; designed layers; failure morphology; performance evaluation

Citation: Deng, J.; Jiang, F.; Zha, X.; Zhang, T.; Yao, H.; Zhu, D.; Zhu, H.; Xie, H.; Wang, F.; Wu, X.; et al. Impact Resistance of CVD Multi-Coatings with Designed Layers. *Coatings* **2023**, *13*, 815. https://doi.org/10.3390/coatings13050815

Academic Editor: Zohra Benzarti

Received: 30 March 2023
Revised: 20 April 2023
Accepted: 21 April 2023
Published: 22 April 2023

Copyright: © 2023 by the authors. Licensee MDPI, Basel, Switzerland. This article is an open access article distributed under the terms and conditions of the Creative Commons Attribution (CC BY) license (https:// creativecommons.org/licenses/by/ 4.0/).

1. Introduction

Coated cutting tools have been widely used in the manufacturing industry, where the parts require high geometrical precision, machining efficiency, and surface quality. Tool coatings have demonstrated excellent static properties such as heat resistance [1], hardness [?], and anti adhesion [3]. However, cutting is a dynamic process that involves complex stress conditions applied on the tool coatings. Therefore, coatings with excellent static properties may not perform well when suffering cyclic and impact loads [4]. The coating surface of cutting tools suffers cyclic fatigue impacts with a frequency ranging from 1 to 1 kHz due to the contact and separation of the tool and workpiece during milling, which can lead to coating damage [5], while the generation of serrated chips during the turning process of difficult-to-machine materials can result in high-frequency cyclic fatigue impacts above 1 kHz, leading to coating chipping. The response of the coating to these cyclic fatigue impacts determines its cutting performance; therefore, the fatigue resistance of coating tools is regarded as an important index to evaluate the performance of coating tools [6].

However, cutting processes contain friction processes, heating processes, deformation processes, as well as impact processes [7], so the coated tool's wear after cutting experiments may not reflect the fatigue impact resistance of cutting tool coatings individually [8]. Impact tests are an effective method to evaluate the performance of cutting tool coatings [9].

Lamri et al. [10,11] utilized electromagnets to drive a rigid indenter to repeatedly impact the sample material with constant acceleration. The initial position of the indenter was adjusted to control the impact velocity, which was detected by a laser displacement sensor. The impact frequency of this impact test device was 10 Hz, and the impact speed could reach 3 m/s. The device controlled the impact force through a piezoelectric actuator, so it could produce impact loads of various modes, frequencies, and durations. Beake [12] et al. developed a low-frequency impact experimental device to study the fatigue characteristics of tetrahedral amorphous carbon coatings and analyzed the influence of different impact loads on the fracture probability of coatings. In the actual cutting process, the tool coating is often impacted by intermediate-frequency loads. The cutting parameters used by the tool in the actual processing directly affect the frequency of the continuous fatigue impact. The milling speed of ordinary materials is generally 40–200 m/min [13]. According to the speed of the milling cutter, the frequency of the fatigue impact of the tool can be calculated to be about 10–50 Hz. However, when milling aerospace titanium alloy materials [14,15], the milling speed is higher. Skordaris et al. [16] analyzed the impact fatigue behavior of tool coatings with different structures and layers through impact tests. The results show that the indentation depth of single-layer TiAlN coatings changes sharply with the increase in impact times, while the multi-layer coatings show more stable fatigue resistance. In addition, in the process of machining difficult materials, the impact frequency of the tool can reach more than 1 kHz due to the production of sawtooth chips [17,18].

Based on the existing literature and research, several experts have developed impact test devices using different principles. For instance, the relationship between high-frequency impact and impact resistance of PVD (Physical Vapor Deposition) coatings during turning and between medium-frequency impact and impact resistance of PVD coatings during milling was established. Zha et al. [19–22] developed high-frequency impact test equipment using an ultrasonic generator. The relationship between high-frequency impact and fatigue impact resistance of PVD coatings during turning was established. Wang [23] et al. also simulated the damage to coating tools in the cutting process through high-frequency impact. However, there are few studies on the tool damage of intermediate-frequency impact and CVD (Chemical Vapor Deposition) coating tools during milling.

In this study, a cyclic impact test device face on milling process CVD coatings was developed based on the vibration of a piezoelectric actuator. The device utilizes the inverse piezoelectric effect, which applies an electric field to the piezoelectric ceramics to extend or shorten the material and carry out high-frequency impacts on coating samples. The impact frequency was set at 10 Hz to 100 Hz. Four different composite layer coatings were used for impact tests, and the results were compared and analyzed. Through the damage situation of various coated tools obtained from this study, we look forward to selecting the most high-performance coated tools, and hope to have some experience and summary on the impact of coating thickness on tool performance. This will help to further change the layer thickness parameters of CVD tools and improve their performance in the future.

2. The Solution for Cyclic Impact Tests

2.1. Design of CVD Coatings

In this study, coating tools were prepared using the CVD method. Four different CVD coating tools, labeled as A, B, C, and D, respectively, were prepared by varying the deposition thickness of three coatings: TiCN, Al_2O_3, and TiN. The impact test and cutting test were conducted on these four tools. Table 1 shows the layer parameters of the four coating tools, and Figure 1 depicts the structure of the prepared coating tools. The substrate was made of a hard alloy material produced by Xiamen Golden Egret Special Alloy Co., Ltd. (Xiamen, China), brand GU20. Its chemical composition was 10.3% Co and 89.7% WC, with a grain size of 0.8 μm. First, coating A was prepared as a TiN layer with a thickness of 1.47 μm. The thickness of the Al_2O_3 layer was 4.33 μm and the TiCN layer thickness was 9.45 μm. Coating B was formed by increasing the thickness of each layer of coating A by 30%; coating C only had two layers, with 5.08 μm of Al_2O_3 and 9.96 μm of TiCN; coating D reduced the thickness of the TiN layer and Al_2O_3 layer, but increased the thickness of the TiCN layer, making its total thickness consistent with that of coating A.

Table 1. Coating parameters of different samples.

Sample	Outer Layer TiN (μm)	Middle Layer Al_2O_3 (μm)	Inner Lying TiCN (μm)	Film Thickness (μm)
A	1.47	4.33	9.45	15.6
B	2.33	5.51	12.5	20.2
C	/	5.08	9.96	15.1
D	0.973	2.26	12.1	15.6

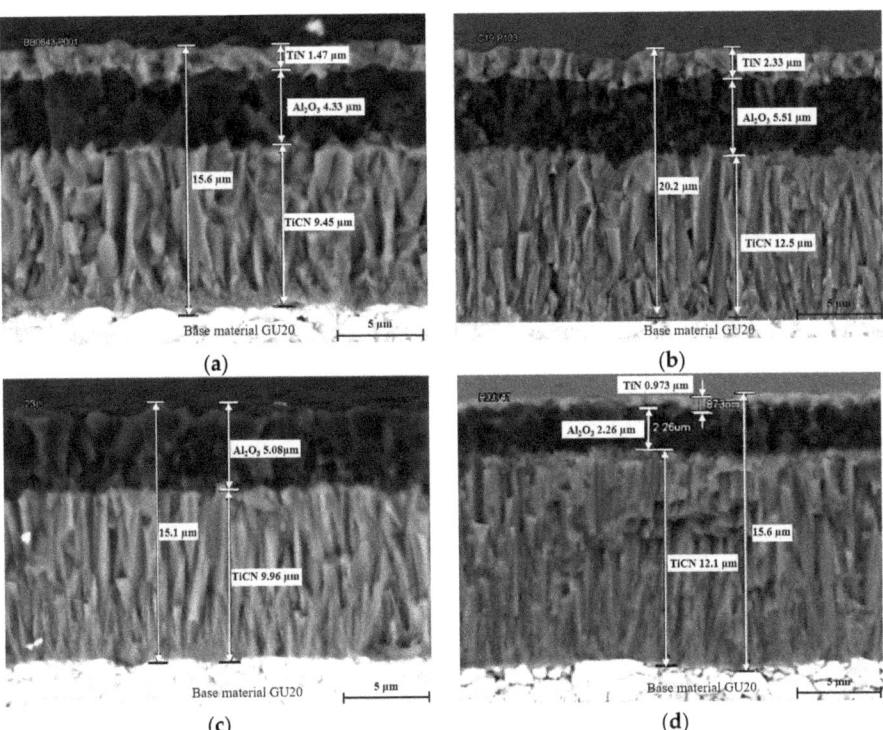

Figure 1. The structure of the prepared coating tools. (a) Coating A, (b) Coating B, (c) Coating C, (d) Coating D.

2.2. Test Principle

The mechanical motion module of the cyclic impact test device was composed of a linear motor and a two-dimensional slide. The linear motor adopted Kollmorgen's linear motor rotor (model IC22150A2TRP1) and linear motor stator (model MC1500512001), supplemented by a marble base, guide rail drag chain, roof, and grating ruler. The overall representation was in the form of a linear motor, as shown in Figure 2. The linear motor actuator and the linear motor stator rod, as well as the guide rail drag chain, were used to achieve the relative motion of the fixed part of the electric motor. The grating ruler was used to provide accurate position information, and the roof was placed on the linear motor actuator to load the "piezoelectric actuator" onto the linear motor. The linear motor was driven by Kollmogen's AC (alternating current) driver (model AKD-P01206-NBEC-0000), thus achieving linear motion of the piezoelectric actuator. The schematic diagram of the entire device is shown in Figure 2, and the main parameters of the linear motor device are shown in Table 2.

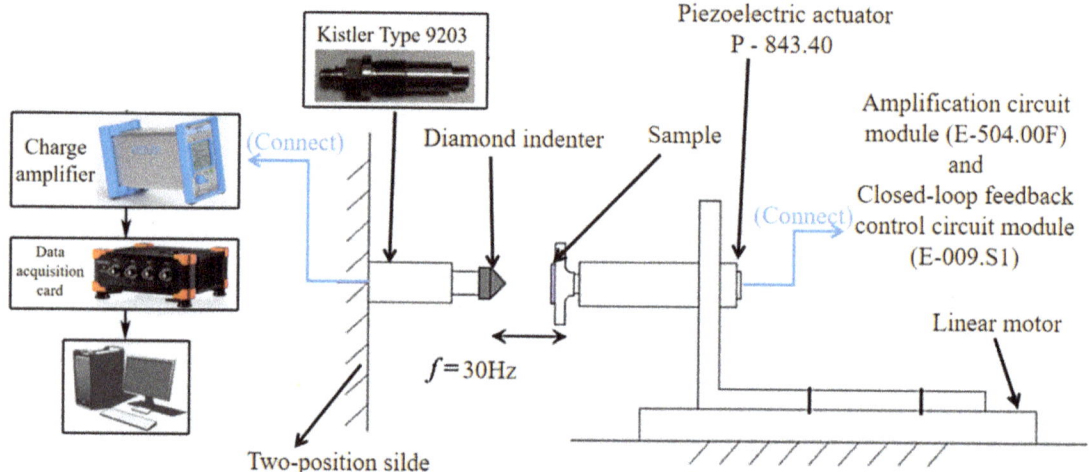

Figure 2. Schematic diagram of the overall device.

Table 2. Main parameters of the linear motor.

Electric Machine Parameter	Data
Rated thrust (N)	67
Peak thrust (N)	202
Top speed (m/s)	4
Peak current duration (s)	<1
Raster resolution (μm)	0.1

The sample was connected to the vibration output end of a piezoelectric actuator (PhysikInstrumente's P-843.40), and its cyclic vibration signal was modulated and output by a signal generator. Then, the signals were amplified and processed by Kistler's amplifier circuit module (E-504.00F) and closed-loop feedback control circuit module (E-509.S1), respectively. Finally, the signal was output to a piezoelectric actuator, which generated cyclic vibration under the action of an electric field. The cyclic impact between the sample and the diamond indenter was generated by the cyclic vibration of the piezoelectric actuator. In [24], the sample was attached to a two-dimensional slide, while in this platform, the sample was attached to the sample stage and connected to a piezoelectric actuator with a thread. At the same time, the pressure head was connected to the force measuring instrument and fixed on the two-dimensional slide. This improvement makes it more efficient to replace the sample in practical applications.

The data acquisition module comprises a dynamometer sensor, a charge amplifier, and a data acquisition card. The dynamometer sensor was a Kistler type 9203 single component micro force sensor that was fixed to the two-dimensional slider with a thread. It operated based on the principle of piezoelectric measurement, where the force acting on the highly sensitive measuring element generates a proportional electric charge at the signal output, which is then converted into an evaluable process signal or curve via the charge amplifier. The force measuring instrument has a measuring range of −500–500 N. The charge amplifier used was a Type 5018A single-channel charge amplifier produced by the same company as the dynamometer sensor. The data acquisition card utilized was the SIRIUS MINI data acquisition system, which enabled the collected force signals to be analyzed and processed on a computer via its supporting software.

Previous studies have indicated that the coating surface of a tool undergoes cyclic fatigue impact in the range of 1–1000 Hz during milling processes. In this paper, we focus on investigating the cyclic fatigue impact that occurs during critical contact between the coating tool and the workpiece.

During the impact test, the coating sample was brought into critical contact with the indenter. The coating sample was attached to the end of the ultrasonic horn and installed on the linear motor. The linear motor then slowly moved the coating sample towards the indenter until contact was made, as indicated by the contact force signal between the coating sample and the indenter reaching 0.01 N.

As shown in Figure 3, once critical contact state was achieved, the power to the signal generator was turned on. The piezoelectric actuator drove the coating sample cyclically, causing the coating sample to be cyclically impacted by the indenter. The number of impact cycles was controlled by setting the impact time. Once the working time of the piezoelectric actuator reached the set value, it automatically stopped, and the motor system was controlled to withdraw the indenter, separating it from the coating sample.

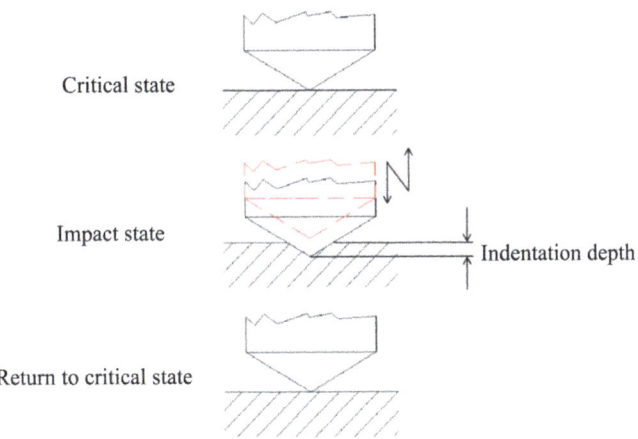

Figure 3. Schematic diagram of the impact test.

2.3. Design of the Impact Test

The indenter used was a Rockwell conical diamond indenter with a fillet radius of 0.05 mm. One end of the dynamometer was attached to the indenter, while the other end was connected to a charge amplifier. The coating sample was secured onto the vibration output end of the piezoceramic actuator, which was then fixed onto the linear motor as a whole via threading, allowing it to move horizontally.

In the cyclic impact test, the impact time was set to 240 s. During the milling process, the speed of the milling cutter is usually 1800 r/min, and each cutting tooth is impacted 30 times per second. Therefore, the impact frequency was set to 30 Hz to simulate the impact on the cutting tooth at a milling speed of 1800 r/min. In addition, the load range of the cutting tool during milling is 1–5 GPa. Therefore, different impact depths were set in the early stage of this experiment to obtain different loads. In order to ensure that the test results did not have contingency, three points at different positions on the sample were randomly selected for a fatigue test in each group of tests, and three control experiments were carried out. Through observation, it was found that the results of the three groups of experiments were consistent. The impact depth H of this experiment was set to 20/25/30/35 μm (the piezoelectric actuator used in the experiment was a P-843.40, produced by PhysikInstrumente. The amplitude of the piezoelectric actuator was controlled by controlling the voltage output from the signal generator. The amplitudes of 20/25/30/35 μm correspond to voltage signals of 3.33/17/5.83 V, respectively), and these loads were in the range of 1–5 GPa. Through the above setup, the impact of the coating tool was finally set to approximate the milling process. The parameter settings for the cyclic impact test are shown in Table 3.

Table 3. The parameter settings of the cyclic impact test.

Frequency of Impact (Hz)	Time of Impact (s)	Depth of Impact H (μm)
		25
30	240	30
		35

3. Results and Discussion

3.1. Cyclic Contact Stress Analysis

Figure 4 shows the impact load of coating A after cyclic impact at different impact depths. As can be seen from the figure, each coating first has a large initial impact load when it is impacted, then quickly enters the fatigue impact stage, and the initial impact load and fatigue impact load increase with the increase in impact depth (taking coating A as an example, when the impact depth is 25 μm the fatigue impact load is 9.42 N, and when the impact depth is 35 μm the fatigue impact load is 13.21 N). By observing the impact value of the whole impact process, it can be found that the coating slowly fails in the impact process due to the long-time impact, which can be proved by the slow decline of the impact value of the fatigue impact cut-off energy. In addition, the impact force at the initial impact stage is very unstable, which is a phenomenon of the first impact under various impact conditions. By comparing the force signal graphs of each coating at different impact depths, it was found that there is no significant difference in the force conditions of each coating.

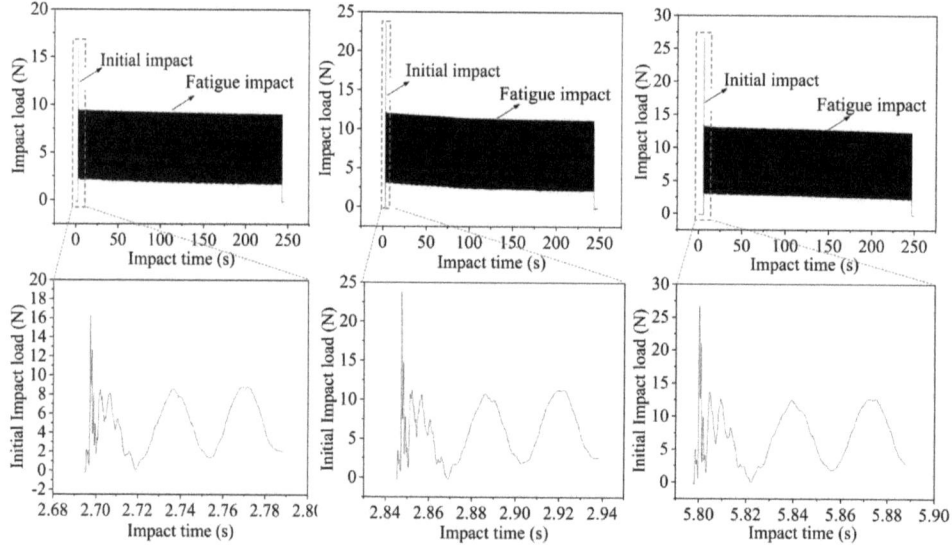

Figure 4. The impact load of a cyclic impact at different impact depths for coating A.

In addition to the cyclic impact loads of coating A at different impact depths, shown in Figure 4, the other three coating tools were also subjected to cyclic impact tests. The cyclic impact loads of each coating tool during the impact process are shown in Table 4 below.

Table 4. The fatigue impact force F.

	25 μm	30 μm	35 μm
Coating A	9.42 N	12.06 N	13.21 N
Coating B	8.72 N	11.12 N	13.09 N
Coating C	8.84 N	10.33 N	12.63 N
Coating D	9.22 N	11.15 N	13.77 N

By observing the impact loads of different coatings at different impact depths, it can be concluded that the coatings have a large initial impact load at the first time of impact. After that, they enter the stable fatigue impact stage, where the impact load increases with the depth of impact. At the same depth of impact, the initial impact load and fatigue impact load of the four coatings are comparable.

3.2. SEM Investigation

A scanning electron microscope was used to observe and study the surface morphology of the coatings after the cyclic impact test. Figure 5 illustrates the surface topography of the tool coatings after cyclic impact at different impact depths. It can be seen from the figure that the surface failure of each coating sample increases with the increase in impact depth, and the failure forms of sample A, sample B, and sample D are mainly radial crack, while the failure form of sample B is mainly annular crack. The damage of four samples under the same test parameters is most significant in sample D, where the surface layer of sample D completely fell off at an impact depth of 35μm. Sample B was slightly more damaged than sample A.

From a macroscopic perspective, the impact depth H increased from 25 μm to 35 μm, resulting in an increase in the degree of damage for each coating. Upon observing the damage of the different coatings, it is evident that the degree of damage for samples B and D is significantly greater than that of sample A. Furthermore, the damage observed in all three coatings is primarily radial cracking.

3.3. Discussion

At the same depth of impact, coating D exhibited the most significant damage. Numerous radial cracks were observed at an impact depth H of 25 μm, while surface spalling occurred at a depth H of 35 μm. Upon examining the impact load F at 35 μm, it was found that coating D experienced an impact for 47 s at that depth. The impact load F rapidly decreased from 13.46 N to 13.06 N within 12 s, which was slightly steeper than the slope observed during the initial and fatigue impact stages. It is inferred that the TiN coating on the surface of coating D spalled during this period. When comparing coating A and coating C, it was observed that coating A had relatively fewer ring cracks, while coating C had more. These key damage characteristics reflect different failure mechanisms during impact.

The impact condition of the tool samples during the impact process was analyzed. Using the experiment on tool sample A at an impact depth of 25 μm as an example, it was observed that the tool first experienced an initial impact with a large load, followed by entering the fatigue impact stage. Based on this observation, it was concluded that

$$\sigma = \frac{N}{S}$$

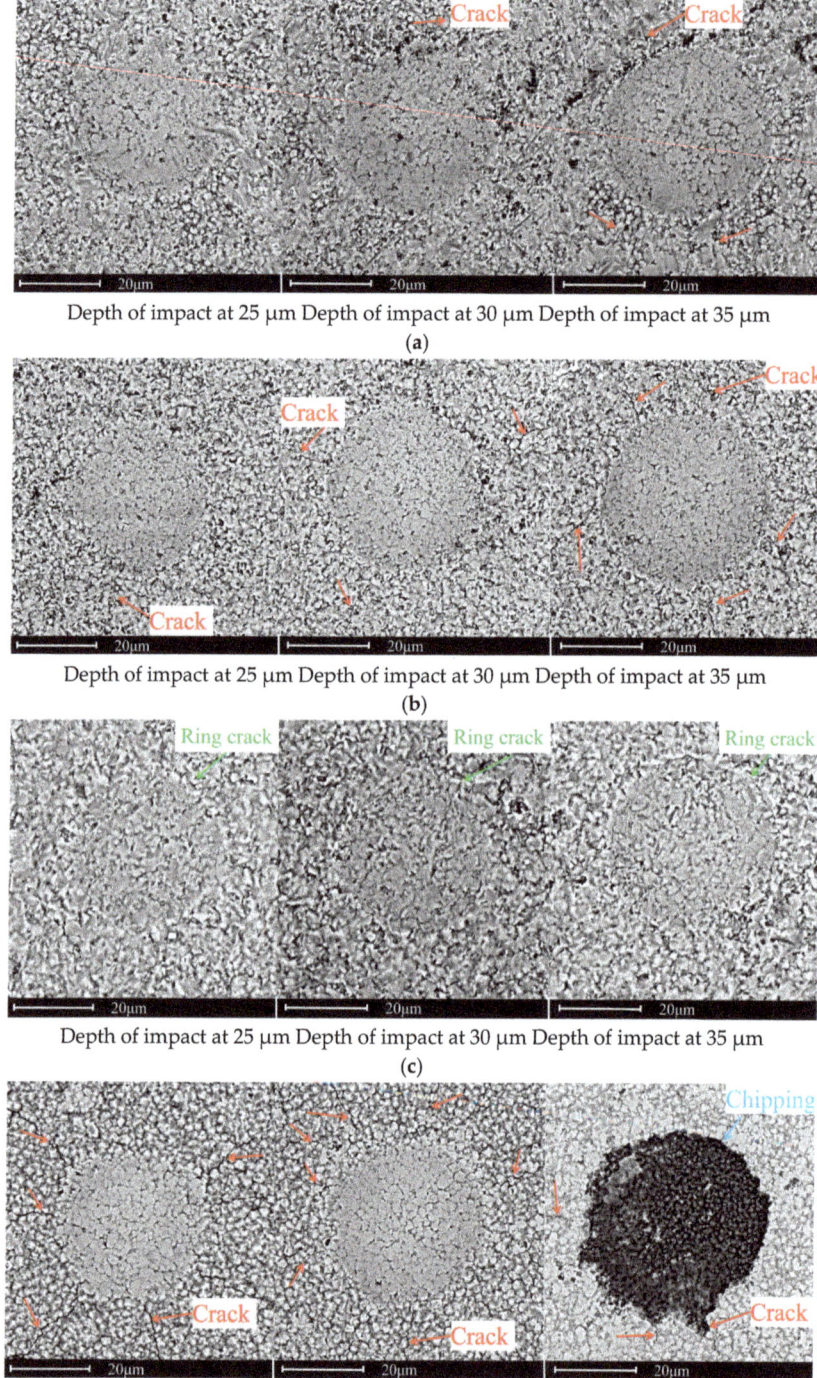

Figure 5. The surface topography of the tool coating after cyclic impact. (**a**) Coating A; (**b**) Coating B; (**c**) Coating C; (**d**) Coating D.

Based on the above formula, the impact stress of each coating tool in the process of cyclic impact was calculated. As shown in Table 5, the impact stress of the coating tool in the cyclic impact process was basically close to 1 GPa, which accords with the impact force of the milling tool in the process of milling.

Table 5. The mean fatigue compression stress.

	25 μm	30 μm	35 μm
Coating A	1.109 GPa	1.105 GPa	0.993 GPa
Coating B	1.023 GPa	1.031 GPa	0.982 GPa
Coating C	1.094 GPa	1.008 GPa	1.013 GPa
Coating D	1.092 GPa	1.030 GPa	1.038 GPa

The contact stress of the coating can be calculated using the formula above, which can also be applied to calculate the coating's stress during the fatigue impact stage. As Figure 6 shows, due to the prolonged impact test time, the coating gradually deteriorated under sustained impact, as evidenced by the gradual but minor reduction in the impact force during the fatigue stage. The results showed that the initial impact compressive stress of each coating sample, corresponding to the impact depth, was similar, and it was also similar to the subsequent fatigue impact stress for all four samples, indicating a consistent trend. As the impact progressed, each coating tool sample exhibited some degree of failure. A comparison of the final compressive stress of each sample revealed that the impact stress of each coating sample decreased as the impact depth increased, suggesting that deeper impacts were more likely to cause coating failure.

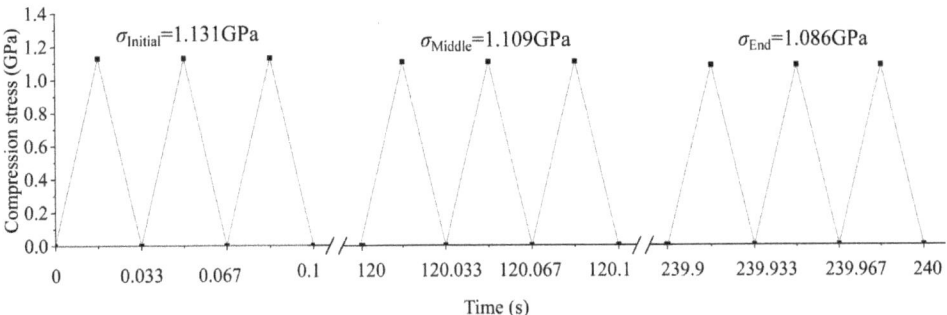

Figure 6. Fatigue compressive stress changes throughout the whole process.

In impact tests, cracks can be caused by both tensile and compressive stresses during cyclic impact. Based on the classical Hertzian crack development characteristics, annular cracks are typically caused by tensile stress, whereas radial cracks are caused by compressive stress.

Upon comparing coatings A and B, it was observed that despite increasing the thickness of TiCN in coating B, its impact resistance was lower than that of coating A. At the same impact depth, the compressive stress of both samples was similar, but the tensile stress of coating A was higher than that of coating B. Upon examining the surface morphology after impact, it was revealed that the main failure mode of both samples was radial cracking. However, the radial cracking of coating B was more severe than that of coating A, which suggests that the thicker bonding layer reduced the coating's impact resistance. This cannot explain the feedback effect of the two samples on tensile stress.

By comparing coating B and coating D, it was found that the TiCN and Al_2O_3 thickness of coating D was reduced. By observing the surface of coating D after impact, it was found that the surface layer of coating D was spalling and the Al_2O_3 layer was completely shed, resulting in lower impact resistance compared with coating A and coating B.

Comparing coating A and coating C, the main difference between them was the absence of a TiN surface layer in coating C. Upon observing the surface after impact, a large number of annular cracks were found on the surface of coating C, and the tensile stress of the two samples was similar, indicating that the tensile strength of coating C without a TiN surface was lower than that of coating A. This suggests that the impact resistance of coating C is lower than that of coating A.

By comparing the damage situation of four coating tools, it was found that they were subjected to similar stress conditions, but their damage was different. The surface morphology of coating tool A was the most complete among the four samples, indicating that coating tool A has the best ability to resist compressive and tensile stresses; the surface damage of coated tool B was slightly increased compared with coated tool A in terms of radial cracks, but no circular cracks appeared yet, indicating that coated tool B has a better ability to resist tensile stress, but its ability to resist compressive stress is slightly lower than that of coated tool A. The surface damage of coating tool C was mainly caused by circular cracks, without obvious radial cracks, indicating that its resistance to compressive stress is slightly stronger than that of coating tools A and B, but its resistance to tensile stress is poor; coating tool D exhibited a large number of radial cracks at an impact depth of 25 μm, and surface delamination occurred at an impact depth of 35 μm, indicating its poor compressive stress ability, but its resistance to tensile stress cannot be determined yet. In summary, coated tool A has the best fatigue resistance, while coated tools B and C are slightly weaker, and coated tool D exhibits the worst fatigue resistance.

4. Conclusions

A cyclic impact system based on a piezoelectric actuator was invented, and the impact frequency can be adjusted. In this experiment, the impact frequency was set to 30 Hz; the cyclic impact load was obtained by controlling the impact depth to simulate cyclic impact during the cutting process. Four coating tools of different layer thicknesses were designed and prepared for this experiment. In this experiment, the difference in impact resistance of coating tools with different layer thicknesses was found:

(1) Coating tool B increased the total layer thickness of coating A by 1.3 times. The study found that under the same impact cyclic load, the surface damage of coated tool B increased by about 10% compared with that of coated tool A, indicating that the fatigue impact resistance of coated tool B, with an increase in total thickness, was slightly lower than that of coated tool A.

(2) Comparing coating A and coating C, the difference between the two coatings is that coating tool C removed the TiN layer and increased the thickness of the Al_2O_3 and TiCN layers, and the overall thickness remained almost unchanged. It was found that the surface crack of coated tool C was slightly more than that of coated tool A, and the cyclic impact resistance of coated tool C was also slightly lower than that of coated tool A.

(3) Compared with coating A and coating D, the difference between them is that coating D reduced the thickness of TiN and Al_2O_3 while increasing the thickness of bonding layer TiCN. However, the experiment found that this thickness change caused more cracks to appear in coated tool D. When the impact depth was 30 μm, the surface cracks of coated tool D increased by more than 50% compared with those of coated tool A. At an impact depth of 35 μm, there was even a phenomenon of surface layer peeling. It shows that the decrease in TiN and Al_2O_3 thickness and the increase in TiN make the cyclic impact resistance of coating tools obviously decrease.

After analyzing the above results, it was found that the presence of TiN coating mainly changes the surface damage form of the coating (radial cracks or circular cracks). The thickness of the TiN surface should not be too small, otherwise surface peeling may occur. In this experiment, an increase in the total thickness of the coating slightly reduced its performance. Therefore, among the four coating tools, coating tool A had the best performance, coating tool D had the worst performance, and coating tool B and coating tool

C had similar performances, both slightly lower than that of coating tool A. In subsequent research, coating tools with different coating parameters can be further prepared to obtain the optimal coating thickness through experiments.

Author Contributions: Conceptualization, J.D., F.J. and H.Y.; investigation, J.D. and T.Z.; funding acquisition, F.J., X.Z., T.Z. and H.Y.; data curation, X.Z.; methodology, D.Z.; supervision, D.Z. and F.W.; resources, H.Z.; validation, H.Z.; software, H.X.; writing—original draft, H.X. and X.W.; writing—review and editing, F.W. and L.Y.; visualization, X.W. and L.Y. All authors have read and agreed to the published version of the manuscript.

Funding: This work was supported by the Integration Project of Industry and Research of the National Natural Science Foundation of China (No. 52275428), the National Natural Science Foundation of China (No. 52205466), the Natural Science Foundation for the Science and Technology Project of Fujian Province (No. 2021J05167), and the Foundation of State Key Laboratory of Digital Manufacturing Equipment and Technology (Grant No. DMETKF2022002).

Institutional Review Board Statement: Not applicable.

Informed Consent Statement: Not applicable.

Data Availability Statement: Data are contained within the article.

Acknowledgments: The authors would like to thank Xiamen Golden Egret Special Alloy Co., Ltd. for their support in preparing coating samples.

Conflicts of Interest: The authors declare no conflict of interest.

References

1. Blinkov, I.V.; Belov, D.S.; Volkhonsky, A.O.; Sergevnin, V.S.; Nizamova, A.N.; Chernogor, A.V.; Kiryukhantsev-Korneev, F.V. Heat Resistance, High-Temperature Tribological Characteristics, and Electrochemical Behavior of Arc-PVD Nanostructural Multilayer Ti–Al–Si–N Coatings. *Prot. Met. Phys. Chem.* **2018**, *54*, 416–424. [CrossRef]
2. Antonyuk, V.S.; Soroka, E.B.; Kalinichenko, V.I. Providing adhesion strength for a substrate-coating system under contact loading. *J. Superhard Mater.* **2008**, *30*, 133–138. [CrossRef]
3. Wang, H.; Lu, H.; Song, X.; Yan, X.; Liu, X.; Nie, Z. Corrosion resistance enhancement of WC cermet coating by carbides alloying. *Corros. Sci.* **2019**, *147*, 372–383. [CrossRef]
4. Sima, M.; Özel, T. Modified material constitutive models for serrated chip formation simulations and experimental validation in machining of titanium alloy Ti–6Al–4V. *Int. J. Mach. Tools Manuf.* **2010**, *50*, 943–960. [CrossRef]
5. Liyao, G.; Minjie, W.; Chunzheng, D. On adiabatic shear localized fracture during serrated chip evolution in high speed machining of hardened AISI 1045 steel. *Int. J. Mech. Sci.* **2013**, *75*, 288–298. [CrossRef]
6. Gassner, M.; Schalk, N.; Tkadletz, M.; Czettl, C.; Mitterer, C. Thermal crack network on CVD TiCN/α-Al_2O_3 coated cemented carbide cutting tools. *Int. J. Refract. Met. Hard Mater.* **2019**, *81*, 1–6. [CrossRef]
7. Zha, X.; Chen, F.; Jiang, F.; Xu, X. Correlation of the fatigue impact resistance of bilayer and nanolayered PVD coatings with their cutting performance in machining Ti_6Al_4V. *Ceram. Int.* **2019**, *45*, 14704–14717. [CrossRef]
8. Wang, T.; Zha, X.; Chen, F.; Wang, J.; Li, Y.; Jiang, F. Mechanical impact test methods for hard coatings of cutting tools: A review. *Int. J. Adv. Manuf. Technol.* **2021**, *115*, 1367–1385. [CrossRef]
9. Beake, B.D.; Bird, A.; Isern, L.; Endrino, J.L.; Jiang, F. Elevated temperature micro-impact testing of TiAlSiN coatings produced by physical vapour deposition. *Thin Solid Films* **2019**, *688*, 137358. [CrossRef]
10. Lamri, S.; Langlade, C.; Kermouche, G. Damage phenomena of thin hard coatings submitted to repeated impacts: Influence of the substrate and film properties. *Mater. Sci. Eng. A* **2013**, *560*, 296–305. [CrossRef]
11. Lamri, S.; Langlade, C.; Kermouche, G. Failure mechanisms of thin hard coatings submitted to repeated impacts: Influence of the film thickness. *Adv. Mat. Res.* **2010**, *112*, 73–82. [CrossRef]
12. Beake, B.D.; Lau, S.P.; Smith, J.F. Evaluating the fracture properties and fatigue wear of tetrahedral amorphous carbon films on silicon by nano-impact testing. *Surf. Coat. Technol.* **2004**, *177*, 611–615. [CrossRef]
13. Zhao, G.; Zhang, X.; Zavalnyi, O.; Liu, Y.; Xiao, W. Extended roughing operations to ISO 14649-11 for milling T-spline surfaces. *Int. J. Adv. Manuf. Technol.* **2019**, *102*, 4319–4335. [CrossRef]
14. Hovsepian, P.E.; Luo, Q.; Robinson, G.; Pittman, M.; Howarth, M.; Doerwald, D.; Tietema, R.; Sim, W.M.; Deeming, A.; Zeus, T. TiAlN/VN superlattice structured PVD coatings: A new alternative in machining of aluminium alloys for aerospace and automotive components. *Surf. Coat. Technol.* **2006**, *201*, 265–272. [CrossRef]
15. Khorasani, A.M.; Gibson, I.; Goldberg, M.; Doeven, E.H.; Littlefair, G. Investigation on the effect of cutting fluid pressure on surface quality measurement in high speed thread milling of brass alloy (C3600) and aluminium alloy (5083). *Measurement* **2016**, *82*, 55–63. [CrossRef]

16. Niu, J.; Huang, C.; Li, C.; Zou, B.; Xu, L.; Wang, J.; Liu, Z. A comprehensive method for selecting cutting tool materials. *Int. J. Adv. Manuf. Technol.* **2020**, *110*, 229–240. [CrossRef]
17. Arriola, I.; Whitenton, E.; Heigel, J.; Arrazola, P.J. Relationship between machinability index and in-process parameters during orthogonal cutting of steels. *CIRP Ann.* **2011**, *60*, 93–96. [CrossRef]
18. Baizeau, T.; Campocasso, S.; Rossi, F.; Poulachon, G.; Hild, F. Cutting force sensor based on digital image correlation for segmented chip formation analysis. *J. Mater. Process. Technol.* **2016**, *238*, 466–473. [CrossRef]
19. Zha, X.; Jiang, F.; Xu, X. Investigation of modelling and stress distribution of a coating/substrate system after an indentation test. *Int. J. Mech. Sci.* **2017**, *134*, 1–14. [CrossRef]
20. Zha, X.; Jiang, F.; Xu, X. Investigating the high frequency fatigue failure mechanisms of mono and multilayer PVD coatings by the cyclic impact tests. *Surf. Coat. Technol.* **2018**, *344*, 689–701. [CrossRef]
21. Zha, X.; Wang, T.; Chen, F.; Wang, J.; Lin, L.; Lin, F.; Xie, H.; Jiang, F. Investigation the fatigue impact behavior and wear mechanisms of bilayer micro-structured and multilayer nano-structured coatings on cemented carbide tools in milling titanium alloy. *Int. J. Refract. Met. Hard Mater.* **2022**, *103*, 105738. [CrossRef]
22. Zha, X.; Wang, T.; Guo, B.; Chen, F.; Lin, L.; Zhang, T.; Jiang, F. Research on the oxidation resistance and ultra-high frequency thermal fatigue shock failure mechanisms of the bilayer and multilayer nano-coatings on cemented carbide tools. *Int. J. Refract. Met. Hard Mater.* **2023**, *110*, 106043. [CrossRef]
23. Wang, T.; Zha, X.; Chen, F.; Wang, J.; Lin, L.; Xie, H.; Lin, F.; Jiang, F. Research on cutting performance of coated cutting tools by a new impact test method considering contact stress condition caused by segmented chips. *J. Manuf. Process.* **2021**, *68*, 1569–1584. [CrossRef]
24. Bouzakis, K.D.; Skordaris, G.; Bouzakis, E.; Michailidis, N. Characterization methods and performance optimization of coated cutting tools. *Ann. Fac. Eng. Hunedoara* **2014**, *12*, 121.

Disclaimer/Publisher's Note: The statements, opinions and data contained in all publications are solely those of the individual author(s) and contributor(s) and not of MDPI and/or the editor(s). MDPI and/or the editor(s) disclaim responsibility for any injury to people or property resulting from any ideas, methods, instructions or products referred to in the content.

Review

A Review of Optical Fiber Sensing Technology Based on Thin Film and Fabry–Perot Cavity

Chaoqun Ma [1,†], Donghong Peng [1,†], Xuanyao Bai [1], Shuangqiang Liu [1,*] and Le Luo [1,2,3,4,*]

1. School of Physics and Astronomy, Sun Yat-Sen University, Zhuhai 519082, China
2. Shenzhen Research Institute of Sun Yat-Sen University, Nanshan, Shenzhen 518087, China
3. State Key Laboratory of Optoelectronic Materials and Technologies, Guangzhou Campus, Sun Yat-Sen University, Guangzhou 510275, China
4. International Quantum Academy, Shenzhen 518048, China
* Correspondence: liushq33@mail.sysu.edu.cn (S.L.); luole5@mail.sysu.edu.cn (L.L.)
† These authors contributed equally to this work.

Abstract: Fiber sensors possess characteristics such as compact structure, simplicity, electromagnetic interference resistance, and reusability, making them widely applicable in various practical engineering applications. Traditional fiber sensors based on different microstructures solely rely on the thermal expansion effect of silica material itself, limiting their usage primarily to temperature or pressure sensing. By employing thin film technology to form Fabry–Perot (FP) cavities on the end-face or inside the fiber, sensitivity to different physical quantities can be achieved using different materials, and this greatly expands the application range of fiber sensing. This paper provides a systematic introduction to the principle of FP cavity fiber optic sensors based on thin film technology and reviews the applications and development trends of this sensor in various measurement fields. Currently, there is a growing need for precise measurements in both scientific research and industrial production. This has led to an increase in the variety of structures and sensing materials used in fiber sensors. The thin film discussed in this paper, suitable for various types of sensing, not only applies to fiber optic FP cavity sensors but also contributes to the research and advancement of other types of fiber sensors.

Keywords: fiber sensor; Fabry–Perot cavity; thin film

1. Introduction

Fiber optic sensing technology utilizes the propagation of light signals in optical fibers to detect external physical quantities. When external physical quantities (temperature, humidity, magnetic field, electric field, etc.) change, the characteristic parameters of the optical signal transmitted through fiber (phase, intensity, wavelength, etc.) also undergo corresponding variations. By establishing the relationship between these characteristic parameters and the physical quantity to be measured, the value of the physical quantity can be derived. Compared to other types of sensors, fiber optic sensors possess advantages such as being compact, lightweight, and resistant to electromagnetic interference. These sensors enable measurements to be conducted in extreme environments, including confined spaces, extreme temperatures, and areas with strong electromagnetic interference. Furthermore, fiber optic sensors demonstrate significant advantages in terms of detection sensitivity, resolution, and signal transmission distance. Based on the signal modulation methods used in fiber optic sensors, they can be classified into several categories, including intensity modulation, wavelength modulation, frequency modulation, polarization modulation, and phase modulation. Interferometric fiber optic sensors based on phase modulation have gained widespread adoption due to their exceptional resolution and sensitivity. Nevertheless, the utilization of interferometric fiber optic sensors based on phase modulation demands high standards for the quality of both the light source and the receiving detection system. Phase modulation-based fiber optic sensors primarily utilize different types of fiber optic interferometers to achieve optical interference. Typical fiber optic interferometers include the Mach–Zehnder (M-Z) interferometer [1,2], Michelson interferometer [3,4], FP interferometer [5,6], and Sagnac interferometer [7,8]. The FP interferometer, which originated in the late 19th century, typically consists of two highly reflective mirrors. The incident light beam undergoes multiple reflections between these mirrors, resulting in multiple-beam interference. Despite the early invention of the FP interferometer, its integration with optical fibers commenced in the 1980s. The fiber optic FP structure, as a crucial component in the development of fiber optic sensing technology, has been subject to in-depth investigations by researchers. Compared to other sensors, the fiber optic FP sensing structure exhibits significant advantages in terms of dynamic range, sensitivity, response speed, and implementation modes. Figure 1 illustrates the annual number of research papers related to FP fiber optic sensors from 1999 to 2023. As observed, there has been a consistent and steady increase in the number of publications in this field in recent years. This upward trend clearly indicates that FP fiber optic sensors have gained significant attention from the academic community, and the research in this area continues to be extensively explored and studied.

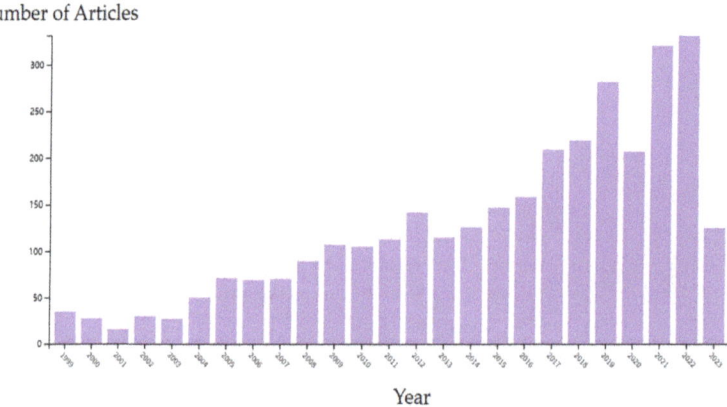

Figure 1. Annual numbers of research papers related to FP fiber optic sensors from 1999 to 2023.

In Mach–Zehnder and Michelson interference optical fiber sensors, the elastic-optic effect of the optical fiber is negligible. Therefore, a lengthy optical fiber is necessary for high sensitivity, which in turn, creates poor thermal stability and increased vulnerability to vibration. Additionally, the phase noise produced by the light source has a significant impact on the interferometer structure, requiring a highly coherent light source for optimal sensor performance [9]. Additionally, in the sensing experiment, fiber optic FP interferometric sensors can effectively mitigate optical power fading caused by polarization. Due to the birefringent nature of optical fibers, Mach–Zehnder and Michelson interferometers split a light beam into two independently propagating beams, each with varying polarization states in a random manner. This leads to a decrease in the efficiency of interference when the two beams recombine, resulting in reduced contrast of interference fringes [9]. Sagnac interferometers can employ polarization control techniques or high birefringent fibers to maintain the polarization characteristics of transmitted light, but this undoubtedly adds complexity and cost to the system [10,11]. However, FP interferometers, due to their short cavity length or propagation in air, can effectively ignore the issue of optical power fading [9]. Fiber optic sensors based on FP interferometers have unique structures and can be fabricated through a variety of methods to meet specific sensing requirements. The sensitivity of the sensor can be enhanced by filling the FP cavity with specific materials. Moreover, an open FP cavity can be effectively employed to sense gas pressure or measure the refractive index of solutions. These properties have greatly contributed to the progress of fiber optic sensing technology. Thin film technology plays a crucial role in fiber optic sensors by coating reflective surfaces inside the FP cavity or on the fiber surface, enabling the measurement of specific physical quantities [12]. For instance, temperature sensing can be achieved by adding a temperature-sensitive film on the reflecting surface [13]. The choice of materials and fabrication methods for the film constituting the FP cavity can be tailored according to the specific requirements of the physical quantity sensor.

Reference [14] summarized the main methods of constructing Fabry–Perot (FP) cavities within optical fibers. These methods can be broadly categorized as non-splicing and splicing techniques. The authors classified important literature on fiber optic FP sensors from 1981 to 2014 based on fabrication methods and sensing applications. They also discussed the development trends in this research direction and provided an overview of successfully implemented fiber optic FP sensors in industrial settings. Reference [15] reported on the research progress of fiber optic sensors based on FP interference. The authors conducted theoretical and experimental analyses of FP sensors in sensing applications such as temperature, displacement, and the refractive indices of liquids and solids. They also provided prospects for future applications of FP sensors in other fields. In contrast to previous review articles, this paper begins with a comprehensive introduction elucidating the principles and sensing mechanisms of fiber sensors employing the FP cavity with thin film. A meticulous analysis is conducted to explore the factors influencing the sensitivity of these sensors. Subsequently, a comprehensive survey is presented, highlighting the research progress of these sensors across diverse domains, encompassing pressure, magnetic field, refractive index, humidity, gas, temperature, as well as biological or medical sensing. Furthermore, an overview is provided, elucidating the diverse thin film materials and FP cavity structures utilized to accomplish these sensing applications.

2. Principle

The Fabry–Perot cavity mainly plays a role in changing the spectrum in the fiber optic sensor technology, changing the external parameters. In conventional fiber optic FP interferometers, one of the cavity mirrors is typically formed by a cleaved single-mode fiber (SMF), while the other mirror consists of a mirror parallel to the fiber end-face. This parallel mirror can be a thin film or another section of the fiber end-face. Upon the emission of light from the fiber core, it traverses the fiber end-face and encounters the mirror, leading to multiple reflections and interference. The optical path difference between consecutive reflections is determined as 2nd, and the corresponding phase difference can be expressed as follows:

$$\varphi = \frac{4\pi n d}{\lambda} \tag{1}$$

where n represents the refractive index of the medium between the two cavity mirrors, d denotes the separation distance between the two mirrors, and λ signifies the wavelength of the incident light wave.

Considering an incident light wave with an amplitude of E and an initial phase factor of 0, the fiber end-face is characterized by transmission coefficient t_1 and reflection coefficient r_1. In the counter-propagating direction, the transmission coefficient is denoted as $t_1{'}$, and the reflection coefficient is represented as $r_1{'}$. Additionally, the mirror surface exhibits a reflection coefficient denoted as r_2. Within the fiber core, all reflected light waves experience interference, resulting in a complex amplitude expressed as [16]:

$$\begin{aligned} E_r &= r_1 E e^0 + t_1 t_1' r_2 E e^{-i\varphi} + t_1 t_1' r_1' r_2^2 E e^{-i2\varphi} + \cdots + t_1 t_1' r_1'^{m-1} r_2^m E e^{-im\varphi} \\ &= \left[r_1 E + t_1 t_1' r_2 E e^{-i\varphi} \left(1 + r_1' r_2 e^{-i\varphi} + \cdots + r_1'^{m-1} r_2^{m-1} e^{-i(m-1)\varphi} \right) \right] \\ &= r_1 E + t_1 t_1' r_2 E e^{-i\varphi} \frac{1 - r_1'^m r_2^m e^{-im\varphi}}{1 - r_1' r_2 \eta e^{-i\varphi}} \end{aligned} \tag{2}$$

where m represents the order of the reflected light. The parentheses in the equation denote the sum of a geometric series. When m tends to infinity, we obtain:

$$E_r = r_1 E + t_1 t'_1 r_2 E e^{-i\varphi} \frac{1}{1 - r'_1 r_2 e^{-i\varphi}} = \frac{r_1 E + (t_1 t'_1 - r'_1 r_1) r_2 E e^{-i\varphi}}{1 - r'_1 r_2 e^{-i\varphi}} \quad (3)$$

according to Fresnel equations:

$$\begin{aligned} r_1^2 &= r'^2_1 = R \\ t_1 t'_1 &= 1 - r_1^2 = T \end{aligned} \quad (4)$$

the total optical intensity of the fiber-reflected FPI is:

$$I_R = E_r \cdot E_r^* = I_0 \cdot \frac{R_1 + R_2 - 2\sqrt{R_1 R_2} \cos\varphi}{1 + R_1 R_2 - 2\sqrt{R_1 R_2} \cos\varphi} \quad (5)$$

In the scenario where the reflectivity R is extremely low, Equation (5) can be simplified to the following form:

$$I_R = 2R(1 - \cos\varphi) I_0 \quad (6)$$

the transmitted light intensity is given by the following expression:

$$I_T = I_0 - I_R = I_0 [1 - 2R(1 - \cos\varphi)] \quad (7)$$

It can be observed that the magnitude of the above expression primarily depends on the phase difference φ. When:

$$\varphi = \frac{4\pi n d}{\lambda_k} = 2k\pi \quad (8)$$

The interference of the reflected light results in destructive interference, characterized by the positive integer k representing the order of resonant wavelengths. Hence, we can express it as follows:

$$\lambda_k = \frac{2nd}{k} \quad (9)$$

differentiating both sides of Equation (9), we obtain:

$$\Delta\lambda_k = \lambda_k \left(\frac{\Delta n}{n} + \frac{\Delta d}{d} \right) \quad (10)$$

Herein, $\Delta\lambda_k$ represents the variation in resonant wavelength, λ_k signifies the resonant wavelength, Δn denotes the change in the refractive index of the medium, and Δd represents the alteration in cavity length. Consequently, modifications in the cavity length d or refractive index n within the cavity induce shifts in the wavelength where destructive interference transpires, thus causing alterations in the interference spectrum. Perturbations in external environmental factors, such as temperature, humidity, or magnetic field, prompt deformations in the thin film, ultimately influencing the FP cavity length or the refractive index of the medium within the cavity. Consequently, the resonant wavelength experiences changes. By measuring the shift in resonant wavelength, the corresponding variation in the external environmental parameter can be determined. Through experimental investigations that establish the relationship between the shift in resonant wavelength and the corresponding environmental parameter, the detection of wavelength shifts enables the inference of changes in the environmental parameter. Thus, the sensing objective can be accomplished.

The aforementioned formula applies to cases where the thin film is extremely thin, rendering its thickness negligible. In this scenario, only two reflecting surfaces need to be considered. When a thin film of a certain thickness is used as the reflecting film in the FP cavity, the consideration of three reflecting surfaces is necessary, as shown in Figure 2. The interfaces M_1 and M_2 form FP_1, while M_2 and M_3 form FP_2. The interface reflectance (R) is given by the following equation:

$$R_1 = \left(\frac{n_1-n_2}{n_1+n_2}\right)^2$$
$$R_2 = R_3 = \left(\frac{n_3-n_2}{n_3+n_2}\right)^2 \tag{11}$$

where n_1, n_2, and n_3 represent the refractive indices of the fiber, air, and the thin film, respectively.

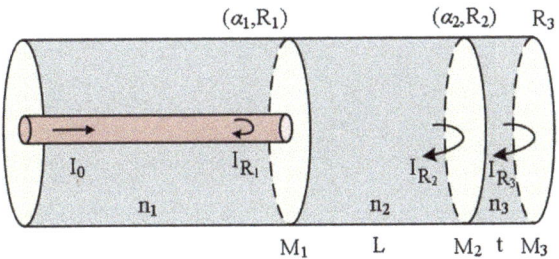

Figure 2. Description of the Optical Field in Thin Film-Based Fiber FP Cavity.

In this analysis, we account for the interference of three reflected beams within the SMF while disregarding higher-order terms. By considering the electric field amplitudes and phase factors, the corresponding expression can be derived as follows:

$$\begin{aligned} E_r = &\sqrt{R_1}E_0 + (1-\alpha_1)(1-R_1)\sqrt{R_2}E_0 e^{-j2\varphi_1} \\ &+ (1-\alpha_2)(1-\alpha_1)(1-R_2)(1-R_1)\sqrt{R_3}E_0 e^{-j2(\varphi_1+\varphi_2)} \end{aligned} \tag{12}$$

where E_r represents the total amplitude of the interference field, E_0 represents the initial amplitude of the incident field, and α represents the optical transmission loss coefficient. The phase factors φ_1 and φ_2 are given by:

$$\varphi_1 = \frac{4\pi n_2 L}{\lambda}, \varphi_2 = \frac{4\pi n_3 t}{\lambda} \tag{13}$$

By representing the power of the FP reflected light as the ratio between the reflected electric field intensity and the incident electric field intensity, the following expression can be derived [17]:

$$\begin{aligned} I_r(\lambda) = &|E_r/E_0|^2 \\ = &R_1 + (1-\alpha_1)^2(1-R_1)^2 R_2 \\ &+ (1-\alpha_1)^2(1-\alpha_2)^2 \times (1-R_1)^2(1-R_2)^2 R_3 \\ &+ 2\sqrt{R_1 R_2}(1-\alpha_1)(1-R_1)\cos(\varphi_1) \\ &+ 2\sqrt{R_2 R_3}\left[\begin{array}{c}(1-\alpha_1)^2(1-\alpha_2)(1-R_1)^2(1-R_2)\times \cos(\varphi_2) \\ +2\sqrt{R_1 R_3}(1-\alpha_1)(1-\alpha_2)(1-R_1)\times(1-R_2)\cos(\varphi_1+\varphi_2)\end{array}\right] \end{aligned} \tag{14}$$

It can be observed that the reflected power in the above equation is primarily composed of three cosine functions that are linearly superimposed. Considering that M1 and M2 form FP1, M2 and M3 form FP2, and M1 and M3 form FP3, each cosine term corresponds to the optical path difference of FP1, FP2, and FP3 cavities, respectively. By employing bandpass filtering techniques, the signals corresponding to different FP cavities can be extracted from the overall spectrum. Typically, when external parameters such as temperature or pressure change, the deformation of the thin film causes a variation in the length of FP1, leading to a spectral shift in the interference pattern associated with FP1. By measuring the displacement of the resonant wavelength, the corresponding change in the environmental parameter can be determined.

When evaluating the performance of a sensor, it is important to consider not only sensitivity but also other parameters, such as the Figure of Merit and Limit of detection. The Figure of Merit (FOM) is an important metric for directly quantifying sensor performance. It is defined as the ratio of the sensor's sensitivity and FWHM, where FWHM is full width at half-maximum of the interference spectrum [18]. As for Fabry–Perot interference, its full width at half maximum (FWHM) is defined as follows [19]:

$$FWHM = \frac{2(1-R)}{\sqrt{R}} \qquad (15)$$

Here, R is the film reflectance. It is evident that to enhance the FOM of a sensor, we can either improve its sensitivity or increase thin film reflectance. The Limit of Detection (LOD) for a sensor is the minimum detectable change in the measured physical quantity. When using spectral shift, the LOD of a sensor is calculated by dividing the wavelength resolution by the sensor's sensitivity [20]. In order to enhance LOD, it is recommended to utilize a spectrometer with greater spectral resolution, opt for a consistent light source, and mitigate environmental disturbances.

It should be noted that in addition to the conventional Fabry–Pérot (FP) cavity mentioned above, there are other FP-like structures in optical fibers, such as the anti-resonant reflecting optical waveguide (ARROW). Proposed by Duguya et al. in 1986 [21], ARROW emerged to address the issue of electromagnetic wave energy leakage caused by the significant difference in refractive index between traditional optical waveguides and silicon-based waveguides. Similar to the FP cavity, ARROW employs a higher refractive index layer between the waveguide and the substrate to confine the propagation of light beams in the waveguide layer. The ARROW structure not only reduces the energy leakage but also elevates the percentage of the energy distribution of the evanescent field in the material being evaluated, enhancing the sensitivity of the waveguide to changes in external physical quantities. Initially, ARROW's application was focused on the field of silicon-based waveguides, with little cross-section with the optical fiber. However, as the photonic crystal fiber preparation technology has matured and improved, ARROW has gradually entered the research scope of fiber sensing. Reference [22] has conducted theoretical analysis and discussion on the anti-resonant mechanism in photonic crystal waveguides, proving the anti-resonant interference phenomenon of light in hollow-core photonic crystal fibers. This provides significant insight for the future application of ARROW in photonic crystal fibers and also opens up a new path for the integration of single-layer quartz tubes and fiber sensing areas. Hence, for integrated waveguide optical sensing, ARROW offers excellent sensing performance and application prospects [23,24].

3. Research Progress of Optical Fiber Fabry–Perot Cavity Sensors

Fiber optic FP cavity sensing is a critical research field that has witnessed significant advancements in recent years. This technology utilizes the FP cavity structure within an optical fiber to achieve high-sensitivity detection of environmental parameters by monitoring the phase variation of the light signal. These sensors offer advantages such as simple structure, high sensitivity, and rapid response, making them widely applicable in fields such as pressure, magnetic field, refractive index, humidity, gas detection, temperature, and biomedical applications. By optimizing the FP cavity structure and adjusting the types of thin film employed, the performances of these sensors can be further enhanced. The research on fiber optic FP cavity sensors and thin film provides essential technical support for achieving high-precision and high-sensitivity monitoring of environmental parameters, opening up new possibilities for scientific research and engineering applications across various domains.

3.1. Pressure Sensor

Pressure measurement is crucial in diverse domains, including oceanography, geology, and medical applications. Notably, significant progress has been made in recent years with the development of various fiber optic pressure sensors. These sensors employ different principles, such as Fiber Bragg Gratings (FBGs) [25,26], Surface Plasmon Resonance (SPR) [27,28], Multimode Interference (MMI) [29,30], and Mach–Zehnder interferometer (MZI) [31,32]. This section primarily focuses on recent advancements in fiber optic FP pressure sensors utilizing thin film technology.

Silicon dioxide (SiO_2) thin diaphragms possess ideal attributes such as high temperature resistance, structural stability, and chemical inertness [33], rendering them suitable for measurements in extreme environments. In [34], a compact fiber optic FP pressure sensor based on a silicon diaphragm was proposed, as shown in Figure 3. The authors utilized a carbon dioxide laser to weld an SMF to a capillary tube, creating a sealed FP cavity between the end-face of the SMF and the silicon diaphragm on the capillary tube's end-face. When the external pressure changes, the silicon diaphragm undergoes deformation, altering the cavity length of the FP cavity. This change in the optical path difference between the two end-faces results in a shift in the interference spectrum, achieving a sensitivity of 9.48 pm/kPa. Additionally, an alternative approach was proposed by [35,36]. This method entailed etching a groove at the fiber end using hydrofluoric acid, followed by bonding the silicon diaphragm to the fiber end through the application of heat. Consequently, FP cavities were formed, exhibiting sensitivities of 11 nm/kPa and 12.4 nm/kPa, respectively.

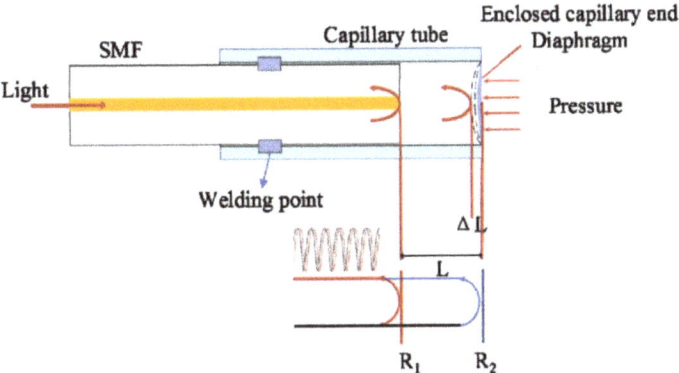

Figure 3. Formation of an FP cavity using an SMF and a capillary tube [34].

However, both silicon and silicon dioxide possess a high Young's modulus [37]. Typically, optical fiber dimensions are around one hundred micrometers, and even with the optimal thickness of the silicon diaphragm, achieving high sensitivity in such dimensions remains challenging. In recent years, considerable research interest has been directed toward the utilization of polymer film for pressure sensing. This film offer advantages such as a low Young's modulus, cost-effectiveness, and flexibility [38,39]. Among the various options, polydimethylsiloxane (PDMS) stands out as a polymeric organosilicon compound. Solid PDMS represents a transparent elastomer that can be easily and swiftly processed. It is cost-effective, optically transparent, and readily bonded with different materials at room temperature [40]. The remarkable elasticity of PDMS can be attributed to its low Young's modulus [41], rendering it an excellent choice as a thin film material for pressure-sensing applications. Reference [42] proposed a fiber optic pressure sensor utilizing PDMS as the reflective film for the FP cavity, as depicted in Figure 4. The authors fused a glass tube to the end of an SMF, followed by the deposition of a PDMS film layer at the end of the glass tube. The incident light beam undergoes the first reflection at R_1 (end-face of the SMF) and the second reflection at R_2 (left side of the PDMS film) after passing through the FP gas chamber. Subsequently, it undergoes the third reflection at R_3 (the right side of the PDMS film). These three surface-reflected beams interfere inside the SMF. When external pressure increases, the polymer film undergoes deformation, resulting in a change in the FP cavity length and a shift in the interference spectrum. Compared to similar research works on PDMS-based pressure sensors, the authors significantly reduced the thickness of the film. This reduction not only enhances the film's response to external pressure but also effectively mitigates the impact of temperature on the pressure sensor. Experimental data show that the pressure sensitivity of the sensor reaches 100 pm/kPa. Similar studies on PDMS-based pressure sensors include [43], which filled a hollow-core fiber with PDMS film at the fiber end to form an FP cavity for measuring gas pressure, achieving a pressure sensitivity of 52.143 nm/MPa. In references [44,45], PDMS film was filled in a capillary tube to form an FP cavity with the end of an SMF for gas pressure measurement, yielding pressure sensitivities of 20.63 nm/MPa and 55 pm/mBar, respectively.

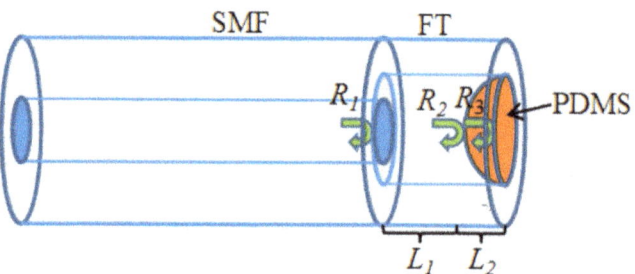

Figure 4. Fiber FP pressure sensors based on PDMS thin film [42].

The fabrication of PDMS film possesses challenges in controlling its thickness and achieving uniform symmetry. Ultraviolet curable polymers are a promising candidate for pressure-sensing film due to their ability to solidify rapidly under ultraviolet (UV) light exposure without requiring high-temperature curing [46,47]. In reference [48], UV-curable polymer film was coated on the end of a hollow core silica tube, forming an FP cavity. The film deforms under external pressure, resulting in a change in cavity length and a corresponding shift in the interference spectrum. Experimental results demonstrated a pressure sensitivity of approximately 396 pm/kPa within the range of 0 to 30 kPa. The thickness of the polymer film is also a crucial factor influencing the sensing performance. To address this, the authors employed a suspension curing method to facilitate precise control over the formation of the UV-curable polymer layer, enabling easy regulation of the film thickness. Additionally, reference [49] proposed a fiber optic FP sensor for perceiving sound pressure using a UV adhesive diaphragm. In this sensor configuration, an FP cavity is formed between the end of an SMF and the UV adhesive diaphragm. The sensing principle is similar to the previous approach. The authors obtained the sound pressure signal by demodulating the changes in reflected optical intensity. Experimental results indicate a sensitivity of 57.3 mV/Pa and a detection range of 21.4 mPa to 3.56 Pa for the sound pressure sensor at a frequency of 1000 Hz. In reference [50], the author employed a welding technique to attach a quartz glass tube to the end of an SMF. Subsequently, a film ultra-thin microbubble structure was created on a liquid-state AB epoxy glue using a bubble-blowing method. Once the bubble solidified, it was transferred to the end of the quartz glass tube, thereby establishing an FP cavity in conjunction with the SMF. Based on the experimental data, the pressure sensor exhibited a sensitivity of 263.15 pm/kPa within the pressure range of 100.0 kPa to 400.0 kPa.

In addition to the devices that utilize polymer as reflective film, there is another approach where polymers are applied externally to the fiber to enhance its pressure-sensing properties. Reference [51] presented an optical fiber FP sensor for measuring water pressure variations. The pressure sensor consisted of two SMFs and an interposed portion of a photopolymerizable resin between their end-faces, forming an FP cavity. When the external water pressure changes, the resin undergoes strain (the study's analysis suggests that the physical length change of the polymer contributes more significantly than the refractive index change), leading to a variation in the FP cavity length and a corresponding shift in the interference spectrum. The study indicates that with an increase in the number of polymer resin layers coated on the outside of the FP cavity, the response of the FP cavity length to strain becomes larger. Consequently, the static hydrostatic pressure sensitivity of the fiber FP pressure sensor increases from 102 to 475 pm/MPa, and its detection range spans from 0 to 20 MPa.

Using thinner film layers can effectively enhance the sensitivity of sensors. Graphene, the thinnest material discovered to date, with a thickness as low as around 0.3 nm [52], exhibits exceptionally high mechanical strength [53]. Utilizing graphene as a thin film can greatly improve the pressure sensitivity of sensors. Reference [54] presented a setup where a capillary tube was fusion-spliced to the end of an SMF, and a graphene film with a thickness of approximately 0.71 nm was transferred to the other end of the capillary tube, forming an FP cavity with the end-face of the SMF. Experimental results demonstrated an average pressure sensitivity of 39.4 nm/kPa within the range of 0–5 kPa. However, the authors observed slight gas leakage into the cavity during the experiments, leading to some measurement errors. Reference [55] proposed a similar pressure sensor based on graphene film. In this setup, the end-face of an SMF was etched to create a groove, and a graphene film was transferred onto the fiber end-face. Experimental results showed a response of 1.28 nm/mmHg for pressure variations in the range of 0–100 mmHg. The authors also provided theoretical analysis for such sensors and proposed a critical thickness for the film. When the film thickness is below this threshold, the sensitivity of the sensor is no longer influenced by the film's thickness and elastic properties, and the performance can be enhanced through FP cavity design. Similarly, in reference [56], an FP cavity formed by the end-face of an SMF and graphene film is utilized for sensing changes in sound pressure, achieving a sensitivity of 43.5 dB re 1 V/Pa at 60 Hz. In reference [57], an FP cavity is formed by the end-face of a multimode fiber and a graphene film for pressure sensing, achieving a sensitivity of 79.956 nm/kPa.

Compared to the aforementioned materials, metal film offers unique advantages due to its high reflectivity and ideal properties, such as being lightweight, physically stable, and easy to manufacture. In reference [58], the authors employed a ceramic ring as a cavity and constructed an FP interferometer using SMF end-faces and a silver film at both ends. When the silver-coated sensing membrane is deformed by external pressure, it causes a change in the length of the FP cavity. The experimental result presented by the authors in the paper is the response of cavity length to external pressure. The sensitivity of cavity length variation for this sensor within the range of 0–10 mPa is reported to be 1677 nm/MPa, with a stress measurement resolution of 60 Pa. In a similar manner, reference [59] also proposed an FP structure enhanced with a silver film with a static pressure sensitivity of 1.6 nm/kPa.

Gold film can also be employed as pressure-sensitive membranes, offering greater chemical stability compared to silver film. Reference [60] proposed a fiber optic FP pressure sensor with a gold film. In this study, the authors placed an SMF and a gold film at the ends of a ceramic ferrule to form the FP cavity, following the same sensing principle as the silver film. Experimental results demonstrated that within the range of 0–100 kPa, this method exhibits a static pressure sensitivity of approximately 19.5 nm/kPa. FP fiber optic pressure sensors have been commercially applied in various examples. For instance, Roctest Company offers the FOP series (FOP-F, FOP-C, and FOP-P) fiber optic piezometers based on stainless steel diaphragms, primarily used in civil engineering. These sensors provide accurate and reliable measurements unaffected by electromagnetic, RF, and lightning interference. Additionally, FISO's FOP-M sensor is designed for pressure measurement under high-temperature conditions, such as in aerospace and automotive research and development, where harsh and hazardous environments are common. The FOP-M pressure sensor exhibits resistance to EMI/RFI/MW, compact size, and reliable measurements in adverse conditions. It boasts good operating conditions, high accuracy, and resistance to corrosive environments. The performances of fiber optic FP pressure sensors based on different thin films, as discussed in this section, are summarized in Table 1.

Table 1. Pressure sensor performance based on different materials.

Material	Sensitivity	Test Range	Reference
Silicon	9.48 pm/kPa	0~200 kPa	[34]
	11 nm/kPa	0~100 kPa	[35]
	12.4 nm/kPa	6.9~48.3 kPa	[36]
PDMS	100 pm/kPa	100~175 kPa	[42]
	52.143 nm/Mpa	0.1~0.7 Mpa	[43]
	20.63 nm/Mpa	0~2 MPa	[44]
	55 pm/mBar	0~50 mBar	[45]
UV	395 pm/kPa	0~30 kPa	[48]
	57.3 mV/Pa	21.4 mPa~3.56 Pa	[49]
AB epoxy glue	263.15 pm/kPa	100.0 ~400.0 kPa	[50]
Graphene	39.4 nm/kPa	0~5 kPa	[54]
	1.28 nm/mmHg	0~100 mmHg	[55]
Silver	1.6 nm/kPa	0~50 psi	[59]
Gold	19.5 nm/kPa	0~100 kPa	[60]

3.2. Magnetic Field Sensor

Magnetic field measurement is very important in various fields, such as scientific research and industrial production. Due to the excellent electromagnetic interference resistance of optical fibers, they have proven to be outstanding magnetic field sensors [61]. Currently, the commonly used material for fiber optic magnetic field sensing is magnetic fluid, which is a composite material formed by suspending magnetic particles in a liquid. These magnetic particles are typically made of ferrite or metallic materials such as iron, nickel, and cobalt, while the liquid base can be water or organic solvents. Under the influence of an external magnetic field, the magnetic fluid particles are guided by magnetic forces, forming chain-like, cluster-like, or ordered structures, thereby altering the morphology and properties of the magnetic fluid. When the external magnetic field disappears, the magnetic fluid particles return to a freely suspended state [62,63]. In recent years, there has been significant development of fiber optic sensors utilizing the properties of magnetic fluids, such as magnetic field sensors based on the MZI [64,65], the Michelson interferometer [66,67], and the multimode interference [68,69]. A wide range of fiber optic magnetic field sensors based on FP Interferometers have been proposed. In reference [70], a reflective FP magnetic field sensor based on the Magneto-Volume Effect of Magnetic Fluid was presented. The authors constructed a fiber optic FP cavity by enclosing two segments of single-mode fibers (SMFs) within a capillary. Subsequently, a precisely measured quantity of magnetic fluid was introduced into the cavity, while a specific section of the cavity was intentionally left empty. As the incident light sequentially traversed the SMF and the empty cavity, it eventually reached the interface of the magnetic fluid (MF), encountering two reflective surfaces: the SMF interface and the MF interface. The reflected light from these two interfaces interfered with the SMF. When an axial magnetic field was applied to the sensor, the magnetic fluid underwent deformation, resulting in changes in the lengths of the empty cavity which served as the FP cavity. Experimental results demonstrated that within the range of 15.5–139.7 G, the sensitivity of this magnetic field sensor reached 268.81 pm/G. Under a step-change magnetic field, the sensor exhibited a rapid response time of 0.2 s. Another similar FP cavity based on the Magneto-Volume Effect of Magnetic Fluid was reported in reference [71]. The authors fusion-spliced an SMF with an HCF and filled the HCF with a controlled amount of magnetic fluid, leaving a section of air column. The interference principle is the same as described above. Experimental results showed that within the range of 109.6–125.8 G, the sensitivity of this sensor was −4219.15 pm/G. However, within the broader range of 0–125.8 G, the linearity of the sensor's response was not ideal, leading to a limited sensing range.

Apart from experiencing changes in volume, magnetic fluid also demonstrates alterations in refractive index due to fluctuations in the magnetic field. Reference [72] proposed a fiber optic FP magnetic field sensor based on the refractive index effect of magnetic fluid. The sensor's FP cavity consisted of an SMF, a capillary glass tube, and the SMF end-face coated with a gold film. The authors inserted the SMF into the capillary tube, leaving a section of air column in the middle, and then encapsulated it with another SMF coated with a gold film. Magnetic fluid was filled in between the capillary tube and the SMF. Due to capillary action [73], the surface of MF became curved and appeared concave. Moreover, the length of the cavity containing the magnetic fluid was very small, making the impact of volume changes negligible. Under the influence of an external magnetic field, the refractive index of the magnetic fluid changed, causing the interference spectrum to shift. Experimental results demonstrated that within the range of 118.768~166.261 G, the sensitivity of the sensor can reach 1.02602 nm/G.

Due to the high viscosity of the magnetic fluid, its response time to magnetic fields is slow, and it is prone to leakage. Researchers have begun investigating the utilization of polymer materials filled with magnetic particles as an alternative to magnetic fluid for sensing applications. Magnetic-Based Polydimethylsiloxane is a magnetic-sensitive polymer formed by filling ferromagnetic material (Mn_3O_4 nanocrystals) into PDMS. In reference [74], the authors used this material as a thin film inside the FP cavity for dual parameter sensing of temperature and magnetic fields. The magnetic-sensitive polymer is coated at the end of an SMF, which forms an FP cavity with another SMF end-face. When there are changes in the external magnetic field or temperature, the magnetic-sensitive material undergoes deformation, causing a change in the FP cavity length and leading to a shift in the interference spectrum. The linear sensitivity of this sensor within the linear magnetic field intensity range of 0–4 mT is 563.2 pm/mT. Reference [75] proposed a fiber optic FP magnetic field sensor based on a magnetic alloy, as shown in Figure 5a. The FP cavity was composed of Fiber Bragg Gratings (FBGs) in two segments of the SMFs, which were filled with silicone between the two SMFs. The device was then bonded to a magnetic alloy called iron–cobalt–vanadium Supermendur using UV glue. This magnetic alloy has a high saturation magnetic flux density (2.4 T) and extremely low magnetic field sensitivity, making it suitable for sensing in a wide range but not for situations with small magnetic fields. When there was a change in the external magnetic field, the magnetic alloy underwent deformation. Due to the lower Young's modulus of silicone compared to silica, the length of the FP cavity changed while the FBG experienced no strain. With the help of this FP structure, its performance could be significantly improved, enabling measurements in the presence of small magnetic fields. The experimental results indicated that the sensitivity of this sensor within the range of 0–70 mT is −34.83 pm/mT. Reference [76] proposed a magnetic field sensor based on magnetostrictive materials. The FP cavity was formed by fusion splicing a segment of hollow-core fiber between two segments of SMFs, and it was fixed onto a magnetostrictive material called Terfenol-D. This material underwent strain when the magnetic field changed, resulting in a change in the length of the FP cavity. However, the response of this material to the magnetic field was nonlinear within a wide range, and Terfenol-D ceased to respond when the magnetic field intensity exceeded 60 mT. Experimental data showed that the sensitivity within the linear range of 10–30 mT is 14.6 pm/mT. Similar approaches using Terfenol-D for magnetic field sensing are also described in [77], where the authors' fusion-spliced a segment of hollow core fiber between an SMF and a hollow silica capillary, and Terfenol-D was attached, as shown in Figure 5b. Experimental data showed a sensitivity of −7.53 nm/mT within the range of 4–10 mT. The performance of the magnetic field sensors described in this section is summarized in Table 2.

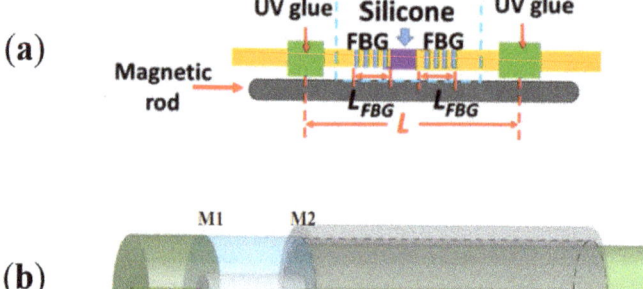

Figure 5. (**a**) Magnetic field sensors based on magnetic alloys [75] (**b**) Magnetic field sensors based on Terferol-D [77].

Table 2. Magnetic field sensor performance based on different materials.

Material	Sensitivity (pm/mT)	Test Range (mT)	Reference
MF	2688.1	1.55~13.97	[70]
	−42,191.5	10.96~12.58	[71]
	10,260.2	11.8768~16.6261	[72]
Mn_3O_4-PDMS	563.2	0~4	[74]
Magnetic alloy	−34.83	0~70	[75]
Terfenol-D	14.6	10~30	[76]
	−7530	4~10	[77]

3.3. Refractive Index Sensing

In the fields of biology, chemistry, and materials science, refractive index measurement plays a crucial role. Various optical fiber sensors are employed for refractive index measurement, including those based on the whispering gallery mode (WGM) interference [78,79], MZI [80,81], surface plasmon resonance (SPR) [82,83], and Bragg fiber gratings [84,85]. To measure the refractive index of a solution, a semi-open Fabry–Perot (FP) cavity can be employed, with a thin film coated on the end-face of the cavity to enhance reflectivity and ensure a high contrast of interference fringes. Reference [86] proposed an optical fiber FP refractive index sensor. A cavity was etched on the end-face of an SMF using laser ablation to create the FP cavity. SiO/TiO film was coated on the reflective interface of the FP cavity as mirrors. A microchannel was present at the fiber end, serving as an inlet for the tested liquid or gas. It is necessary for the microchannel to be sufficiently small to maintain the optimal performance of the FP cavity. When the liquid or gas entered through the small hole, it changed the refractive index in the FP cavity, resulting in a shift in the interference spectrum. The experimental results demonstrated a high sensitivity of 1130.887 nm/RI, indicating the sensor's capability for accurate refractive index measurements. Furthermore, by adjusting the cavity length or enhancing the reflectivity, the sensor response and optical performance can be effectively optimized. A similar refractive index sensor was proposed in reference [87]. In this case, a silver film was used as the reflective coating. The measured refractive index sensitivity is 1025 nm/RIU. Reference [88] suggested the use of ZnO coated on both ends of an SMF as a thin film to form the FP cavity for refractive index measurement of chemical substances. Reference [89] introduced a fiber refractive index sensor, where two SMF ends formed an FP cavity, which was subsequently filled with UV glue. This sensor responded to changes in the refractive index of the liquid based on the evanescent field effect [90], with a refractive index sensitivity of 156.8 nm/RIU. Reference [91] presented a simple fiber FP sensor with a structure similar to the previous one. The authors filled the FP cavity with three different polymers, UV88, NOA68, and Loctite 3525. The refractive index sensing performance of these three materials is shown in the table below, along with the refractive index sensing performance of other refractive index sensors mentioned in this section (Table 3).

Table 3. Refractive index sensor performance based on different materials.

Material	Sensitivity (nm/RIU)	Reference
SiO/TiO	1130.887	[86]
Silver	1025	[87]
UV	156.8	[89]
UV88	24.678	[91]
NOA68	81.096	[91]
Loctite 3525	34.395	[91]

3.4. Humidity Sensor

With increasing attention to meteorology, agriculture, industry, and indoor environments, the importance of humidity sensors has become more prominent. Fiber optic humidity sensors have been proposed, such as humidity sensors based on FBG [92,93], long-period fiber gratings (LPFG) [94,95], MZI [96,97], and Michelson interferometers [98,99]. Polymethyl methacrylate (PMMA) is a commonly used synthetic polymer, also known as acrylic or organic glass, which can absorb or release moisture, causing volume changes. When the humidity increases, PMMA absorbs water molecules and expands, resulting in an increase in volume. When the humidity decreases, PMMA releases water molecules and contracts. This property can be utilized to detect humidity changes and is less susceptible to external environmental interference [100]. Reference [101] proposed an optical fiber humidity sensor based on an external FP interferometer structure. The authors fusion-spliced a section of coreless fiber (CLF) to the end of an SMF to enlarge the beam diameter and improve the stability of the extinction ratio of the interference spectrum. The end-faces of CLF and another SMF formed an FP cavity, which was filled with PMMA. When the external humidity changed, PMMA underwent deformation, causing a change in the FP cavity length and resulting in a shift in the interference spectrum. Experimental results showed that within the range of 25% to 80% relative humidity (RH), the RH sensitivity was 0.1747 nm/%RH, with a response time of 4.5 min, and the sensor was unaffected by temperature change in the range of 30–55 °C. Reference [102] utilized a multimode fiber end-face and an SMF end-face to form an FP cavity and filled the cavity with PMMA for humidity sensing. Within the range of 35% to 85% RH, the sensor achieved a sensitivity of 127 pm/%RH. Reference [103] wrapped PMMA material around the end of an SMF to form an FP cavity for humidity sensing. Within the range of 10% to 70% RH, the sensor achieved a sensitivity of 0.4172 nm/%RH.

There are other kinds of polymers that can be used for humidity sensing. Reference [104] proposed an open FP cavity, which was filled with nanocomposite polyacrylamide (PAM). PAM exhibits a change in refractive index when it absorbs water molecules [105]. A change in external humidity causes a shift in the interference spectrum output by the sensor. Experimental results showed that the relative humidity (RH) sensitivity of this sensor was approximately 0.1 nm/%RH in the range of 38% to 78% RH, and approximately 5.868 nm/%RH in the range of 88% to 98% RH. Reference [106] suggested using two Bragg fiber gratings as an FP cavity and filling the gap between the two Bragg fiber gratings with agarose gel. This sensor exploited the property of agarose gel to expand and change the refractive index with increasing humidity [107]. Moreover, FBGs are sensitive to temperature, which effectively solves the problem of cross-sensitivity between temperature and humidity. Within the range of 43% to 63% RH, the average sensitivity of this sensor was 22.5 pm/%RH. Reference [108] proposed filling the gap between two SMF end-faces with poly(N-isopropyl acrylamide) (PNIPAM) hydrogel, as shown in Figure 6. This sensor utilized the refractive index change property of PNIPAM with humidity variation [109]. In the relative humidity range of 45% to 75%, the measured relative humidity sensitivity was 1.634 nm/%RH, with good repeatability.

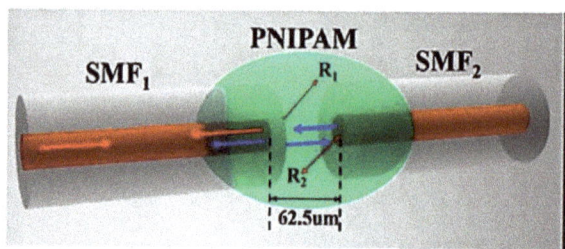

Figure 6. Humidity sensors based on the use of PNIPAM [108].

Using polymers as reflective film in the FP cavity for humidity sensing is one of the main techniques used today. In reference [110], a sensor for simultaneous measurement of temperature and relative humidity was proposed, consisting of an FBG and an FP interferometer. The author fusion-spliced an SMF with a hollow capillary tube and coated the end of the capillary tube with a layer of polyimide film. The film and the end-face of the SMF formed the FP cavity, and an FBG was cascaded to eliminate the influence of temperature changes. When the external humidity changed, the polyimide underwent deformation [111], resulting in a change in the length of the FP cavity and a shift in the interference spectrum. Experimental results indicated that the sensitivity of this sensor in the range of 20% to 90% relative humidity (RH) was 22.07 pm/%RH. Reference [112] used polyvinylidene fluoride (PVDF) as the humidity-sensing film. The structure and sensing principle was similar to the previous approach. The sensor exhibited a sensitivity of 32.54 pm/%RH at a constant temperature. Furthermore, reference [113] proposed a humidity sensor based on polyvinyl alcohol (PVA) film. The author fusion-spliced a section of hollow-core fiber between two segments of SMFs and coated the other end of the second segment of SMF with PVA film. The end-faces of the two aligned SMFs formed FP_1, and the two ends of the second segment of SMF formed FP_2. When the external humidity changed, the PVA film underwent deformation [114], causing a change in the length of the FP_2 cavity and resulting in a spectral shift. The cascaded FP cavities generated a vernier effect, enhancing the sensitivity of the sensor [115]. The author also mentioned that the envelope spectrum of the sensing cavity and the two FP cavities can be separated using a bandpass filter, effectively addressing the issue of cross-sensitivity between temperature and humidity. Experimental results indicated that when the humidity increased from 35% RH to 85% RH, the sensitivity to humidity reached 1.454 nm/%RH.

Besides polymers for humidity sensing, there are other materials that can be used in fiber optic FP cavity humidity sensors. Reference [116] proposed a fiber optic FP humidity sensor based on porous anodic alumina (PAA) film. The author used UV adhesive to attach the porous PAA film to the end of an SMF and the capillary structure of the film caused changes in the effective refractive index due to the condensation of water molecules [117]. Experimental results showed that the sensitivity in the range of 20% to 90% RH could reach 0.31 nm/%RH. Reference [118] presented an FP cavity utilizing the hydrophilic properties of graphene oxide. The author fusion-spliced a capillary tube to the end of an SMF and added a graphene oxide diaphragm at the end of the capillary tube. Due to the absorption and desorption behavior of water molecules by graphene oxide, it underwent expansion or contraction [119], resulting in a change in the length of the FP cavity and a shift in the interference spectrum. Experimental results demonstrated that this humidity sensor exhibited an average wavelength variation of 0.2 nm/%RH in the range of 10% to 90% RH. The performance of fiber optic FP humidity sensors based on different film is summarized in the Table 4 in this section.

Table 4. Humidity sensor performance based on different materials.

Material	Sensitivity (nm/%RH)	Range (%RH)	Reference
PMMA	0.1747	25~80	[101]
	0.127	35~85	[102]
	0.4172	10~70	[102]
PAM	0.1	38~78	[104]
	5.868	88~98	[104]
Agarose gel	0.0225	43~63	[106]
PNIPAM	1.634	45~75	[108]
POLYIMIDE	0.02207	20~90	[110]
PVDF	0.03254	20~80	[112]
PVA	0.001454	35~85	[113]
PAA	0.31	20~90	[116]
Graphene oxide	0.2	10~90	[118]

3.5. Gas Detection

Gas detection plays a crucial role in various aspects of everyday life, such as ensuring the safety of spaces, environmental protection, industrial production, health preservation, and disaster response. In recent years, fiber optic sensing has achieved significant advancements in gas detection. Examples include gas sensors based on fiber Bragg gratings [120,121], Sagnac interferometry [122,123], Mach–Zehnder interferometry [124,125], and Michelson interferometry [126,127]. This section will introduce fiber optic FP Interferometer (FPI) gas detection sensors based on different thin films.

Carbon dioxide (CO_2) is one of the major greenhouse gases closely associated with global climate change. Monitoring CO_2 concentration provides valuable information about the accumulation of greenhouse gases in the atmosphere. This is crucial for assessing the impact of climate change, formulating mitigation and adaptation measures, and promoting sustainable development. In reference [128], a CO_2 sensor based on a polyethyleneimine/poly (vinyl alcohol) (PEI/PVA) coating was proposed. The authors coated the end-face of an SMF with PEI/PVA, forming an FP cavity between the end-faces of the SMF and the PEI/PVA coating. When the PEI/PVA film is exposed to varying CO_2 environments, it can alter the optical path difference of the FPI [129]. Consequently, the CO_2 gas concentration can be measured through wavelength shifts in the interference fringe pattern. The proposed FPI sensor exhibited high sensitivity to changes in CO_2 concentration, with a sensitivity of 0.281 nm/% in the range of 7.6% to 86.9%. In reference [130], a CO_2 sensor utilizing polyhexamethylene biguanide (PHMB) as the thin film was presented. The authors coated the end of an SMF with PHMB film as the reflective coating for the FP cavity. Absorption and release of CO_2 gas molecules caused variations in the refractive index of PHMB, leading to spectral shifts in the interference pattern [131]. Experimental results demonstrated a sensitivity of 1.22 pm/ppm in the CO_2 concentration range of 0–700 ppm. However, both above-mentioned sensors exhibited a response to temperature, necessitating further improvements to address cross-sensitivity. Techniques such as incorporating FBG into the SMF can be employed to solve this issue.

As is well known, carbon monoxide (CO) is an explosive and toxic gas that, when inhaled, can cause oxygen deprivation in the blood, leading to harm to the heart and nervous system and even coma or death. However, its colorless and odorless nature makes it difficult to detect, highlighting the necessity of highly sensitive carbon monoxide sensors. Reference [132] proposed and fabricated an optical fiber FP interferometric carbon monoxide gas sensor based on a polyaniline/Co_3O_4 (PANI/Co_3O_4) film. The sensor consisted of a segment of SMF fusion-spliced with an endlessly photonic crystal fiber (EPCF), with the PANI/Co_3O_4 thin film coated on the end of the EPCF. This film adsorbed CO molecules, thereby altering the optical path of light transmission within the film.

Experimental results showed that within the range of 0–70 ppm, the sensor achieved a sensitivity of 21.61 pm/ppm for detecting CO. The authors further immersed the sensor in air containing various gases such as nitrogen, carbon dioxide, and oxygen, which are commonly present in the atmosphere. They observed that the interference spectrum did not show any significant changes, indicating the sensor's high selectivity towards carbon monoxide. However, it was noted that the sensor exhibited relatively long response and recovery times, which could potentially be attributed to the characteristics of the materials used in the sensor design. Further investigation is needed to improve the response and recovery times of the sensor.

Ammonia (NH_3) detection has broad practical significance, including industrial safety, agricultural applications, indoor air quality, and biological research. Reference [133] compared fiber optic FPI gas sensors based on porous graphene film (G-FPI) and Fe_3O_4-graphene nanocomposite film (FG-FPI) for the detection of ammonia gas at room temperature. The mechanism of these two sensors was based on the change in the refractive index of the thin film under different NH_3 gas concentrations. The sensor structure was simple, with a sensing film coated on the end of an SMF. However, both sensors initially exhibited good linearity at low NH_3 concentrations, but as the concentration increased, a nonlinear relationship appeared due to saturation effects. The sensors were tested in the range of 1.5 ppm to 150 ppm, and the experimental results showed that the G-FPI sensor achieved a sensitivity of ~25 pm/ppm at 150 ppm, while the FG-FPI sensor achieved a sensitivity of ~36 pm/ppm, the lowest detection limit of both sensor probes could be in the range of around 10 ppb and 7 ppb, respectively. Throughout the measurement process, the FG-FPI sensor outperformed the G-FPI sensor. Reference [134] investigated the influence of the thickness of ITO and SnO_2 on the sensitivity of microstructured optical fiber FP sensors. The authors connected a section of four-bridge double-Y-shape core microstructured optical fiber (MOF) to the end of an SMF to form an FP cavity and deposited metal oxides inside the holes of the MOF. The sensing mechanism was based on the change in the refractive index of the metal oxide after gas molecule adsorption [135]. However, both oxide-based sensors exhibited relatively long response and recovery times. The performance of fiber optic FP gas sensors based on different thin films is summarized in Table 5 in the original paper.

Table 5. Gas detection sensor performance based on different materials.

Material	Sensitivity (pm/ppm)	Range (ppm)	Reference
PEI/PVA	0.281	76,000~869,000(CO_2)	[128]
PHMB	1.22	0–700 (CO_2)	[130]
PCG	21.61	0~70 (CO)	[132]
G-FPI	25	0~150 (NH_3)	[133]
FG-FPI	36	0~150 (NH_3)	[133]

3.6. Temperature Sensor

Temperature measurement has important applications in various fields, such as industrial production, construction, healthcare, and nuclear energy. However, existing electronic temperature sensors are difficult to use in the presence of strong electromagnetic fields. In contrast, optical fibers can overcome this interference, as demonstrated by temperature sensors based on FBG [136,137], WGM [138,139], MZI [140,141], and Michelson interferometry [142,143]. This section will focus on the recent achievements of fiber FP devices based on different thin films for temperature sensing.

A fiber FP sensor based on silicon thin film was proposed in [144] for simultaneous temperature and pressure measurements. The author inserted an SMF into a glass ferrule, followed by the adhesion of heat-resistant glass and silicon wafer. The silicon wafer had a silicon film at its end to sense external temperature and pressure changes, while the heat-resistant glass only responded to temperature. By using bandpass filtering, the signals provided by the two sensing elements could be extracted separately, addressing the issue of cross-sensitivity between temperature and pressure. Experimental results showed a sensitivity of 142.02 nm/°C within the range of −20 to 70 °C for this sensor. Another temperature sensor based on a silicon thin film was presented in [145], as shown in Figure 7. The author inserted a multimode fiber into a glass ferrule and fused an empty cavity at the end, which was then bonded with a layer of silicon film. This silicon film divided the cavity into two chambers, with the chamber closer to the multimode fiber serving as the FP cavity and the other side as the sensing chamber. As the external temperature increased, the thermal expansion of the air inside the sensing chamber caused a pressure difference between the two chambers, leading to the deformation of the silicon film and resulting in a change in the cavity length of the FP cavity. Experimental results demonstrated a temperature sensitivity of 6.07 nm/°C within the range of −50 to 100 °C, and the response time of the sensor within the range of 28 to 100 °C was approximately 1.3 s.

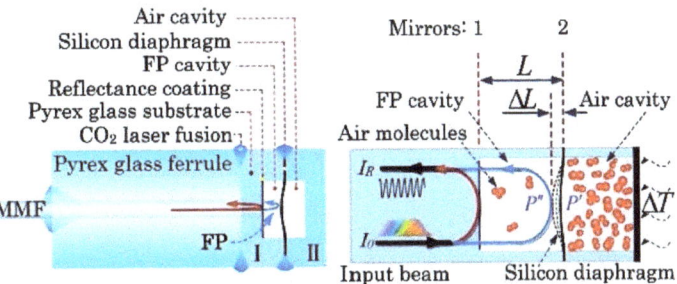

Figure 7. Differential-pressure-based fiber optic temperature sensor [145].

In addition to silicon thin film, alternative materials are also employed for temperature sensing in fiber FP sensors. An optic fiber FP temperature sensor based on PDMS thin film was proposed in reference [146]. The sensor consisted of two sensing devices, with each FP cavity formed by splicing a section of silica tube to the end of an SMF. One cavity was filled with a PDMS thin film, denoted as FP_1, which was sensitive to both temperature and pressure, while the other cavity was filled with a layer of UV adhesive thin film, denoted as FP_2, which was only sensitive to pressure. Like the previous approach, the use of dual FP cavities generated a vernier effect to enhance the sensing sensitivity. By employing bandpass filtering to extract the signals from each FP cavity separately, the issue of cross-sensitivity between temperature and pressure could be resolved. The temperature sensitivity of this sensor within the range of 44 to 49 °C was measured to be 10.29 nm/°C. An FPI based on a Polyimide tube for measuring seawater temperature and pressure was proposed in reference [147], as shown in Figure 8a. In the FP cavity, variations in external temperature and pressure caused the deformation of the PI tube, leading to a change in the FP cavity length. To address the issue of temperature and pressure cross-sensitivity, a cascaded FBG was employed. The higher thermal expansion coefficient of the polymer compared to silicon-based devices contributed to improved sensor sensitivity. Experimental results indicated a temperature sensitivity of 18.910 nm/°C within the range of 24 to 43 °C. In [148], PDMS was deposited on the end of a standard SMF, and it included a layer of carbon nanoparticles (CNPs). The optical-thermal effect of carbon nanoparticles [149] was utilized to create a microbubble within the coating using a laser, enhancing the sensitivity of the sensor. The structure of this sensor is depicted in Figure 8b. The temperature sensitivity of this sensor reached up to 790 pm/°C within the temperature range of 27 to 40 °C. In reference [150], the authors employed a rapid immersion method of multimode fiber into molten tellurite glass, which alleviated the challenge of material fusion. Upon solidification of the tellurite glass, a microcavity is formed at the end of the multimode fiber, serving as an FP cavity. The external temperature variations induce changes in the refractive index and shape of the tellurite glass, resulting in the shift of the interference spectrum. The temperature sensitivity of the sensor reaches 62 pm/°C within a wide measurement range of 20 °C to 170 °C. Furthermore, this sensor demonstrates fast response and excellent stability.

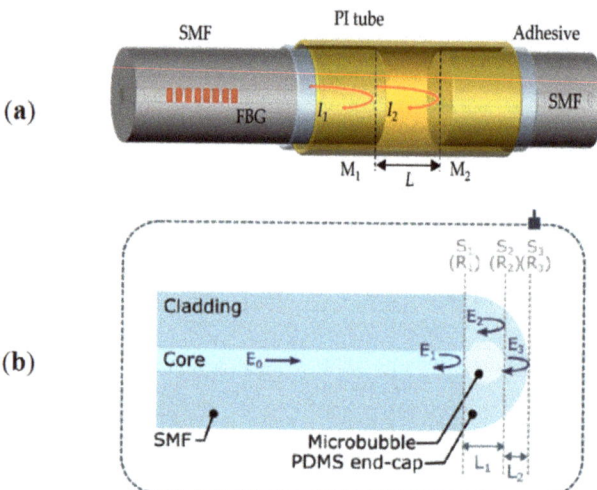

Figure 8. (a) Temperature sensors based on PI (polyimide) tube [147] (b) Temperature sensors based on PDMS film [148].

Solid materials often exhibit uneven thermal stress during the sensing process, which can increase the response time of the sensor [151]. On the other hand, liquid materials rarely experience thermal stress imbalance, allowing for faster sensor response. In reference [152], the authors constructed an FP sensing cavity by connecting an SMF, a thick-core capillary, a thin-core capillary, and another SMF in sequence. The capillary was filled with ethanol solution. When the temperature rises, the ethanol solution expands, causing a change in its refractive index [153], which leads to a shift in the interference spectrum. Since the fusion splicing process involves discharging, some of the ethanol solution vaporizes at high temperatures, resulting in an air column. The article points out that this sealed space increases the boiling point of ethanol, so the air column should not be too large. However, it should not be too small either, as it would restrict the expansion of ethanol, affecting the sensitivity of the sensor. Experimental data show that when the volume ratio of ethanol to air is 2.89, the temperature sensitivity of the sensor reaches −497.6 pm/°C within the range of 20 to 180 °C, and the maximum detectable temperature can reach 220 °C. Furthermore, the proposed fiber temperature sensor exhibits fast response (less than 1 s for temperature variations of about 90 °C) and good repeatability.

The previously mentioned fiber temperature sensors have a maximum temperature limit of only 220 °C, which is insufficient for high-temperature measurements. In [154], a fiber FP temperature sensor using sapphire material was proposed. The authors fixed a sapphire chip on the end-face of a sapphire fiber as the FP cavity. When the temperature changed, the thickness and refractive index of the sapphire chip changed, resulting in a shift in the interference spectrum. The sensor was tested within the range of 25 to 1550 °C, and the sensitivities at temperatures of 500, 1000, and 1550 °C were 20.63, 26.25, and 32.45 pm/°C, respectively. Another sapphire fiber FP structure for high-temperature measurement was proposed in reference [155]. The FP cavity was formed by the end-face of the sapphire fiber and the end-face of a sapphire rod. The sensor was experimentally tested within the range of 0 to 1500 °C, but the data showed that the sensor's response to temperature was nonlinear, and the sensitivity increased with temperature.

In the field of low-temperature measurements, there are three commonly used instruments: helium vapor pressure thermometers, gas thermometers, and platinum resistance thermometers. At low temperatures, the thermal expansion coefficient of silica (SiO_2) is extremely low, rendering bare fiber Bragg gratings (FBGs) unsuitable for temperature sensing below 40 K [156]. Therefore, additional materials are needed to enable fiber-based temperature sensing at low temperatures, such as the application of metal coatings on FBGs. Among them, lead-coated FBGs are currently the most sensitive, exhibiting a sensitivity of 8.7 pm/K at 5 K [157]. In reference [158], a fiber FP sensor designed for low-temperature measurements was proposed. The sensor featured a simple structure, where two SMFs were enclosed within ceramic rings, forming an FP cavity with the fiber end-faces. The assembly was then inserted into a copper sleeve. The working principle was straightforward: as the external temperature changed, both the ceramic rings and the copper sleeve underwent deformations proportional to the temperature variations, causing a shift in the interference spectrum. The copper sleeve was crucial for the sensor's low-temperature sensing range of 5–75 K since copper maintained a considerable thermal expansion coefficient within this range, whereas the thermal expansion coefficients of the ceramic rings (alumina) and silica started to increase gradually from near-zero to around 75 K. The study also conducted a further analysis of the sensor and suggested that zinc, with a thermal expansion coefficient better than copper below 50 K, could potentially be employed as a substitute to enhance sensing performance. In summary, this paper presents a rare investigation of fiber FP sensors in low-temperature conditions. Roctest Company offers two FP-based fiber optic temperature sensors, namely FOT-F and FOT-N. These sensors are designed based on highly stable glass materials that exhibit thermal expansion. They are suitable for highly precise, stable, and repeatable measurements. FOT-F can be used in vacuum environments, high-pressure applications, or high-voltage environments, while FOT-N is primarily used for embedding in concrete structures or air. The performance of the aforementioned fiber FP temperature sensors for temperature measurement is summarized in Table 6.

Table 6. Temperature sensor performance based on different materials.

Material	Sensitivity	Test Range	Reference
Silicon	142.02 nm/°C	−20~70 °C	[144]
	6.07 nm/°C	−50~100 °C	[145]
PDMS	10.29 nm/°C	44~49 °C	[146]
	62 pm/°C	20~170 °C	[148]
PI	18.910 nm/°C	24~43 °C	[147]
Ethanol	−497.6 pm/°C	20~180 °C	[152]
Sapphire	20.63 pm/°C (at 500 °C)	25~1550 °C	[154]
	26.25 pm/°C (at 1000 °C)	25~1550 °C	[154]
	32.45 pm/°C (at 1550 °C)	25~1550 °C	[154]
Cu/Al_2O_3	2.10 nm/K	5.367~15.069 K	[158]
	1.95 nm/K	15~50 K	[158]
	7.73 nm/K	96.5~142.69 K	[158]
	5.33 nm/K	150.19~200.36 K	[158]
	4.35 nm/K	250.18~290.98 K	[158]

3.7. Biological or Medical Sensor

In recent years, fiber optic sensing has attracted increasing attention in the fields of biology and medicine, such as the use of LPG-based biosensors [159,160], Mach–Zehnder Interferometer (MZI)-based medical sensors [161,162], and SPR-based biosensors [163,164]. This section introduces some research on fiber optic FP sensors in the field of biology and medicine. Reference [165] proposed a fiber optic FP sensor for detecting Microcystin-LR (MCT). The authors coated the end-face of an SMF with a Molecularly Imprinted Polymer to form the FP cavity, which exhibits affinity for the selected "template molecule". When the sensor meets the MCT solution, the refractive index changes, leading to a shift in the interference spectrum. Experimental results showed that the sensor has a sensitivity of approximately 12.4 nm L/µg in the concentration range of 0.3–1.4 µg/L for MCT. Reference [166] presented a fiber optic FP interferometric immunosensor based on chitosan/polystyrene sulfonate film, as shown in Figure 9. Chitosan is a biopolymer that can preserve the biological properties of proteins, making it suitable for immobilizing immunoglobulin G (IgG). When anti-immunoglobulin G (anti-IgG) is adsorbed onto the chitosan substrate, the adsorbed protein can be considered to be forming a membrane, which increases the thickness of the sensing film and adjusts its refractive index, resulting in the shift of the output interference spectrum. The focus of the experiment was on the power variation in the spectrum. The authors determined protein binding events by demodulating the changes in effective optical length. The sensitivity of the sensor is 0.033 m/(pg/mm^2). Reference [167] proposed a fiber optic FP sensor based on antibody–antigen-specific binding. The authors prepared the FP interferometer by splicing the ends of a short-section hollow-core photonic crystal fiber to an SMF and cutting the SMF pigtail to an appropriate length. Then, goat anti-rabbit immunoglobulin G (IgG) was covalently immobilized and fixed onto the salinization-modified end of an SMF to detect the specific rabbit IgG. The specific binding of anti-rabbit IgG and rabbit IgG changed the thickness of the sensing layer, causing a change in light absorption by the reflection surface fixed by the biomolecular layer and a change in the effective cavity length of the FP sensor, resulting in a change in the optical intensity and a shift in the interference spectrum. Experimental results showed that during the antigen binding process, the wavelength increased by 190 pm, and the fringe contrast decreased by 2.15 dB, indicating the successful binding of the antibody and antigen, thereby confirming the feasibility of the proposed immunosensor.

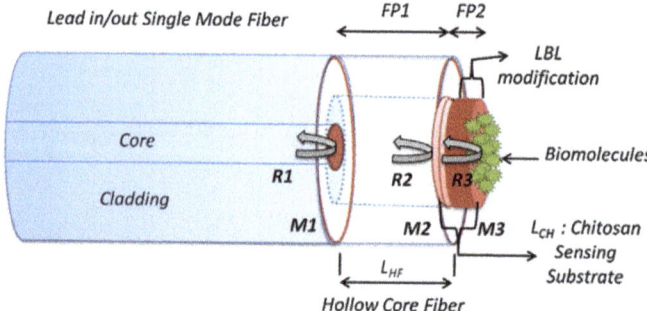

Figure 9. F-P interferometric immunosensor based on a chitosan/polystyrene sulfonate film [166].

Reference [168] presented a fiber optic FP sensor based on polypropylene film for respiratory detection in humans. This thin film had a low Young's modulus, which enhanced pressure-sensing sensitivity. The sensor structure was also simple, with a polypropylene film prepared at the end of a nut and an SMF inserted at the other end to form the FP cavity. Experimental data showed a sensitivity of −0.581 nm/Pa for the respiratory sensor. However, this sensor was influenced by humidity, particularly in high humidity conditions, which limited its operating conditions. Reference [169] proposed a fiber optic FP sensor for monitoring human respiration using humidity sensing. The sensing mechanism utilized the adsorption of water molecules by chitosan polymers, which caused changes in refractive index and volume [170]. The authors fusion-spliced a section of MOF (Microstructured Optical Fiber) to the end of an SMF to form the FP cavity and filled the cavity with chitosan polymer. Experimental results showed that within the range of 70% to 95% relative humidity (RH), the sensor has a sensitivity of 68.55 pm/%RH. The impact of temperature within the range of 20 to 70 °C is minimal. This sensor exhibits fast response times, with a rise time of 80 ms and a recovery time of 70 ms. Reference [171] proposed a fiber optic sensor for DNA detection. Two sections of C-type optical fibers were inserted into an SMF to form two FP cavities. One FP cavity was used for DNA detection, where the binding of pDNA and cDNA in the solution adds a biological layer on the cross-section of the SMF, resulting in a refractive index change and a shift in the interference spectrum. The other FP cavity was filled with PDMS to monitor environmental temperature changes and prevent interference during the detection process.

Reference [172] presented a fiber optic FP sensor for measuring blood temperature. The authors fusion-spliced a short section of multimode fiber to the end of an SMF and etched a groove to place borosilicate glass, forming the FP cavity. Due to the high coefficient of thermal expansion of borosilicate glass, the FP cavity length changed when the external temperature varied, resulting in a shift in the interference spectrum. Experimental results showed that the sensor has a sensitivity of 0.0103 nm/°C within the range of 38 to 40 °C. This sensor is flexible for temperature measurement in various locations of the body, and it exhibits a very fast response time of less than 1 s. Reference [173] proposed a fiber optic FP sensor for monitoring blood pressure change. The authors fusion-spliced a section of multimode fiber to the end of an SMF and etched a groove at the end of the multimode fiber. They then bonded a thin film of silicon dioxide to the end for pressure sensing. When the external pressure changed, the deformation of the thin film led to a change in cavity length, and the blood pressure could be determined based on the spectral shift. Experimental tests were conducted in pig arteries, and the sensor demonstrated a sensitivity of 0.035 mV/mmHg. Reference [174] introduced a fiber optic FP sensor for monitoring heart rate. The authors inserted two sections of SMFs into a capillary to form the FP cavity and bonded the SMFs to the capillary using ethyl cyanoacrylate (EtCNA). The device was then placed within a circular frame. The material EtCNA has a low Young's modulus, making it suitable for detecting low-frequency vibrations. When external vibrations occurred, the circular frame underwent deformation, resulting in a change in the cavity length of the interferometric device. Experimental results demonstrated that the sensor has a strain sensitivity of 2.57 pm/μN and exhibited good response to low-frequency vibrations at 1~3 Hz. Currently, there are several fiber optic FP sensors available on the market for medical applications. For instance, the FOP-M260(FISO Technologies, Inc., Quebec, QC, Canada) fiber optic pressure sensor is specifically designed for the medical field. It is a small-sized, high-precision sensor that can be used for monitoring left ventricular pressure, arterial blood pressure, intracranial pressure, and other parameters. The sensor ensures high stability and does not pose any harm to the human body.

4. Conclusions

This paper presents a comprehensive overview and analysis of fiber FP sensors based on different thin films. In various sensing applications, it is crucial to select thin film materials that exhibit sensitivity to the specific physical quantities being measured, either as the core components of the sensor or to enhance its sensitivity. Challenges encountered in practical implementation include achieving secure integration of thin film with optical fibers and precise control over film size and thickness to meet the requirements of high-performance sensors, thereby enabling the fabrication of stable and reliable fiber FP sensors. Additionally, addressing the issue of mutual interference among multiple parameters and cross-sensitivity, where a particular material is influenced by more than one physical quantity, represents an important research direction for fiber FP cavity sensors. In summary, thin film technology and FP cavities, as essential components of fiber optic sensing, offer novel avenues to enhance the sensitivity, selectivity, and reliability of sensors. With continued technological advancements and expanded applications, this fiber optic sensing technology holds great promise for playing a more significant role in various fields and providing innovative solutions to scientific and engineering problems.

Author Contributions: Conceptualization, S.L.; methodology, C.M. and D.P.; formal analysis, X.B.; writing—original draft, C.M. and D.P.; writing—review and editing, S.L.; supervision, S.L. and L.L. All authors have read and agreed to the published version of the manuscript.

Funding: National Key Research and Development Program (Grant No. 2022YFC2204402), Guangdong Science and Technology Project (Grant No. 20220505020011), Shenzhen Science and Technology Program (Grant No. 2021Szvup172), and Shenzhen Science and Technology Program (Grant No. JCYJ20220818102003006).

Institutional Review Board Statement: Not applicable.

Informed Consent Statement: Not applicable.

Data Availability Statement: Data-sharing is not applicable to this article.

Conflicts of Interest: The authors declare no conflict of interest.

References

1. Lu, P.; Men, L.; Sooley, K.; Chen, Q. Tapered fiber Mach-Zehnder interferometer for simultaneous measurement of refractive index and temperature. *Appl. Phys. Lett.* **2009**, *94*, 131110. [CrossRef]
2. Tian, Z.; Yam, S.S.H.; Barnes, J.; Bock, W.; Greig, P.; Fraser, J.M.; Loock, H.-P.; Oleschuk, R.D. Refractive index sensing with Mach-Zehnder interferometer based on concatenating two single-mode fiber tapers. *IEEE Photonics Technol. Lett.* **2008**, *20*, 626–628. [CrossRef]
3. Tian, Z.; Yam, S.S.H.; Loock, H.-P. Refractive index sensor based on an abrupt taper Michelson interferometer in a single-mode fiber. *Opt. Lett.* **2008**, *33*, 1105–1107. [CrossRef] [PubMed]
4. Tian, Z.; Yam, S.S.H.; Loock, H.-P. Single-mode fiber refractive index sensor based on core-offset attenuators. *IEEE Photonics Technol. Lett.* **2008**, *20*, 1387–1389. [CrossRef]
5. Choi, H.Y.; Park, K.S.; Park, S.J.; Paek, U.-C.; Lee, B.H.; Choi, E.S. Miniature fiber-optic high temperature sensor based on a hybrid structured Fabry-Perot interferometer. *Opt. Lett.* **2008**, *33*, 2455–2457. [CrossRef]
6. Choi, H.Y.; Mudhana, G.; Park, K.S.; Paek, U.-C.; Lee, B.H. Cross-talk free and ultra-compact fiber optic sensor for simultaneous measurement of temperature and refractive index. *Opt. Express* **2010**, *18*, 141–149. [CrossRef]
7. Dong, X.; Tam, H.Y.; Shum, P. Temperature-insensitive strain sensor with polarization-maintaining photonic crystal fiber based Sagnac interferometer. *Appl. Phys. Lett.* **2007**, *90*, 151113. [CrossRef]
8. Fu, H.Y.; Tam, H.Y.; Shao, L.-Y.; Dong, X.; Wai, P.K.A.; Lu, C.; Khijwania, S.K. Pressure sensor realized with polarization-maintaining photonic crystal fiber-based Sagnac interferometer. *Appl. Opt.* **2008**, *47*, 2835–2839. [CrossRef]
9. Wang, W. Fabry-Perot Interference Fiber Acoustic Wave Sensor Based on Laser Welding All-Silica Glass. *Materials* **2022**, *15*, 2484. [CrossRef]
10. Ruan, J. High sensitivity Sagnac interferometric strain sensor based polarization maintaining fibre enhanced coupling. *IET Optoelectron.* **2021**, *15*, 48–51. [CrossRef]
11. Tian, J.; Zuo, Y.; Hou, M.; Jiang, Y. Magnetic field measurement based on a fiber laser oscillation circuit merged with a polarization-maintaining fiber Sagnac interference structure. *Opt. Express* **2021**, *29*, 8763–8769. [CrossRef] [PubMed]
12. Wu, B.; Zhao, C.; Kang, J.; Wang, D. Characteristic study on volatile organic compounds optical fiber sensor with zeolite thin film-coated spherical end. *Opt. Fiber Technol.* **2017**, *34*, 91–97. [CrossRef]
13. Li, J.-X.; Tong, Z.-R.; Jing, L.; Zhang, W.-H.; Qin, J.; Liu, J.-W. Fiber temperature and humidity sensor based on photonic crystal fiber coated with graphene oxide. *Opt. Commun.* **2020**, *467*, 134–139. [CrossRef]
14. Islam, M.R.; Ali, M.M.; Lai, M.-H.; Lim, K.-S.; Ahmad, H. Chronology of Fabry-Perot Interferometer Fiber-Optic Sensors and Their Applications: A Review. *Sensors* **2014**, *14*, 7451–7488. [CrossRef]
15. Huang, Y.W.; Tao, J.; Huang, X.G. Research Progress on F-P InterferenceBased Fiber-Optic Sensors. *Sensors* **2016**, *16*, 1424. [CrossRef]
16. Rao, Y.-J.; Ran, Z.-L.; Gong, Y. *Fiber-Optic Fabry-Perot Sensors: An Introduction*; CRC Press: Boca Raton, FL, USA, 2017.
17. Wang, X.; Jiang, J.; Wang, S.; Liu, K.; Liu, T. All-silicon dual-cavity fiber-optic pressure sensor with ultralow pressure-temperature cross-sensitivity and wide working temperature range. *Photonics Res.* **2021**, *9*, 521–529. [CrossRef]
18. Zeng, L.; Chen, M.; Yan, W.; Li, Z.; Yang, F. Si-grating-assisted SPR sensor with high figure of merit based on Fabry-Perot cavity. *Opt. Commun.* **2020**, *457*, 124641. [CrossRef]
19. Born, M.; Wolf, E. *Principles of Optics: Electromagnetic Theory of Propagation, Interference and Diffraction of Light*; Elsevier: Amsterdam, The Netherlands, 2013.
20. Li, Y.; Li, Y.; Liu, Y.; Li, Y.; Qu, S. Detection limit analysis of optical fiber sensors based on interferometers with the Vernier-effect. *Opt. Express* **2022**, *30*, 35734–35748. [CrossRef]
21. Duguay, M.A.; Kokubun, Y.; Koch, T.L.; Pfeiffer, L. Antiresonant reflecting optical waveguides in SiO_2-Si multilayer structures. *Appl. Phys. Lett.* **1986**, *49*, 13–15. [CrossRef]
22. Litchinitser, N.M.; Abeeluck, A.K.; Headley, C.; Eggleton, B.J. Antiresonant reflecting photonic crystal optical waveguides. *Opt. Lett.* **2002**, *27*, 1592–1594. [CrossRef]
23. Hou, M.; Zhu, F.; Wang, Y.; Wang, Y.; Liao, C.; Liu, S.; Lu, P. Antiresonant reflecting guidance mechanism in hollow-core fiber for gas pressure sensing. *Opt. Express* **2016**, *24*, 27890–27898. [CrossRef]
24. Gao, R.; Lu, D.-F.; Cheng, J.; Jiang, Y.; Jiang, L.; Qi, Z.-M. Humidity sensor based on power leakage at resonance wavelengths of a hollow core fiber coated with reduced graphene oxide. *Sens. Actuators B Chem.* **2016**, *222*, 618–624. [CrossRef]
25. Rosolem, J.B.; Penze, R.S.; Floridia, C.; Bassan, F.R.; Peres, R.; de Costa, E.F.; de Araujo Silva, A.; Coral, A.D.; Junior, J.R.N.; Vasconcelos, D.; et al. Dynamic Effects of Temperature on FBG Pressure Sensors Used in Combustion Engines. *IEEE Sens. J.* **2021**, *21*, 3020–3027. [CrossRef]
26. Vaddadi, V.S.C.S.; Parne, S.R.; Afzulpurkar, S.; Desai, S.P.; Parambil, V.V. Design and development of pressure sensor based on Fiber Bragg Grating (FBG) for ocean applications. *Eur. Phys. J. Appl. Phys.* **2020**, *90*, 30501. [CrossRef]

27. Lu, Y.; Tian, F.; Chen, Y.; Han, Z.; Zeng, Z.; Liu, C.; Yang, X.; Li, L.; Zhang, J. Characteristics of a capillary single core fiber based on SPR for hydraulic pressure sensing. *Opt. Commun.* **2023**, *530*, 129125. [CrossRef]
28. Zhao, Y.; Wu, Q.-L.; Zhang, Y.-N. Simultaneous measurement of salinity, temperature and pressure in seawater using optical fiber SPR sensor. *Measurement* **2019**, *148*, 106792. [CrossRef]
29. Reja, M.I.; Nguyen, L.V.; Ebendorff-Heidepriem, H.; Warren-Smith, S.C. Multipoint pressure sensing at up to 900 °C using a fiber optic multimode interferometer. *Opt. Fiber Technol.* **2023**, *75*, 103157. [CrossRef]
30. Pang, Y.-N.; Liu, B.; Liu, J.; Wan, S.-P.; Wu, T.; Yuan, J.; Xin, X.; He, X.-D.; Wu, Q. Singlemode-Multimode-Singlemode Optical Fiber Sensor for Accurate Blood Pressure Monitoring. *J. Light. Technol.* **2022**, *40*, 4443–4450. [CrossRef]
31. Zhao, Y.; Li, H.; Li, J.; Zhou, A. Cascaded fiber MZIs for simultaneous measurement of pressure and temperature. *Opt. Fiber Technol.* **2021**, *66*, 102629. [CrossRef]
32. Lei, X.; Dong, X.; Lu, C.; Sun, T.; Grattan, K.T.V. Underwater Pressure and Temperature Sensor Based on a Special Dual-Mode Optical Fiber. *IEEE Access* **2020**, *8*, 146463–146471. [CrossRef]
33. Tada, H.; Kumpel, A.E.; Lathrop, R.E.; Slanina, J.B.; Nieva, P.; Zavracky, P.; Miaoulis, I.N.; Wong, P.Y. Thermal expansion coefficient of polycrystalline silicon and silicon dioxide thin films at high temperatures. *J. Appl. Phys.* **2000**, *87*, 4189–4193. [CrossRef]
34. Li, H.; Luo, X.; Zhang, H.; Dong, M.; Zhu, L. All-Silica Diaphragm-based Optical Fiber Fabry-Perot Pressure Sensor Fabricated by CO_2 Laser Melting Capillary End Face. *Optik* **2023**, *287*, 170994. [CrossRef]
35. Wang, W.; Wu, N.; Tian, Y.; Niezrecki, C.; Wang, X. Miniature all-silica optical fiber pressure sensor with an ultrathin uniform diaphragm. *Opt. Express* **2010**, *18*, 9006–9014. [CrossRef]
36. Guo, X.; Zhou, J.; Du, C.; Wang, X. Highly Sensitive Miniature All-Silica Fiber Tip Fabry-Perot Pressure Sensor. *IEEE Photonics Technol. Lett.* **2019**, *31*, 689–692. [CrossRef]
37. Hopcroft, M.A.; Nix, W.D.; Kenny, T.W. What is the Young's Modulus of Silicon? *J. Microelectromech. Syst.* **2010**, *19*, 229–238. [CrossRef]
38. Spitalsky, Z.; Tasis, D.; Papagelis, K.; Galiotis, C. Carbon nanotube-polymer composites: Chemistry, processing, mechanical and electrical properties. *Prog. Polym. Sci.* **2010**, *35*, 357–401. [CrossRef]
39. Tjong, S.C. Structural and mechanical properties of polymer nanocomposites. *Mater. Sci. Eng. R Rep.* **2006**, *53*, 73–197. [CrossRef]
40. Schneider, F.; Draheirn, J.; Kamberger, R.; Wallrabe, U. Process and material properties of polydimethylsiloxane (PDMS) for Optical MEMS. *Sens. Actuators A Phys.* **2009**, *151*, 95–99. [CrossRef]
41. Liu, M.; Sun, J.; Chen, Q. Influences of heating temperature on mechanical properties of polydimethylsiloxane. *Sens. Actuators A Phys.* **2009**, *151*, 42–45. [CrossRef]
42. Luo, C.; Liu, X.; Liu, J.; Shen, J.; Li, H.; Zhang, S.; Hu, J.; Zhang, Q.; Wang, G.; Huang, M. An Optimized PDMS Thin Film Immersed Fabry-Perot Fiber Optic Pressure Sensor for Sensitivity Enhancement. *Coatings* **2019**, *9*, 290. [CrossRef]
43. Wei, X.; Song, X.; Li, C.; Hou, L.; Li, Z.; Li, Y.; Ran, L. Optical Fiber Gas Pressure Sensor Based on Polydimethylsiloxane Microcavity. *J. Light. Technol.* **2021**, *39*, 2988–2993. [CrossRef]
44. Fu, D.; Liu, X.; Shang, J.; Sun, W.; Liu, Y. A Simple, Highly Sensitive Fiber Sensor for Simultaneous Measurement of Pressure and Temperature. *IEEE Photonics Technol. Lett.* **2020**, *32*, 747–750. [CrossRef]
45. Ranjbar-Naeini, O.; Barandak, A.; Tahmasebi, M.; Pooladmast, A.; Latifi, H. Characterization the Effect of Pressure and Concentration of Acetone Gas on Micro Polymeric Curved Diaphragm Fabry Perot Optical Fiber Sensor. In Proceedings of the Optical Fiber Sensors, Lausanne, Switzerland, 24–28 September 2018; p. WF10.
46. Endruweit, A.; Johnson, M.S.; Long, A.C. Curing of composite components by ultraviolet radiation: A review. *Polym. Compos.* **2006**, *27*, 119–128. [CrossRef]
47. Priola, A.; Gozzelino, G.; Ferrero, F.; Malucelli, G. Properties of polymeric films obtained from u.v. cured poly(ethylene glycol) diacrylates. *Polymer* **1993**, *34*, 3653–3657. [CrossRef]
48. Chen, Y.; Zheng, Y.; Liang, D.; Zhang, Y.; Guo, J.; Lian, S.; Yu, Y.; Du, C.; Ruan, S. Fiber-Tip Fabry-Perot Cavity Pressure Sensor with UV-Curable Polymer Film Based on Suspension Curing Method. *IEEE Sens. J.* **2022**, *22*, 6651–6660. [CrossRef]
49. Liu, L.; Lu, P.; Wang, S.; Fu, X.; Sun, Y.; Liu, D.; Zhang, J.; Xu, H.; Yao, Q. UV Adhesive Diaphragm-Based FPI Sensor for Very-Low-Frequency Acoustic Sensing. *IEEE Photonics J.* **2016**, *8*, 6800709. [CrossRef]
50. Zhang, S.; Lei, Q.; Hu, J.; Zhao, Y.; Gao, H.; Shen, J.; Li, C. An optical fiber pressure sensor with ultra-thin epoxy film and high sensitivity characteristics based on blowing bubble method. *IEEE Photonics J.* **2021**, *13*, 6800510. [CrossRef]
51. Oliveira, R.; Bilro, L.; Nogueira, R.; Rocha, A.M. Adhesive Based Fabry-Perot Hydrostatic Pressure Sensor with Improved and Controlled Sensitivity. *J. Light. Technol.* **2019**, *37*, 1909–1915. [CrossRef]
52. Ni, Z.H.; Wang, H.M.; Kasim, J.; Fan, H.M.; Yu, T.; Wu, Y.H.; Feng, Y.P.; Shen, Z.X. Graphene thickness determination using reflection and contrast spectroscopy. *Nano Lett.* **2007**, *7*, 2758–2763. [CrossRef]
53. Lee, C.; Wei, X.; Kysar, J.W.; Hone, J. Measurement of the elastic properties and intrinsic strength of monolayer graphene. *Science* **2008**, *321*, 385–388. [CrossRef]
54. Ma, J.; Jin, W.; Ho, H.L.; Dai, J.Y. High-sensitivity fiber-tip pressure sensor with graphene diaphragm. *Opt. Lett.* **2012**, *37*, 2493–2495. [CrossRef] [PubMed]
55. Cui, Q.; Thakur, P.; Rablau, C.; Avrutsky, I.; Cheng, M.M.-C. Miniature Optical Fiber Pressure Sensor with Exfoliated Graphene Diaphragm. *IEEE Sens. J.* **2019**, *19*, 5621–5631. [CrossRef]

56. Ge, Y.; Shen, L.; Sun, M. Temperature Compensation for Optical Fiber Graphene Micro-Pressure Sensor Using Genetic Wavelet Neural Networks. *IEEE Sens. J.* **2021**, *21*, 24195–24201. [CrossRef]
57. Ni, W.; Lu, P.; Liu, D.; Zhang, J. Graphene Diaphragm-based Extrinsic Fabry-Perot Interferometer for Low Frequency Acoustic Sensing. In Proceedings of the Conference on Lasers and Electro-Optics Pacific Rim (CLEO-PR), Singapore, 31 July–4 August 2017.
58. Cai, J.; Wang, G.; Wu, Y.; Gao, Z.; Qiu, Z. Extrinsic Optical Fiber Pressure Sensor Based on FP Cavity. *Instrum. Mes. Métrol.* **2020**, *19*, 43–49.
59. Guo, F.; Fink, T.; Han, M.; Koester, L.; Turner, J.; Huang, J. High-sensitivity, high-frequency extrinsic Fabry-Perot interferometric fiber-tip sensor based on a thin silver diaphragm. *Opt. Lett.* **2012**, *37*, 1505–1507. [CrossRef]
60. Huang, Q.; Deng, S.; Li, M.; Wen, X.; Lu, H. Fabry-Perot acoustic sensor based on a thin gold diaphragm. *Opt. Eng.* **2020**, *59*, 064105. [CrossRef]
61. Zhang, Z.; Wang, Y.; Feng, Z.; Shi, J.; Wang, C.; Li, D.; Yao, H.; Li, C.; Kang, R.; Li, L.; et al. Preliminary Strain Measurement in High Field Superconducting Magnets with Fiber Bragg Grating. *IEEE Trans. Appl. Supercond.* **2022**, *32*, 9001105. [CrossRef]
62. Odenbach, S. Recent progress in magnetic fluid research. *J. Phys. Condens. Matter* **2004**, *16*, R1135. [CrossRef]
63. Philip, J. Magnetic nanofluids: Recent advances, applications, challenges, and future directions. *Adv. Colloid Interface Sci.* **2022**, *311*, 102810. [CrossRef]
64. Zhang, R.; Pu, S.; Li, Y.; Zhao, Y.; Jia, Z.; Yao, J.; Li, Y. Mach-Zehnder interferometer cascaded with FBG for simultaneous measurement of magnetic field and temperature. *IEEE Sens. J.* **2019**, *19*, 4079–4083. [CrossRef]
65. Zeng, L.; Sun, X.; Zhang, L.; Hu, Y.; Duan, J.A. High sensitivity magnetic field sensor based on a Mach-Zehnder interferometer and magnetic fluid. *Optik* **2022**, *249*, 168234. [CrossRef]
66. Zhang, Y.; Guo, Y.; Zhu, F.; Qi, K. In-line Michelson interferometer based on microspherical structure for magnetic field applications. In Proceedings of the Optical and Quantum Sensing and Precision Metrology, Online Only, 5 March 2021; pp. 146–151.
67. Xiong, Z.; Guan, C.; Duan, Z.; Cheng, T.; Ye, P.; Yang, J.; Shi, J.; Yang, J.; Yuan, L.; Grattan, K.T.V. All-optical vector magnetic field sensor based on a side-polished two-core fiber Michelson interferometer. *Opt. Express* **2022**, *30*, 22746–22754. [CrossRef]
68. Huang, Y.; Wang, T.; Deng, C.; Zhang, X.; Pang, F.; Bai, X.; Dong, W.; Wang, L.; Chen, Z. A Highly Sensitive Intensity-Modulated Optical Fiber Magnetic Field Sensor Based on the Magnetic Fluid and Multimode Interference. *J. Sens.* **2017**, *2017*, 9573061. [CrossRef]
69. Ning, Y.; Zhang, Y.; Guo, H.; Zhang, M.; Zhang, Y.; Li, S.; Liu, Z.; Zhang, J.; Yang, X.; Yuan, L. Optical Fiber Magnetic Field Sensor Based on Silk Fibroin Hydrogel. *IEEE Sens. J.* **2022**, *22*, 14878–14882. [CrossRef]
70. Zhao, Y.; Wang, X.-X.; Lv, R.-Q.; Zheng, H.-K.; Zhou, Y.-F.; Chen, M.-Q. Reflective highly sensitive Fabry–Pérot magnetic field sensor based on magneto-volume effect of magnetic fluid. *IEEE Trans. Instrum. Meas.* **2021**, *70*, 7003506. [CrossRef]
71. Wang, X.-X.; Zhao, Y.; Lv, R.-Q.; Zheng, H.-K.; Cai, L. Magnetic Field Measurement Method Based on the Magneto-Volume Effect of Hollow Core Fiber Filled with Magnetic Fluid. *IEEE Trans. Instrum. Meas.* **2021**, *70*, 9513708. [CrossRef]
72. Zhao, Y.; Wang, X.-X.; Lv, R.-Q.; Li, G.-L.; Zheng, H.-K.; Zhou, Y.-F. Highly Sensitive Reflective Fabry-Perot Magnetic Field Sensor Using Magnetic Fluid Based on Vernier Effect. *IEEE Trans. Instrum. Meas.* **2021**, *70*, 7000808. [CrossRef]
73. Hocking, L.M.; Rivers, A.D. The spreading of a drop by capillary action. *J. Fluid Mech.* **1982**, *121*, 425–442. [CrossRef]
74. Sun, B.; Bai, M.; Ma, X.; Wang, X.; Zhang, Z.; Zhang, L. Magnetic-Based Polydimethylsiloxane Cap for Simultaneous Measurement of Magnetic Field and Temperature. *J. Light. Technol.* **2022**, *40*, 2625–2630. [CrossRef]
75. Zhou, B.; Lu, C.; Mao, B.-M.; Tam, H.-Y.; He, S. Magnetic field sensor of enhanced sensitivity and temperature self-calibration based on silica fiber Fabry-Perot resonator with silicone cavity. *Opt. Express* **2017**, *25*, 8108–8114. [CrossRef]
76. Xu, J.; Huang, K.; Zheng, J.; Li, J.; Pei, L.; You, H.; Ning, T. Sensitivity enhanced magnetic field sensor based on hollow core fiber Fabry-Perot interferometer and vernier effect. *IEEE Photonics J.* **2022**, *14*, 6841205. [CrossRef]
77. Wang, Z.; Jiang, S.; Yang, P.; Wei, W.; Bao, W.; Peng, B. High-sensitivity and high extinction ratio fiber strain sensor with temperature insensitivity by cascaded MZI and FPI. *Opt. Express* **2023**, *31*, 7073–7089. [CrossRef] [PubMed]
78. Liu, G.; Li, K.; Hao, P.; Zhou, W.; Wu, Y.; Xuan, M. Bent optical fiber taper for refractive index detection with a high sensitivity. *Sens. Actuators A Phys.* **2013**, *201*, 352–356. [CrossRef]
79. Chen, Y.; Han, Q.; Liu, T.; Liu, F.; Yao, Y. Simultaneous measurement of refractive index and temperature using a cascaded FBG/droplet-like fiber structure. *IEEE Sens. J.* **2015**, *15*, 6432–6436. [CrossRef]
80. Wang, Y.; Liu, B.; Wu, Y.; Mao, Y.; Zhao, L.; Sun, T.; Nan, T.; Han, Y. Temperature insensitive fiber Fabry-Perot/Mach-Zehnder hybrid interferometer based on photonic crystal fiber for transverse load and refractive index measurement. *Opt. Fiber Technol.* **2020**, *56*, 102163. [CrossRef]
81. Xia, F.; Zhao, Y. RI sensing system with high sensitivity and large measurement range using a microfiber MZI and a photonic crystal fiber MZI. *Measurement* **2020**, *156*, 107603. [CrossRef]
82. Jiang, X.; Wang, Q. Refractive index sensitivity enhancement of optical fiber SPR sensor utilizing layer of MWCNT/PtNPs composite. *Opt. Fiber Technol.* **2019**, *51*, 118–124. [CrossRef]
83. Teng, C.; Shao, P.; Li, S.; Li, S.; Liu, H.; Deng, H.; Chen, M.; Yuan, L.; Deng, S. Double-side polished U-shape plastic optical fiber based SPR sensor for the simultaneous measurement of refractive index and temperature. *Opt. Commun.* **2022**, *525*, 128844. [CrossRef]
84. Liang, W.; Huang, Y.; Xu, Y.; Lee, R.K.; Yariv, A. Highly sensitive fiber Bragg grating refractive index sensors. *Appl. Phys. Lett.* **2005**, *86*, 151122. [CrossRef]

85. Upadhyay, C.; Dhawan, D. Fiber Bragg grating refractive index sensor based on double D-shaped fiber. *Opt. Quantum Electron.* **2023**, *55*, 271. [CrossRef]
86. Ran, Z.; Rao, Y.; Zhang, J.; Liu, Z.; Xu, B. A miniature fiber-optic refractive-index sensor based on laser-machined Fabry–Perot interferometer tip. *J. Light. Technol.* **2009**, *27*, 5426–5429.
87. Quan, M.; Lu, Z.; Tian, J.; Yao, Y. Refractive index Fabry–Perot interferometric fiber sensor based on a microporous silver diaphragm and silica tube. In Proceedings of the Asia Communications and Photonics Conference, Hong Kong, China, 19–23 November 2015; p. ASu2A.47.
88. Majchrowicz, D.; Hirsch, M.; Wierzba, P.; Bechelany, M.; Viter, R.; Jedrzejewska-Szczerska, M. Application of Thin ZnO ALD Layers in Fiber-Optic Fabry-Perot Sensing Interferometers. *Sensors* **2016**, *16*, 416. [CrossRef]
89. Zhang, T.; Liu, Y.; Yang, D.; Wang, Y.; Fu, H.; Jia, Z.; Gao, H. Constructed fiber-optic FPI-based multi-parameters sensor for simultaneous measurement of pressure and temperature, refractive index and temperature. *Opt. Fiber Technol.* **2019**, *49*, 64–70. [CrossRef]
90. Ujah, E.; Lai, M.; Slaughter, G. Ultrasensitive tapered optical fiber refractive index glucose sensor. *Sci. Rep.* **2023**, *13*, 4495. [CrossRef] [PubMed]
91. Liu, Z.-D.; Liu, B.; Liu, J.; Fu, Y.; Wang, M.; Wan, S.-P.; He, X.; Yuan, J.; Chan, H.-P.; Wu, Q. Sensing Characteristics of Fiber Fabry-Perot Sensors Based on Polymer Materials. *IEEE Access* **2020**, *8*, 171316–171324. [CrossRef]
92. Guo, J.-Y.; Shi, B.; Sun, M.-Y.; Zhang, C.-C.; Wei, G.-Q.; Liu, J. Characterization of an ORMOCER®-coated FBG sensor for relative humidity sensing. *Measurement* **2021**, *171*, 108851. [CrossRef]
93. Riza, M.A.; Go, Y.I.; Harun, S.W.; Anas, S.B.A. Optimal etching process and cladding dimension for improved coating of porous hemispherical ZnO nanostructure on FBG humidity sensor. *Laser Phys.* **2023**, *33*, 075901. [CrossRef]
94. Wang, Y.; Liu, Y.; Zou, F.; Jiang, C.; Mou, C.; Wang, T. Humidity Sensor Based on a Long-Period Fiber Grating Coated with Polymer Composite Film. *Sensors* **2019**, *19*, 2263. [CrossRef]
95. Yan, J.; Feng, J.; Ge, J.; Chen, J.; Wang, F.; Xiang, C.; Wang, D.; Yu, Q.; Zeng, H. Highly sensitive humidity sensor based on a GO/Co-MOF-74 coated long period fiber grating. *IEEE Photonics Technol. Lett.* **2021**, *34*, 77–80. [CrossRef]
96. Zhang, J.; Tong, Z.; Zhang, W.; Zhao, Y.; Liu, Y. Research on NCF-PCF-NCF Structure Interference Characteristic for Temperature and Relative Humidity Measurement. *IEEE Photonics J.* **2021**, *13*, 7100805. [CrossRef]
97. Gao, P.; Zheng, X.; Liu, Y.; Wang, Z. Monitoring relative humidity using a Mach-Zehnder interferometer (MZI) senor based upon a photonic crystal fiber (PCF) coated with polyvinyl alcohol (PVA). *Instrum. Sci. Technol.* **2023**, *51*, 198–208. [CrossRef]
98. Hu, P.; Dong, X.; Ni, K.; Chen, L.H.; Wong, W.C.; Chan, C.C. Sensitivity-enhanced Michelson interferometric humidity sensor with waist-enlarged fiber bitaper. *Sens. Actuators B Chem.* **2014**, *194*, 180–184. [CrossRef]
99. Shao, M.; Sun, H.; Liang, J.; Han, L.; Feng, D. In-fiber Michelson interferometer in photonic crystal fiber for humidity measurement. *IEEE Sens. J.* **2020**, *21*, 1561–1567. [CrossRef]
100. Ali, U.; Abd Karim, K.J.B.; Buang, N.A. A Review of the Properties and Applications of Poly (Methyl Methacrylate) (PMMA). *Polym. Rev.* **2015**, *55*, 678–705. [CrossRef]
101. Cui, J.; Chen, G.; Li, J. PMMA-coated SMF-CLF-SMF-cascaded fiber structure and its humidity sensing characteristics. *Appl. Phys. B* **2022**, *128*, 36. [CrossRef]
102. Li, M.; Ma, C.; Li, D.; Bao, S.; Jin, J.; Zhang, Y.; Liu, Q.; Liu, M.; Zhang, Y.; Li, T.; et al. Dual-parameter optical fiber sensor for temperature and humidity based on PMMA-microsphere and FBG composite structure. *Opt. Fiber Technol.* **2023**, *78*, 103292. [CrossRef]
103. Meng, J.; Ma, J.-N.; Li, J.; Yan, H.; Meng, F. Humidity sensing and temperature response performance of polymer gel cold-spliced optical fiber Fabry-Perot interferometer. *Opt. Fiber Technol.* **2022**, *68*, 102823. [CrossRef]
104. Yao, J.; Zhu, T.; Duan, D.-W.; Deng, M. Nanocomposite polyacrylamide based open cavity fiber Fabry-Perot humidity sensor. *Appl. Opt.* **2012**, *51*, 7643–7647. [CrossRef]
105. Chen, H.; You, T.; Xu, G.; Gao, Y.; Zhang, C.; Yang, N.; Yin, P. Humidity-responsive nanocomposite of gold nanoparticles and polyacrylamide brushes grafted on Ag film: Synthesis and application as plasmonic nanosensor. *Sci. China Mater.* **2018**, *61*, 1201–1208. [CrossRef]
106. Wang, C.; Zhou, B.; Jiang, H.; He, S. Agarose Filled Fabry-Perot Cavity for Temperature Self-Calibration Humidity Sensing. *IEEE Photonics Technol. Lett.* **2016**, *28*, 2027–2030. [CrossRef]
107. Mathew, J.; Semenova, Y.; Farrell, G. Effect of coating thickness on the sensitivity of a humidity sensor based on an Agarose coated photonic crystal fiber interferometer. *Opt. Express* **2013**, *21*, 6313–6320. [CrossRef] [PubMed]
108. Xia, R.-L.; Liu, J.; Shi, J.; He, X.-D.; Yuan, J.; Pike, A.R.; Chu, L.; Wu, Q.; Liu, B. Compact fiber Fabry-Perot sensors filled with PNIPAM hydrogel for highly sensitive relative humidity measurement. *Measurement* **2022**, *201*, 111781. [CrossRef]
109. Li, S.; Hernandez, S.; Salazar, N. Biopolymer-Based Hydrogels for Harvesting Water from Humid Air: A Review. *Sustainability* **2023**, *15*, 848. [CrossRef]
110. Wang, Y.; Huang, Q.; Zhu, W.; Yang, M. Simultaneous Measurement of Temperature and Relative Humidity Based on FBG and FP Interferometer. *IEEE Photonics Technol. Lett.* **2018**, *30*, 833–836. [CrossRef]
111. Xu, Y.; Zhao, X.; Li, Y.; Qin, Z.; Pang, Y.; Liu, Z. Simultaneous measurement of relative humidity and temperature based on forward Brillouin scattering in polyimide-overlaid fiber. *Sens. Actuators B Chem.* **2021**, *348*, 130702. [CrossRef]

112. Vaz, A.; Barroca, N.; Ribeiro, M.; Pereira, A.; Frazao, O. Optical Fiber Humidity Sensor Based on Polyvinylidene Fluoride Fabry-Perot. *IEEE Photonics Technol. Lett.* **2019**, *31*, 549–552. [CrossRef]
113. Yu, C.; Gong, H.; Zhang, Z.; Ni, K.; Zhao, C. Temperature-Compensated High-Sensitivity Relative Humidity Sensor Based on Band-Pass Filtering and Vernier Effect. *IEEE Trans. Instrum. Meas.* **2022**, *71*, 7001808. [CrossRef]
114. Lakouraj, M.M.; Tajbakhsh, M.; Mokhtary, M. Synthesis and swelling characterization of cross-linked PVP/PVA hydrogels. *Iran. Polym. J.* **2005**, *14*, 1022–1030.
115. Gomes, A.D.; Bartelt, H.; Frazao, O. Optical Vernier Effect: Recent Advances and Developments. *Laser Photonics Rev.* **2021**, *15*, 2000588. [CrossRef]
116. Huang, C.; Xie, W.; Yang, M.; Dai, J.; Zhang, B. Optical Fiber Fabry-Perot Humidity Sensor Based on Porous Al_2O_3 Film. *IEEE Photonics Technol. Lett.* **2015**, *27*, 2127–2130. [CrossRef]
117. Peng, J.; Wang, W.; Qu, Y.; Sun, T.; Lv, D.; Dai, J.; Yang, M. Thin films based one-dimensional photonic crystal for humidity detection. *Sens. Actuators A Phys.* **2017**, *263*, 209–215. [CrossRef]
118. Li, C.; Yu, X.; Zhou, W.; Cui, Y.; Liu, J.; Fan, S. Ultrafast miniature fiber-tip Fabry-Perot humidity sensor with thin graphene oxide diaphragm. *Opt. Lett.* **2018**, *43*, 4719–4722. [CrossRef] [PubMed]
119. Yao, L.; Zheng, M.; Li, H.; Ma, L.; Shen, W. High-performance humidity sensors based on high-field anodized porous alumina films. *Nanotechnology* **2009**, *20*, 395501. [CrossRef] [PubMed]
120. Yan, G.; Zhang, A.P.; Ma, G.; Wang, B.; Kim, B.; Im, J.; He, S.; Chung, Y. Fiber-Optic Acetylene Gas Sensor Based on Microstructured Optical Fiber Bragg Gratings. *IEEE Photonics Technol. Lett.* **2011**, *23*, 1588–1590. [CrossRef]
121. Zhou, Z.; Xu, Y.; Qiao, C.; Liu, L.; Jia, Y. A novel low-cost gas sensor for CO2 detection using polymer-coated fiber Bragg grating. *Sens. Actuators B Chem.* **2021**, *332*, 129482. [CrossRef]
122. Yang, F.; Jin, W.; Lin, Y.; Wang, C.; Lut, H.; Tan, Y. Hollow-core microstructured optical fiber gas sensors. *J. Light. Technol.* **2017**, *35*, 3413–3424. [CrossRef]
123. Liu, C.; Chen, H.; Chen, Q.; Zheng, Y.; Gao, Z.; Fan, X.; Wu, B.; Shum, P.P. An ultra-high sensitivity methane gas sensor based on Vernier effect in two parallel optical fiber Sagnac loops. *Opt. Commun.* **2023**, *540*, 129509. [CrossRef]
124. Hao, T.; Chiang, K.S. Graphene-Based Ammonia-Gas Sensor Using In-Fiber Mach-Zehnder Interferometer. *IEEE Photonics Technol. Lett.* **2017**, *29*, 2035–2038. [CrossRef]
125. Feng, X.; Feng, W.; Tao, C.; Deng, D.; Qin, X.; Chen, R. Hydrogen sulfide gas sensor based on graphene-coated tapered photonic crystal fiber interferometer. *Sens. Actuators B Chem.* **2017**, *247*, 540–545. [CrossRef]
126. He, S.; Liu, Y.; Feng, W.; Li, B.; Yang, X.; Huang, X. Carbon monoxide gas sensor based on an α-Fe_2O_3/reduced graphene oxide quantum dots composite film integrated Michelson interferometer. *Meas. Sci. Technol.* **2021**, *33*, 035102. [CrossRef]
127. Zhou, J.; Huang, X.; Feng, W. Carbon monoxide gas sensor based on Co/Ni-MOF-74 coated no-core-fiber Michelson interferometer. *Phys. Scr.* **2023**, *98*, 015012. [CrossRef]
128. Ma, W.; Wang, R.; Rong, Q.; Shao, Z.; Zhang, W.; Guo, T.; Wang, J.; Qiao, X. CO_2 Gas Sensing Using Optical Fiber Fabry-Perot Interferometer Based on Polyethyleneimine/Poly (Vinyl Alcohol) Coating. *IEEE Photonics J.* **2017**, *9*, 6802808. [CrossRef]
129. Helberg, R.M.L.; Dai, Z.; Ansaloni, L.; Deng, L. PVA/PVP blend polymer matrix for hosting carriers in facilitated transport membranes: Synergistic enhancement of CO_2 separation performance. *Green Energy Environ.* **2020**, *5*, 59–68. [CrossRef]
130. Ma, W.; Xing, J.; Wang, R.; Rong, Q.; Zhang, W.; Li, Y.; Zhang, J.; Qiao, X. Optical fiber Fabry–Perot interferometric CO_2 gas sensor using guanidine derivative polymer functionalized layer. *IEEE Sens. J.* **2018**, *18*, 1924–1929. [CrossRef]
131. Kazanskiy, N.L.; Butt, M.A.; Khonina, S.N. Carbon Dioxide Gas Sensor Based on Polyhexamethylene Biguanide Polymer Deposited on Silicon Nano-Cylinders Metasurface. *Sensors* **2021**, *21*, 378. [CrossRef]
132. Peng, J.; Feng, W.; Yang, X.; Huang, G.; Liu, S. Dual Fabry-Perot Interferometric Carbon Monoxide Sensor Based on the PANI/Co_3O_4 Sensitive Membrane-Coated Fibre Tip. *Z. Fur Naturforsch. Sect. A* **2019**, *74*, 101–107. [CrossRef]
133. Pawar, D.; Rao, B.V.B.; Kale, S.N. Fe_3O_4-decorated graphene assembled porous carbon nanocomposite for ammonia sensing: Study using an optical fiber Fabry-Perot interferometer. *Analyst* **2018**, *143*, 1890–1898. [CrossRef]
134. Lopez-Torres, D.; Lopez-Aldaba, A.; Elostia Aguado, C.; Auguste, J.-L.; Jamier, R.; Roy, P.; Lopez-Amo, M.; Arregui, F.J. Sensitivity Optimization of a Microstructured Optical Fiber Ammonia Gas Sensor by Means of Tuning the Thickness of a Metal Oxide Nano-Coating. *IEEE Sens. J.* **2019**, *19*, 4982–4991. [CrossRef]
135. Bochenkov, V.; Sergeev, G. Sensitivity, selectivity, and stability of gas-sensitive metal-oxide nanostructures. *Met. Oxide Nanostruct. Their Appl.* **2010**, *3*, 31–52.
136. Feng, Z.; Cheng, Y.; Chen, M.; Yuan, L.; Hong, D.; Li, L. Temperature-Compensated Multi-Point Strain Sensing Based on Cascaded FBG and Optical FMCW Interferometry. *Sensors* **2022**, *22*, 3970. [CrossRef] [PubMed]
137. Eid, M.M.; Seliem, A.S.; Rashed, A.Z.; Mohammed, A.E.-N.A.; Ali, M.Y.; Abaza, S.S. High sensitivity sapphire FBG temperature sensors for the signal processing of data communications technology. *Indones. J. Electr. Eng. Comput. Sci.* **2021**, *21*, 1567–1574. [CrossRef]
138. Munoz-Hernandez, T.; Reyes-Vera, E.; Torres, P. Temperature Sensor Based on Whispering Gallery Modes of Metal-Filled Side-Hole Photonic Crystal Fiber Resonators. *IEEE Sens. J.* **2020**, *20*, 9170–9178. [CrossRef]
139. Yang, Y.; Wang, Z.; Zhang, X.; Zhang, Q.; Wang, T. Recent progress of in-fiber WGM microsphere resonator. *Front. Optoelectron.* **2023**, *16*, 10. [CrossRef] [PubMed]

140. Chen, S.; Pan, P.; Xie, T.; Fu, H. Sensitivity enhanced fiber optic temperature sensor based on optical carrier microwave photonic interferometry with harmonic Vernier effect. *Opt. Laser Technol.* **2023**, *160*, 109029. [CrossRef]
141. Li, X.; Tan, J.; Li, W.; Yang, C.; Tan, Q.; Feng, G. A high-sensitivity optical fiber temperature sensor with composite materials. *Opt. Fiber Technol.* **2022**, *68*, 102821. [CrossRef]
142. Shao, M.; Zhang, R.; Gao, H.; Liu, Y.; Qiao, X.; Lin, Y. A High-Sensitivity Low-Temperature Sensor Based on Michelson Interferometer in Seven-Core Fiber. *IEEE Photonics Technol. Lett.* **2021**, *33*, 1293–1296. [CrossRef]
143. Wang, S.; Yang, Y.; Niu, P.; Wu, S.; Liu, S.; Jin, R.-B.; Lu, P.; Hu, X.; Dai, N. Fiber tip Michelson interferometer for temperature sensing based on polymer-filled suspended core fiber. *Optics & Laser Technology* **2021**, *141*, 107147.
144. Yin, J.; Liu, T.; Jiang, J.; Liu, K.; Wang, S.; Qin, Z.; Zou, S. Batch-Producible Fiber-Optic Fabry-Perot Sensor for Simultaneous Pressure and Temperature Sensing. *IEEE Photonics Technol. Lett.* **2014**, *26*, 2070–2073.
145. Liu, T.; Yin, J.; Jiang, J.; Liu, K.; Wang, S.; Zou, S. Differential-pressure-based fiber-optic temperature sensor using Fabry-Perot interferometry. *Opt. Lett.* **2015**, *40*, 1049–1052. [CrossRef]
146. Pan, R.; Yang, W.; Li, L.; Yang, Y.; Zhang, L.; Yu, X.; Fan, J.; Yu, S.; Xiong, Y. A High-Sensitive Fiber-Optic Fabry-Perot Sensor with Parallel Polymer-Air Cavities Based on Vernier Effect for Simultaneous Measurement of Pressure and Temperature. *IEEE Sens. J.* **2021**, *21*, 21577–21585. [CrossRef]
147. Zhang, S.; Mei, Y.; Xia, T.; Cao, Z.; Liu, Z.; Li, Z. Simultaneous measurement of temperature and pressure based on Fabry-Perot Interferometry for marine monitoring. *Sensors* **2022**, *22*, 4979. [CrossRef] [PubMed]
148. Defas-Brucil, R.; Cano-Velazquez, M.S.; Velazquez-Benitez, A.M.; Hernandez-Cordero, J. Microbubble end-capped fiber-optic Fabry-Perot sensors. *Opt. Lett.* **2022**, *47*, 5569–5572. [CrossRef] [PubMed]
149. Dean, L.M.; Ravindra, A.; Guo, A.X.; Yourdkhani, M.; Sottos, N.R. Photothermal Initiation of Frontal Polymerization Using Carbon Nanoparticles. *ACS Appl. Polym. Mater.* **2020**, *2*, 4690–4696. [CrossRef]
150. Zhou, X.; Li, X.; Li, S.; Yan, X.; Zhang, X.; Wang, F.; Suzuki, T.; Ohishi, Y.; Cheng, T. A miniature optical fiber Fabry–Perot interferometer temperature sensor based on tellurite glass. *IEEE Trans. Instrum. Meas.* **2021**, *70*, 7005706. [CrossRef]
151. Mohanna, Y.; Saugrain, J.-M.; Rousseau, J.-C.; Ledoux, P. Relaxation of internal stresses in optical fibers. *J. Light. Technol.* **1990**, *8*, 1799–1802. [CrossRef]
152. Liao, Y.; Liu, Y.; Li, Y.; Lang, C.; Cao, K.; Qu, S. Large-Range, Highly-Sensitive, and Fast-Responsive Optical Fiber Temperature Sensor Based on the Sealed Ethanol in Liquid State Up to its Supercritical Temperature. *IEEE Photonics J.* **2019**, *11*, 6803112. [CrossRef]
153. Scott, T.A., Jr. Refractive index of ethanol-water mixtures and density and refractive index of ethanol-water-ethyl ether mixtures. *J. Phys. Chem.* **1946**, *50*, 406–412. [CrossRef]
154. Wang, B.; Niu, Y.; Zheng, S.; Yin, Y.; Ding, M. A high temperature sensor based on sapphire fiber Fabry-Perot interferometer. *IEEE Photonics Technol. Lett.* **2019**, *32*, 89–92. [CrossRef]
155. Cui, Y.; Jiang, Y.; Zhang, Y.; Feng, X.; Hu, J.; Jiang, L. An all-sapphire fiber temperature sensor for high-temperature measurement. *Meas. Sci. Technol.* **2022**, *33*, 105115. [CrossRef]
156. Lupi, C.; Felli, F.; Ippoliti, L.; Caponero, M.; Ciotti, M.; Nardelli, V.; Paolozzi, A. Metal coating for enhancing the sensitivity of fibre Bragg grating sensors at cryogenic temperature. *Smart Mater. Struct.* **2005**, *14*, N71. [CrossRef]
157. Lupi, C.; Felli, F.; Brotzu, A.; Caponero, M.A.; Paolozzi, A. Improving FBG sensor sensitivity at cryogenic temperature by metal coating. *IEEE Sens. J.* **2008**, *8*, 1299–1304. [CrossRef]
158. Yang, J.; Yin, B.; Dong, X.; Huang, W.; Zou, Y. Ultrasensitive Cryogenic Temperature Sensor Based on a Metalized Optical Fiber Fabry-Perot Interferometer. *J. Light. Technol.* **2022**, *40*, 5729–5735. [CrossRef]
159. Kaushik, S.; Tiwari, U.; Nilima; Prashar, S.; Das, B.; Sinha, R.K. Label-free detection of Escherichia coli bacteria by cascaded chirped long period gratings immunosensor. *Rev. Sci. Instrum.* **2019**, *90*, 025003. [CrossRef] [PubMed]
160. Eftimov, T.; Janik, M.; Koba, M.; Smietana, M.; Mikulic, P.; Bock, W. Long-Period Gratings and Microcavity In-Line Mach Zehnder Interferometers as Highly Sensitive Optical Fiber Platforms for Bacteria Sensing. *Sensors* **2020**, *20*, 3772. [CrossRef]
161. Sirkis, T.; Beiderman, Y.; Agdarov, S.; Beiderman, Y.; Zalevsky, Z. Fiber sensor for non-contact estimation of vital bio-signs. *Opt. Commun.* **2017**, *391*, 63–67. [CrossRef]
162. Yu, C.; Xu, W.; Zhang, N.; Yu, C. Non-invasive smart health monitoring system based on optical fiber interferometers. In Proceedings of the 16th International Conference on Optical Communications and Networks (ICOCN), Wuzhen, China, 7–10 August 2017.
163. Gasparyan, V.K.; Bazukyan, I.L. Lectin sensitized anisotropic silver nanoparticles for detection of some bacteria. *Anal. Chim. Acta* **2013**, *766*, 83–87. [CrossRef]
164. Arcas, A.d.S.; Dutra, F.d.S.; Allil, R.C.; Werneck, M.M. Surface plasmon resonance and bending loss-based U-shaped plastic optical fiber biosensors. *Sensors* **2018**, *18*, 648. [CrossRef]
165. Queiros, R.B.; Silva, S.O.; Noronha, J.P.; Frazao, O.; Jorge, P.; Aguilar, G.; Marques, P.V.S.; Sales, M.G.F. Microcystin-LR detection in water by the Fabry-Perot interferometer using an optical fibre coated with a sol-gel imprinted sensing membrane. *Biosens. Bioelectron.* **2011**, *26*, 3932–3937. [CrossRef]
166. Chen, L.H.; Chan, C.C.; Menon, R.; Balamurali, P.; Wong, W.C.; Ang, X.M.; Hu, P.B.; Shaillender, M.; Neu, B.; Zu, P.; et al. Fabry-Perot fiber-optic immunosensor based on suspended layer-by-layer (chitosan/polystyrene sulfonate) membrane. *Sens. Actuators B Chem.* **2013**, *188*, 185–192. [CrossRef]

167. Liu, X.; Jiang, M.; Dong, T.; Sui, Q.; Geng, X. Label-Free Immunosensor Based on Optical Fiber Fabry-Perot Interferometer. *IEEE Sens. J.* **2016**, *16*, 7515–7520. [CrossRef]
168. Li, Y.; Dong, B.; Chen, E.; Wang, X.; Zhao, Y.; Zhao, W.; Wang, Y. Breathing process monitoring with a biaxially oriented polypropylene film based fiber Fabry-Perot sensor. *Opt. Commun.* **2020**, *475*, 126292. [CrossRef]
169. Shrivastav, A.M.; Gunawardena, D.S.; Liu, Z.; Tam, H.-Y. Microstructured optical fiber based Fabry-Perot interferometer as a humidity sensor utilizing chitosan polymeric matrix for breath monitoring. *Sci. Rep.* **2020**, *10*, 6002. [CrossRef] [PubMed]
170. Sergeev, A.; Voznesenskiy, S. Specific features of chitosan waveguides optical response formation to changes in the values of relative humidity. *Opt. Mater.* **2015**, *43*, 33–35. [CrossRef]
171. Li, X.; Li, F.; Zhou, X.; Zhang, Y.; Linh Viet, N.; Warren-Smith, S.C.; Zhao, Y. Optical Fiber DNA Biosensor with Temperature Monitoring Based on Double Microcavities FabryPerot Interference and Vernier Combined Effect. *IEEE Trans. Instrum. Meas.* **2023**, *72*, 7001208.
172. Zou, X.; Wu, N.; Tian, Y.; Ouyang, J.; Barringhaus, K.; Wang, X. Miniature Fabry–Perot fiber optic sensor for intravascular blood temperature measurements. *IEEE Sens. J.* **2013**, *13*, 2155–2160. [CrossRef]
173. Viphavakit, C.; O'Keeffe, S.; Yang, M.; Andersson-Engels, S.; Lewis, E. Gold Enhanced Hemoglobin Interaction in a Fabry-Perot Based Optical Fiber Sensor for Measurement of Blood Refractive Index. *J. Light. Technol.* **2018**, *36*, 1118–1124. [CrossRef]
174. Li, Y.; Dong, B.; Chen, E.; Wang, X.; Zhao, Y. Heart-Rate Monitoring with an Ethyl Alpha-Cyanoacrylate Based Fiber Fabry-Perot Sensor. *IEEE J. Sel. Top. Quantum Electron.* **2021**, *27*, 5600206. [CrossRef]

Disclaimer/Publisher's Note: The statements, opinions and data contained in all publications are solely those of the individual author(s) and contributor(s) and not of MDPI and/or the editor(s). MDPI and/or the editor(s) disclaim responsibility for any injury to people or property resulting from any ideas, methods, instructions or products referred to in the content.

Article

Enhancing the Performance and Stability of Perovskite Solar Cells via Morpholinium Tetrafluoroborate Additive Engineering: Insights and Implications

Jianxiao Bian [1,2,*], Yingtang Sun [3], Jinchang Guo [1], Xin Liu [1] and Yang Liu [1]

[1] School of Intelligent Manufacturing, Longdong University, Qingyang 745000, China; guojinchang2008@163.com (J.G.); liux0704@163.com (X.L.); xfliuyang-1201@163.com (Y.L.)
[2] Shaanxi Key Laboratory of Non-Traditional Machining, Xi'an Technological University, Xi'an 710021, China
[3] Liaohe Oilfield Qingyang Exploration and Development Branch, Qingyang 745000, China; meng6tang@163.com
* Correspondence: jxbian@ldxy.edu.cn

Abstract: Perovskite solar cells (PSCs), since their inception in 2009, have experienced a meteoric rise in power conversion efficiencies (PCEs), challenging established photovoltaic technologies. However, their commercial deployment is hindered by stability and performance issues related to the presence of defects at the perovskite surface and grain boundaries. This study focused on the exploration of Morpholinium tetrafluoroborate (MOT) as a post-treatment additive to mitigate these challenges. Comprehensive characterization techniques revealed that the synergistic action of Morpholine and BF_4^- ions in MOT substantially improved the quality of the perovskite films and passivates surface and bulk defects, yielding notable enhancements in device PCE and stability. MOT-doped PSCs exhibited a PCE of 23.83% and retain 92% of the initial PCE after 2000 h of continuous illumination under one sun condition. The findings underscore the significance of additive engineering in advancing perovskite solar cell technology, opening up prospects for high-performing and durable perovskite photovoltaic devices.

Keywords: Morpholine Tetrafluoroborate; stability; passivates surface; perovskite solar cells

1. Introduction

Since their advent in 2009, organic–inorganic perovskite solar cells (PSCs) have emerged as a prominent player in the field of photovoltaics, captivating the scientific community with their high power conversion efficiencies (PCEs), tunable band gaps, solution-processability, and cost-effectiveness [1–10]. Over the past decade, the PCEs of PSCs have experienced an extraordinary surge from a mere 3.8% to over 25%, posing a significant challenge to established solar technologies such as crystalline silicon and thin-film solar cells [11–14]. However, despite these remarkable advancements, the path to commercializing PSCs is fraught with critical challenges, primarily centered around device performance and stability. Defects at the perovskite surface and grain boundaries have emerged as major contributors to non-radiative recombination and ion migration. Moreover, the ubiquitous solution spin-coating method for perovskite fabrication tends to introduce numerous surface and bulk defects due to the presence of non coordinated lead iodide octahedra in the solution. As a consequence, the fabricated perovskite films often exhibit numerous iodine vacancies, thereby undermining both device performance and longevity [15–17]. In the pursuit of overcoming these hurdles, intensive research efforts have been directed towards devising strategies to alleviate these defects. One promising approach lies in the post-treatment of the prepared perovskite films with various molecular additives. Such additives have demonstrated their potential in controlling crystal growth, passivating defects, and modulating electronic properties of the perovskite materials [18–22]. The introduction of additives such as Phenethylammonium iodide (PEAI),

which in situ forms films on mixed cation films to improve the quality of perovskite films, has successfully boosted the PCE of PSCs up to 23.32% [23]. Furthermore, multifunctional ionic liquids, possessing a broad range of desirable properties, have been employed as efficient passivating materials to elevate the performance of PSCs, evidenced by the augmentation in PCE to 23.25% and 9.92% upon the integration of 1,3-dimethyl-3-imidazolium hexafluorophosphate (DMIMPF6) and 1-butyl-2,3-dimethylimidazolium chloride ([BMMIm]Cl), respectively [24,25]. In light of these advancements, MOT, due to its unique structure that allows for strong bonding with lead ions and potential replacement of missing iodide ions, has been identified as a promising additive [26–28]. The interactions facilitated by MOT can profoundly influence the crystal growth mechanism and defect passivation, thereby enhancing both the performance and stability of PSCs. However, empirical insights into the role of MOT as a dopant in PSCs are relatively scant in existing literature.

Building upon the existing advancements in the field of perovskite solar cells (PSCs), our study introduced a novel aspect of employing MOT as an overlayer on the perovskite surface, a method not extensively explored in the current literature. The primary objectives of this research were to elucidate the multifaceted influence of MOT in tailoring the morphology, crystallinity, electrical properties, and stability of perovskite films. In the context of this study, MOT was employed as an overlayer on the perovskite surface to mitigate challenges at the interface between the perovskite and Spiro-OMeTAD layers. Although MOT may function as an additive in some applications, in this specific configuration, it is grown directly on the perovskite surface. This unique approach sets our work apart from existing strategies and allows us to unravel new pathways in enhancing both the performance and stability of PSCs. Our findings, marked by a remarkable PCE of 23.83% in MOT-modified perovskite film, not only contribute to a deeper understanding of the perovskite modification techniques but also signify a significant leap in the ongoing pursuit of highly efficient and stable PSCs. This study, therefore, stands as a valuable addition to the scientific community, bridging the gap between theoretical potential and practical implementation of MOT in the realm of PSCs.

2. Results and Discussion

2.1. Structure and Morphology

The synthesis process of the perovskite devices is shown in Figure 1a. Initially, an electron transport layer was fabricated on an FTO substrate through spin coating of SnO_2 solution. Subsequently, a perovskite precursor solution was drop cast onto the FTO/SnO_2 substrate. The prepared substrate was then annealed on a hot plate at 150 °C for 15 min, resulting in an FTO/SnO_2/Perovskite film. Different concentrations of MOT molecules were then dissolved in isopropyl alcohol (IPA) solution and drop cast onto the prepared FTO/SnO_2/perovskite film. This was followed by a second crystallization step involving spin coating. After this step, the films were again annealed at 150 °C for 1 min to solidify the structure. For comparison, control devices were fabricated following the same procedure, but without the addition of MOT molecules on the surface of the perovskite film. The results of these experiments should offer valuable insights into the effects of MOT molecule concentration and annealing time on the performance and stability of perovskite devices. Figure 2b presents a detailed depiction of the three-dimensional structure of MOT molecules. The image illustrates the three-dimensional arrangement of the Morpholine Tetrafluoroborate (MOT) molecule. The MOT molecule consists of a central morpholine ring, which is a six-membered heterocyclic ring containing four carbon atoms and two nitrogen atoms. Attached to the morpholine ring are four fluorine atoms, one at each position where a hydrogen atom would typically be present. The fluorine atoms are positioned tetrahedrally around the morpholine ring, with each fluorine atom bonded to one of the carbon atoms. The tetrafluoroborate anion, BF_4^-, and these elements to interact with the perovskite surface, leading to a reduction in surface defects. Concurrently, the tetrafluoroborate moiety demonstrated its utility in effectively compensating for the deficiency of iodine elements within the perovskite structure, thereby ameliorating associated defects. These observations imply that the structural attributes of MOT molecules may contribute significantly to the optimization of perovskite device performance. Figure 1c showcases the Ultraviolet-Visible (UV-Vis) absorption spectra for perovskite films, comparing samples that were optimized with MOT molecules to those that have not. The UV-Vis absorption spectra revealed that the perovskite films, post-MOT optimization, exhibited enhanced absorption, notably within the 300 to 600 nm wavelength range. This heightened absorption was likely attributable to the introduction of uncoordinated BF_4^- ions, which resulted in the coordination of perovskite crystals, potentially leading to the formation of the fully coordinated $Pb(I + BF_4)X$ complex. These findings suggest that the implementation of MOT optimization strategies can significantly influence the photophysical properties of perovskite films, thus enhancing their performance. Figure 1d–g provides the comprehensive X-ray Photoelectron Spectroscopy (XPS) spectra and narrow scans for O 1 s, C 1 s, B 1 s, and F 1 s of perovskite films, encompassing both KPF-modified and unmodified samples. The XPS data revealed that the perovskite films subjected to MOT ionic liquid modification exhibited distinctive XPS peaks. Specifically, an O 1 s peak was associated with the C-OH bond, and C 1 s spectra presented peaks corresponding to C-C, C-O-C, and C-N bonds, which was indicative of the MOT molecular structure's influence. Moreover, the emergence of B 1 s and F 1 s peaks asserted the successful integration of MOT into the perovskite film structure, a feature absent in the unmodified samples. To better contrast the chemical alterations instigated by MOT optimization, Figure S1 illustrates the complete XPS spectra for perovskite films both before and after MOT modification, clearly highlighting the absence of B and F elemental intensities in the non-optimized films. These results underscore the substantial impact of MOT modification on the chemical properties of perovskite films.

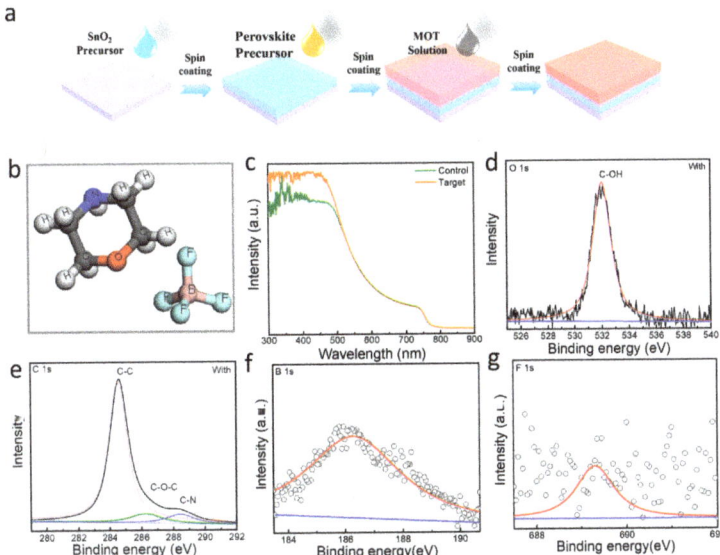

Figure 1. (**a**) Schematic diagram of the fabrication process of perovskite films with MOT. (**b**) 3D chemical structure of MOT. (**c**) UV-Vis absorption spectra of the perovskite solutions without and with MOT. Full XPS spectra and narrow scans of (**d**) O 1 s, (**e**) C 1 s, (**f**) B 1 s, and (**g**) F 1 s for thin films of perovskite without and with Morpholine Tetrafluoroborate (MOT).

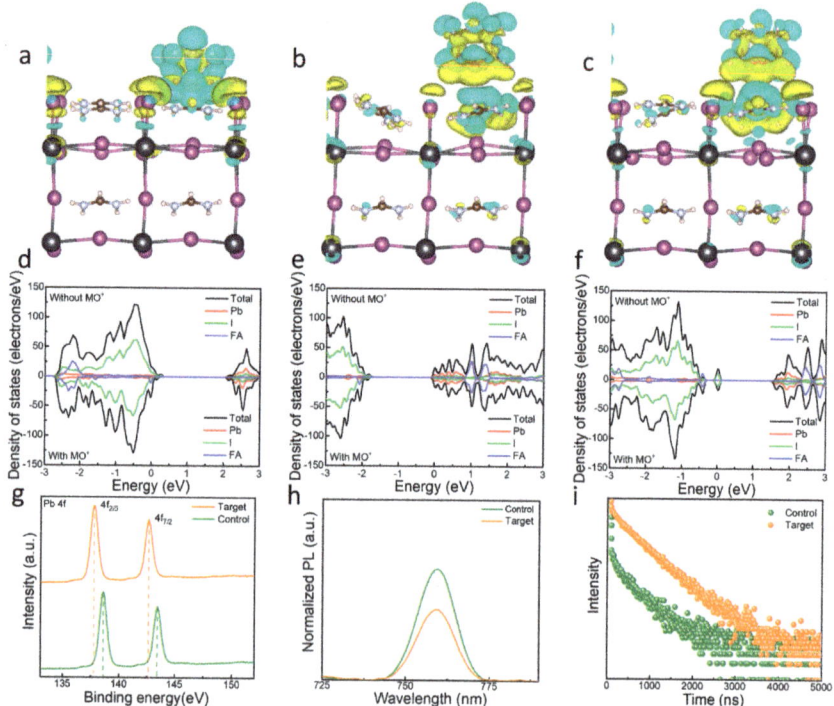

Figure 2. Chemical structure of MOT. Charge density difference of MOT passivated FAI–terminated FAPbI$_3$ surface with V$_{FA}$ (**a**), V$_I$ (**b**) and I$_i$ (**c**). PDOS of passivated FAI–terminated FAPbI$_3$ surface with V$_{FA}$ (**d**), V$_I$ (**e**) and I$_i$ (**f**) in the presence or not of MOT. (**g**) The XPS spectra for Pb 4f from the with and without MOT modified perovskite films. (**h**) PL and (**i**) TRPL spectra of perovskite films from the perovskite films with and without MOT–modification.

Figures 2a and S2a delineate the chemical structure of MOT, utilized for the surface treatment of perovskite in this investigation. Comprising two components—morpholine and tetrafluoroborate anion—the ionic liquid exhibited ease of diffusion and surface coverage owing to its small size, enabling the extensive passivation of perovskite defects. To capture the intricacies of atomic-level interactions between perovskite and MOT, simulations were performed focusing on the action of morpholine and the tetrafluoroborate anion on typical defects (F$_A$ vacancies (V$_{FA}$), I interstitial atoms (I$_i$), and point defect I vacancies (V$_I$) present on an FAI-terminated FAPbI$_3$ perovskite surface with defects. Employing Density Functional Theory (DFT) and MOT as the basic unit, we observed electronic redistribution following bond or atom interactions, as represented by blue areas. Conversely, yellow areas highlighted electrons that were captured by pertinent atoms. Charge transfer, which was witnessed between I atoms on the perovskite surface and oxygen atoms within morpholine, and between the tetrafluoroborate anion and Pb atoms, is suggestive of strong interactions. These interactions restricted the migration and oscillation of I-, induced I vacancies, and upon the introduction of the tetrafluoroborate anion, rectified the lack of I ions, resulting in lattice deformations and structural distortions. Further interactions were observed between FA in perovskite and the oxygen and the tetrafluoroborate anion in MOT, which limited FA migration. To expound the implications of morpholine and tetrafluoroborate passivation on perovskite material properties, we conducted measurements of the partial density of states (PDOS) of perovskite films, both untreated and treated with the ionic modifier (refer to Figure 1d–f and Figure S2c). We identified the introduction of trap states by V$_{FA}$, V$_I$, and I$_i$ at the perovskite surface, serving as non-radiative recombination centers and causing energy loss. DFT predictions suggested a possible reduction in or suppression of these defects within the bandgap following morpholine and tetrafluoroborate passivation, implying that these components acted by compensating for electronic defects at perovskite lattice defect sites. As shown in Figure 2g, the X-ray photoelectron spectra (XPS) of control and MOT-treated perovskite films underscored the effective interaction of MOT, as evidenced by the upshift of Pb^{2+} binding energy peak in the Pb 4f spectra of the treated perovskite films. These results align well with the experimental findings from the XPS study and serve to corroborate the theoretical projections made regarding the interactions within the perovskite films. To analyze the influence of MOT treatment on the optical and photovoltaic properties of perovskite films, we evaluated the steady-state photoluminescence (PL) spectra of these films. Owing to enhanced crystallinity and a reduction in film defects, we observed an augmentation in the PL intensity of the perovskite films post-MOT treatment. As depicted in Figure 2h, the treated perovskite films manifested superior emission intensity at approximately 765 nm. Time-resolved photoluminescence (TRPL) spectra (Figure 2i) further illustrated that the average carrier lifetime (τ) of the MOT-treated perovskite films was notably longer at 50.48 ns compared to the control films, which had a τ value of 13.63 ns. This observation was consistent with decay times $\tau 1$ and $\tau 2$, as detailed in the Supporting Information, Table S1. The escalation in PL intensity and carrier lifetime underscores that defects within the perovskite films can be effectively passivated by MOT molecules. The data presented herein corroborate the promising potential of employing MOT treatment to significantly improve the photophysical and optoelectronic properties of perovskite films, thereby positively impacting their performance in photovoltaic applications.

Figure 3a presents the chemical structure of MOT and a mechanistic explanation for its role in passivating the perovskite. Comprising morpholine and tetrafluoroborate groups, MOT functioned as a multi-faceted agent to improve the quality and stability of perovskite films. The morpholine component promoted superior crystal growth and defect passivation through its strong binding with lead. Concurrently, the tetrafluoroborate group effectively substituted missing iodine ions and restricts the escape of FA, thereby elevating the overall stability of the perovskite. These combined actions significantly boosted the moisture stability of perovskite films, positioning MOT as a promising candidate for enhancing both the power conversion efficiency (PCE) and the durability of perovskite-based photovoltaics. As such, the deployment of MOT provided a vital pathway towards achieving high-performance and long-lasting perovskite solar cells. Figure 3b–e presents SEM (Scanning Electron Microscopy) images of perovskite films that underwent treatment with varying concentrations of MOT. Notably, the discrepancies observed in the film morphology were directly attributed to alterations in the MOT additive concentration, assuming all other parameters involved in film fabrication remained constant. In Figure 3b, the virgin film manifested a rather uneven coverage of the perovskite layer, characterized by tiny pores and remnants of unreacted PbI_2. The grain size of the perovskite material was determined through the analysis of SEM images, allowing us to evaluate the microstructural characteristics of the material. There was no calculation of crystal size using XRD in this particular study. A noticeable enhancement in uniformity and crystalline growth was observable in the MOT-treated film (as shown in Figure 3c), despite the persistence of some PbI_2 residues. Increasing the concentration of MOT led to more noticeable advancements in film morphology (Figure 3d). The absence of discernible pinholes or fissures, the augmentation of grain size, and the elimination of PbI2 residues affirmed the beneficial impact of MOT. However, an excessive concentration of MOT, demonstrated in Figure 3e, negatively impacted the surface coverage of the film. Therefore, these results indicate that the introduction of an optimized amount of MOT significantly fosters grain growth and enhances film quality. An overabundance of MOT, however, impairs the quality of the perovskite layer, emphasizing the importance of finding the balance in additive concentration. The most superior film attributes were demonstrated when the MOT concentration was kept at 1 mg/mL, which is indicative of the essential role of carefully controlled additive concentrations in creating high-performance perovskite solar cells.

In order to further substantiate our conclusions, we conducted X-ray diffraction (XRD) analysis of the perovskite thin films treated with different concentrations of MOT (see Figure 3f). Notably, perovskite films modified by various concentrations of MOT showed a distinctive peak at around 11 degrees. This peak is presumably indicative of a new crystalline phase resulting from the binding of MOT onto the perovskite surface, which appeared to enhance the perovskite's overall stability. Moreover, we analyzed the full width at half maximum (FWHM) of the XRD peaks corresponding to the different MOT concentrations (Figure 3g). The smallest FWHM, signifying the highest crystallinity, was observed for the film with the MOT concentration at 1 mg/mL. Interestingly, the film with an MOT concentration of 2 mg/mL also exhibited a relatively small FWHM, suggesting good crystallinity. However, SEM images for this sample revealed a surface layer comprised of a new phase. A statistical box plot analysis of grain sizes for the perovskite films prepared with varying concentrations of MOT has been presented in Figure 3h. It was observed that the perovskite film exhibited an optimal grain size, and, and the average grain size reached 500 nm when treated with an MOT concentration of 1 mg/mL. This optimal grain size can be attributed to the efficacious role of MOT in regulating the growth kinetics of the perovskite crystals, thereby leading to improved crystallinity. These results substantiate that the addition of MOT can notably refine the microstructure of perovskite thin films, potentially enhancing the overall optoelectronic performance of perovskite-based devices.

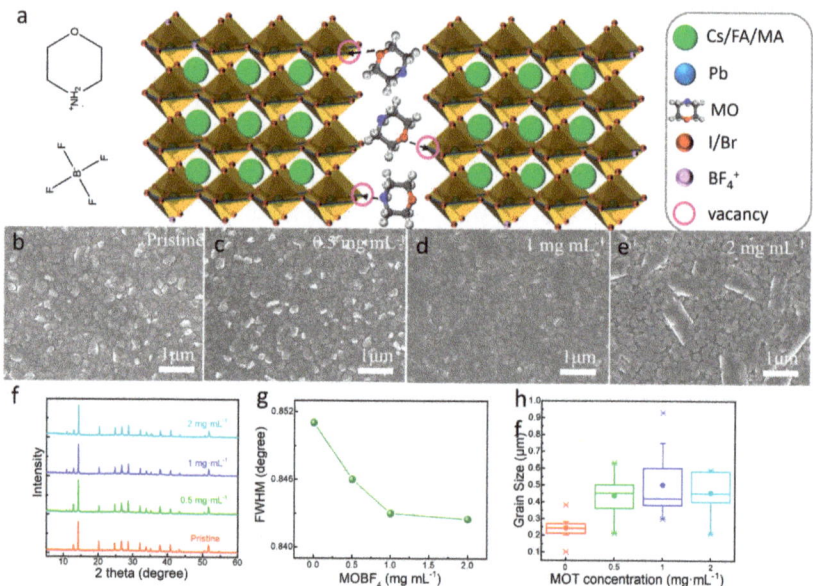

Figure 3. (**a**) Working mechanism and chemical composition characterization of films. (**b**–**e**) SEM image of the films of different concentration treatment. (**f**) XRD patterns (**g**) Corresponding FWHM of perovskite films based on different MOT concentration. (**h**) Statistical graph of grain size of perovskite films based on different MOT concentration.

2.2. Photovoltaic Performance of the PSC

The Kelvin Probe Force Microscopy (KPFM) was employed to further dissect the surface characteristics and electrical properties of perovskite films post Morpholine Tetrafluoroborate (MOT) modification. KPFM is known to offer reliable insights into the local surface potential, impacted by the contact potential difference (CPD)—the disparity in the work function between the microscope tip and sample surface. Our study showed a significant shift in CPD maps (Figure 4a,b) post MOT treatment, providing clues towards understanding the improved open-circuit voltage (V_{OC}) and work function. Figure 4c depicts the local potential distributions from random KPFM scans, where a substantial increase in the average surface potential, from 200.31 mV to 297.63 mV, was observed. This reinforces the idea that MOT modification successfully reduces perovskite defects, fostering enhanced carrier transport within the film and, consequently, a superior Voc. Collectively, our study underscores the promising role of MOT modification in perovskite films to enhance their electrical performance parameters and interface characteristics, as validated by KPFM analysis. This revelation could provide valuable guidance for the future development and optimization of perovskite-based devices. In a quest to delve deeper into the impact of Morpholine Tetrafluoroborate (MOT) on the surface characteristics of perovskite films, we present images acquired from the 3D height sensor Atomic Force Microscopy (AFM) in Figure 4d,e. The MOT-treated perovskite film showed a marked reduction in root mean square roughness, down to 45 nm from the 84 nm observed in the control film. This decrease in surface roughness fosters enhanced contact between the hole transport layer and the perovskite layer, thereby facilitating efficient carrier extraction. Most critically, MOT treatment led to a decrease in film roughness, improved uniformity, mitigation of cracks, and an enhancement in the crystallinity of the perovskite film. These findings collectively underscore the positive influence of MOT in optimizing the surface morphology and overall quality of perovskite films. Such improvements hold significant implications for elevating the performance and stability of perovskite materials in optoelectronic devices, among other applications. To elucidate the reduction in trap-assisted recombination due to MOT modification, we examined the electron trap density (N_t) of perovskite films via JV curves from pure electronic devices. This evaluation was undertaken to quantify the effectiveness of defect passivation (refer to Figure 4f). The J-V curves conventionally encompass three distinct regions: the ohmic, the trap-filled limit (TFL), and space-charge-limited current (SCLC). During low-voltage conditions, ohmic contact transpired within the device. In the TFL region, as the applied bias exceeded the trap-filled limit voltage (V_{TFL})—equivalent to the inflection voltage—a marked escalation in the current signals the culmination of defect filling. Consequently, a diminished fill voltage indicates a reduced defect quantity in the film. The data exhibited V_{TFL} values of 0.32 V for the control and 0.24 V for the MOT-modified perovskite films. This discrepancy underscores the ability of MOT to effectively mitigate defects within the perovskite film, which is consistent with the improvement in carrier lifetime. This work thus highlights the crucial role of MOT in enhancing the quality and stability of perovskite-based devices by attenuating defect-related recombinations.

As represented in Figure 5a and Table 1, the current-voltage (JV) characteristics of the control and MOT-optimized devices manifested distinctive enhancements in photovoltaic performance. The superior device from the control group demonstrated a power conversion efficiency (PCE) of 19.83% in forward scanning (FS), with short-circuit current density (J_{SC}) of 24.05 mA cm^{-2}, open-circuit voltage (V_{OC}) of 1.09 V, and fill factor (FF) of 75.64%. In the reverse scan (RS), it exhibited a PCE of 22.06%, with corresponding J_{SC}, V_{OC}, and FF values of 24.52 mA cm^{-2}, 1.10 V, and 81.15%, respectively. Conversely, the MOT-treated device exhibited substantial enhancements in the RS, achieving a PCE of 23.83% with J_{SC}, V_{OC}, and FF values of 24.78 mA cm^{-2}, 1.17 V, and 82.52%. In FS, the PCE reached 22.67%, with corresponding J_{SC}, V_{OC}, and FF values of 24.68 mA cm^{-2}, 1.17 V, and 78.50%. A significant reduction in hysteresis was evident in the MOT-optimized devices. This is likely attributed to the effective passivation of defects at the perovskite film surface and grain boundaries and to the suppression of ion migration within these regions. These findings underline the influential role of MOT treatment in enhancing perovskite photovoltaic performance. To elucidate the carrier transport dynamics of the perovskite solar cells (PSCs), the devices' Mott–Schottky curves were meticulously examined utilizing an electrochemical workstation (refer to Figure 5b). As a result, the optimized devices with MOT molecular enhancements exhibited a significant improvement in PCE, correlated with the enhancement in Voc, FF, and Jsc, as evidenced by the corresponding external quantum efficiency spectra (Figure S3). This underscores the pronounced enhancement in device performance following the optimization of MOT molecules. The findings depict that the MOT-modified device manifested a steeper gradient in comparison to its control counterpart. This suggests a substantial diminution in trap state density, consequently leading to a reduction in non-radiative charge recombination. In addition, the built-in potential, determined via the intersection point of the Mott–Schottky graph's linear state with the x-axis, was found to be 0.87 V for the control device and was enhanced to 0.99 V for the MOT-treated device. An elevated built-in potential invariably benefits the efficiency of carrier separation, transportation, and subsequent extraction, thus underlining the significance of the MOT modification in improving perovskite solar cell performance. Figure 5c presents the Nyquist plots for the perovskite solar cells. Notably, the MOT-treated device showcased a lesser charge transfer resistance relative to the control device. This could be attributed to the elevation in the crystallinity of the perovskite thin film, which ostensibly had a positive impact on the fill factor (FF). Furthermore, as a result of proficient defect passivation, a marked increase was observed in the recombination resistance following the application of MOT. This mirrors a diminished defect density, which, in turn, was expected to contribute favorably towards the amplification of the open-circuit voltage (V_{OC}). Figure 5d portrays the function between different light intensities and the open-circuit voltage (V_{OC}) to further explicate alterations in electron-hole recombination, incited by defects under varying light intensities in PSCs. It is noteworthy that the slope pertaining to the MOT-treated PSCs was inferior to that of the control devices, which suggests an effective mitigation of the trap-induced carrier recombination. These electrical characterizations provide substantial evidence for the diminishment of trap states associated with carrier recombination following the introduction of MOT, contributing positively towards charge extraction efficiency.

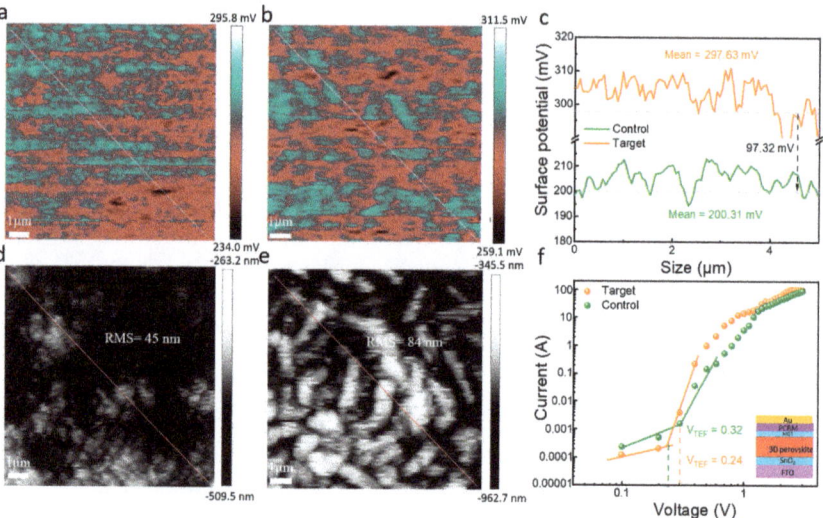

Figure 4. KPFM image of (**a**) with and (**b**) without MOT–modified perovskite films. (**c**) Contact potential difference along the solid white lines drawn in (**a**,**b**). AFM image of (**d**) with and (**e**) without MOT–modified perovskite films. (**f**) The structure (FTO/SnO$_2$/Perovskite/MOT/PCBM/Au) of the electron-only device is shown in the inset.

In a bid to investigate the ramifications of MOT optimization on the enduring stability of devices, we documented the X-ray diffraction (XRD) patterns of unencapsulated perovskite films under stress testing with high relative humidity (RH) at 80 ± 10% (Figure 6c). Contrasted with MOT-optimized films, the control films manifested an intensified PbI$_2$ peak after a 12-day aging period, signifying the decomposition of the perovskite film.

Conversely, the perovskite peaks persisted dominantly in the MOT-optimized films, suggesting the preservation of the perovskite structure post-aging. The control 3D perovskite film exhibited a relatively low water contact angle of 71.2°, whereas the MOT-optimized film displayed a higher angle of 78.3° (Figure 6a,b), an implication of the MOT overlayer functioning as a barrier to water ingress into the perovskite lattice, thereby bolstering moisture stability. Further, after 2000 h at 40% RH, MOT-modified devices evinced no noticeable decomposition, retaining 92% of their original PCE. In our investigation, we identified MOT as a significant agent in enhancing the stability of perovskite materials. By employing MOT as an overlayer on the perovskite surface, we observed improvements in both hydrophobicity and resilience under high humidity conditions. MOT's unique structure, which allows for strong bonding with lead ions and potential replacement of missing iodide ions, plays a vital role in this enhancement. Through its influence on crystal growth and defect passivation, MOT mitigates non-radiative recombination and ion migration, contributing to increased material stability. In contrast, the control PSCs demonstrated a PCE reduction to 45% of their original value (Figure 6d). The superior quality and enduring stability render our advanced devices highly promising for practical implementations.

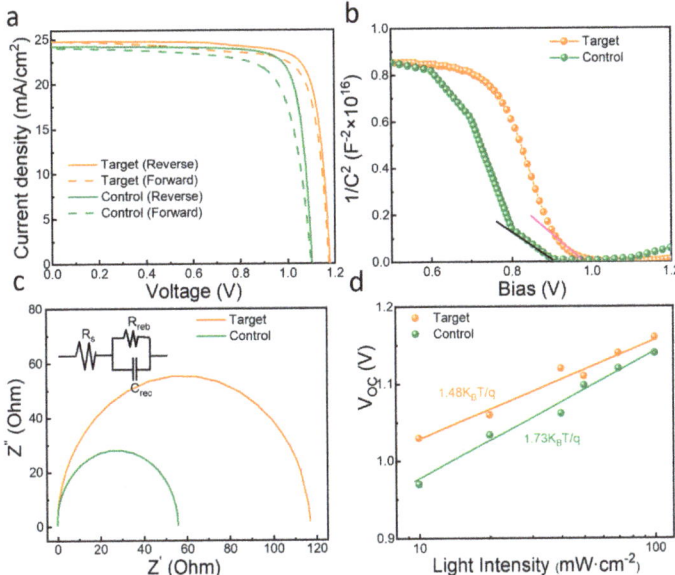

Figure 5. (a) J–V curves of the devices without and with MOT measured in RS and FS. (b) Mott–Schottky plot for control and MOT-modified PSCs. (c) Nyquist plots of the control and MOT-modified PSCs measured in the frequency range from 1 MHz to 1 Hz at 0 V bias under dark. (d) Light-intensity-dependent VOC of the control and MOT modified PSCs.

Table 1. Photovoltaic characteristics of best devices prepared without and with MOT treatment at forward (FS) and reverse (RS) scans.

Samples	Scanning Direction	V_{OC} (V)	J_{SC} (mA·cm^{-2})	FF (%)	PCE (%)
Control	RS	1.10	24.52	81.15	22.06
	FS	1.09	24.05	75.64	19.83
With MOT	RS	1.17	24.78	82.52	23.83
	FS	1.17	24.68	78.50	22.67

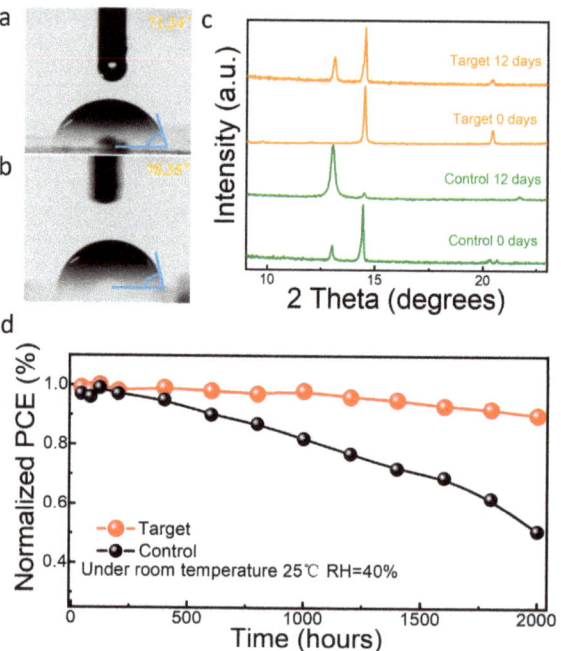

Figure 6. Water-contact-angle measurement of (**a**) control (**b**) with MOT-treated perovskite films. (**c**) XRD patterns of perovskite films stored in high humidity (80 ± 10% RH) for 12 days. (**d**) Normalized PCE stability curves of control and modified devices at 40% RH.

3. Conclusions

Building upon the comprehensive findings of our investigation, we conclude that MOT provides substantial advancements in the optimization of both performance and stability of perovskite solar cells (PSCs). In our experimental paradigm, we observed that MOT, as an effective dopant, not only mediated the growth of perovskite crystals but also passivated defects within the perovskite stratum, which is substantiated by the pronounced morphological transitions and improvements in crystallography evidenced by SEM and XRD analyses, respectively. A specific concentration of MOT, precisely at 1 mg/mL, emerged as the most optimal, inducing the most substantial positive transformations in both film properties and crystallinity. An assessment of the photovoltaic properties of the device highlighted an appreciable escalation in the power conversion efficiency (PCE) of PSCs post-MOT treatment in comparison to the untreated controls, with a substantial mitigation in hysteresis. The substantial decrease in trap states and an enhancement in charge carrier separation and transport in MOT-incorporated devices was particularly noticeable, contributing to their superior performance. Furthermore, the experiment substantiated the pivotal role of MOT in augmenting the longevity and resilience of PSCs, especially under challenging environmental conditions. A significant enhancement in hydrophobicity was observed, evidenced by an increased water contact angle, along with a decreased rate of degradation under high humidity stress tests for the MOT-incorporated films. In light of these findings, our study demonstrates the potential for the integration of MOT in perovskite solar cell fabrication as a strategic approach towards improved device efficiency and stability. These findings pave the way for more practical applications of PSCs and highlight the need for continued exploration and refinement of MOT and analogous dopants to facilitate the development of superior, high-efficiency PSCs.

4. Experiment Section

4.1. DFT Simulation

DFT calculations utilized the VASP package with a 420 eV energy cutoff for plane wave basis sets. The Monkhorst–Pack k-point mesh, measuring $2 \times 2 \times 1$, was selected for Brillouin zone integration. Perovskite surface models employed $(2 \times 2 \times 2)$ supercells with a 15 Å vacuum separation for periodic images. Structure relaxation continued until total energy reached a convergence of 1.0×10^{-5} eV/atom.

4.2. Materials

Chemicals, such as Cesium iodide (CsI), Lead iodide (PbI$_2$), Lead bromide (PbBr$_2$), FAI, MAI, MABr, MOT, Dimethylsulfoxide (DMSO), N,N-Dimethylformamide (DMF), isopropanol (IPA), SnO$_2$, FTO glass, [6,6]-phenyl-C61-butyric acid methyl ester (PCBM), and Spiro-OMeTAD, were sourced from various reputable suppliers and utilized as received.

4.3. Device Fabrication

Our process commenced with the preparation of Fluorine-doped tin oxide (FTO) glass substrates. These substrates were subjected to ultrasonic cleaning in a specific sequence of FTO cleaning solution, deionized water, and ethanol, each for a duration of 20 min. To improve the surface wettability, the cleaned substrates were exposed to UV/ozone for an additional 30 min. Following substrate preparation, we established the electron transport layer using a SnO$_2$ precursor solution (2.67% concentration in water). This solution was spin-coated onto the FTO substrates at a speed of 3500 rpm, lasting for 30 s, and was subsequently annealed at 150 °C for half an hour on a hotplate. In parallel, we prepared the $Cs_{0.05}(FA_{0.83}MA_{0.17})_{0.95}Pb(I_{0.83}Br_{0.17})_3$ precursor solution. This involved dissolving certain quantities of FAI, MABr, PbI$_2$, and PbBr$_2$ into a DMF/DMSO solution at a volume ratio of 4:1. Separately, we dissolved CsI in DMSO and integrated this into the precursor solution. This complex perovskite solution was then spin-coated onto the FTO/SnO$_2$ substrate using two spin speeds and dropwise addition of 8 microliters of chlorobenzene as anti-solvent, and subsequently, samples were exposed to MOT in a nitrogen-filled glovebox. MOT, dissolved in IPA, was then spin-coated onto the perovskite surface at 3000 rpm for 15 s. We then prepared the Spiro-OMeTAD solution (consisting of Spiro-MeOTAD, 4-tert-butylpyridine, and a lithiumbis-(trifluoromethylsulfonyl) imide solution in chlorobenzene) and deposited it onto the perovskite layer. This was achieved through spin-coating at 3000 rpm for 30 s in a nitrogen-rich environment. The final step involved the thermal evaporation of a 100 nm thick Au electrode onto the structure, resulting in a fully formed FTO/SnO$_2$/Perovskite/MOT/Spiro-MeOTAD/Au device. As a comparative reference, an FTO/SnO$_2$/Perovskite/MOT/PCBM/Au device was also fabricated using a solution of [6,6]-phenyl-C61-butyric acid methyl ester (PCBM) in chlorobenzene, serving to estimate electron trap-state densities.

4.4. Device Characterizations

We initiated our examinations with an analysis of the surface morphology of the various perovskite films and a cross-sectional inspection of the PSCs. These investigations were facilitated by the advanced JEOF 7610F emission scanning electron microscope (SEM) (JEOL, Tokyo, Japan). For our X-ray diffraction (XRD) analyses of the perovskite films, we used a cutting-edge Rigaku SmartLab(9Kw)D X-ray diffractometer (Rigaku, Tokyo, Japan). This high-tech instrument utilized a monochromatized Cu Kα target radiation source, running at a scanning rate of 8°/min. We conducted our electrochemical impedance spectroscopy (EIS) tests with the reliable CHI660 electrochemical workstation (Shanghai Chenhua Instrument Co., Ltd., Shanghai, China) allowing us to inspect the impedance characteristics of the perovskite films. The absorption spectra were compiled using a UV-1900 spectrometer from the renowned company (Shimadzu, Kyoto, Japan). Photoluminescence (PL) characterizations were realized through a state-of-the-art fluorescence luminescence spectrometer, enabling us to study the light-emitting properties of the perovskite films.

Current-voltage (J-V) characteristics were evaluated under AM 1.5 G irradiation with the aid of a customized 71S type solar simulator system from SOFN Instruments Co., Ltd. (Beijing, China), while a Keithley 2400 served as a highly precise digital sourcemeter. We ensured the light intensity was properly calibrated through a standard silicon cell. We utilized an Edinburgh Instruments spectrometer (FLS980) (Edinburgh Instruments Ltd., Livingston, UK) to capture both time-resolved PL (TRPL) spectra and steady-state PL spectra under an excitation wavelength of 480 nm. Both the control perovskite film and the N-MOT-treated perovskite film were thoroughly analyzed using an AFM and KPFM (Keysight 5500 scanning probe microscope, Keysight, Santa Rosa, CA, USA). This was carried out in tapping mode for AFM and in contact mode with a bias voltage of 1 V for KPFM. Lastly, our examination encompassed X-ray photoelectron spectroscopy (XPS) conducted on a Thermo ESCALAB 250XI (Thermo Fisher, Waltham, MA, USA), enabling us to study the electronic structure of the perovskite films.

Supplementary Materials: The following supporting information can be downloaded at: https://www.mdpi.com/article/10.3390/coatings13091528/s1, Figure S1: XPS survey spectrum for C 1s, N 1S, Cl 2p, Pb 4f of the different perovskite films.; Figure S2: (a) Chemical structure of MOT-treatment. Charge density difference of MOT passivated FAI-terminated FAPbI$_3$ surface with (a) Ii. PDOS of passivated FAI-terminated FAPbI$_3$ surface with (d) Ii in the presence or not of MOT.; Figure S3: IPCE spectra of the control and target devices.; Table S1: The summary of TRPL curve analysis: short and long lifetime contributions and electron lifetimes.

Author Contributions: Formal analysis: J.B.; investigation: J.B.; project administration: J.B.; writing—original draft preparation: X.L., Y.L. and J.G.; writing—review and editing: J.B., Y.S. and J.G. All authors have read and agreed to the published version of the manuscript.

Funding: This work was supported by the Innovation Fund Project of College Teachers of Gansu Provincial Department of Education (2023B-206) and the Qingyang City Science and Technology Planning Project (QY-STK-2022B-151).

Institutional Review Board Statement: Not applicable.

Informed Consent Statement: Not applicable.

Data Availability Statement: Data is contained within the article.

Conflicts of Interest: The authors declare no conflict of interest.

References

1. Noh, J.H.; Im, S.H.; Heo, J.H.; Mandal, T.N.; Seok, S.I. Chemical management for colorful, efficient, and stable inorganic–organic hybrid nanostructured solar cells. *Nano Lett.* **2013**, *13*, 1764–1769. [CrossRef]
2. Kim, Y.; Cook, S.; Tuladhar, S.M.; Choulis, S.A.; Nelson, J.; Durrant, J.R.; Bradley, D.D.C.; Giles, M.; McCulloch, I.; Ha, C.-S.; et al. A strong regioregularity effect in self-organizing conjugated polymer films and high-efficiency polythiophene: Fullerene solar cells. *Nat. Mater.* **2006**, *5*, 197–203. [CrossRef]
3. Chiou, N.-R.; Lu, C.; Guan, J.; Lee, L.J.; Epstein, A.J. Growth and alignment of polyaniline nanofibres with superhydrophobic, superhydrophilic and other properties. *Nat. Nano* **2007**, *2*, 354–357. [CrossRef] [PubMed]
4. Jeon, N.J.; Noh, J.H.; Yang, W.S.; Kim, Y.C.; Ryu, S.; Seo, J.; Seok, S.I. Compositional engineering of perovskite materials for high-performance solar cells. *Nature* **2015**, *517*, 476–480. [CrossRef]
5. Liu, M.Z.; Johnston, M.B.; Snaith, H.J. Efficient planar heterojunction perovskite solar cells by vapour deposition. *Nature* **2013**, *501*, 395–398. [CrossRef] [PubMed]
6. Ummadisingu, A.; Steier, L.; Seo, J.-Y.; Matsui, T.; Abate, A.; Tress, W.; Grätzel, M. The effect of illumination on the formation of metal halide perovskite films. *Nature* **2017**, *545*, 208–212. [CrossRef]
7. Wu, Y.; Yang, X.; Chen, W.; Yue, Y.; Cai, M.; Xie, F.; Bi, E.; Islam, A.; Han, L. Perovskite solar cells with 18.21% efficiency and area over 1 cm^2 fabricated by heterojunction engineering. *Nat. Energy* **2016**, *1*, 16148. [CrossRef]
8. Liu, D.; Kelly, T.L. Perovskite solar cells with a planar heterojunction structure prepared using room-temperature solution processing techniques. *Nat. Photonics* **2014**, *8*, 133–138. [CrossRef]
9. Bella, F.; Griffini, G.; Correa-Baena, J.-P.; Saracco, G.; Grätzel, M.; Hagfeldt, A.; Turri, S.; Gerbaldi, C. Improving efficiency and stability of perovskite solar cells with photocurable fluoropolymers. *Science* **2016**, *354*, 203–206. [CrossRef]

10. Stranks, S.D.; Eperon, G.E.; Grancini, G.; Menelaou, C.; Alcocer, M.J.P.; Leijtens, T.; Herz, L.M.; Petrozza, A.; Snaith, H.J. Electron-Hole Diffusion Lengths Exceeding 1 Micrometer in an Organometal Trihalide Perovskite Absorber. *Science* **2013**, *342*, 341–344. [CrossRef]
11. Xing, G.; Mathews, N.; Sun, S.; Lim, S.S.; Lam, Y.M.; Grätzel, M.; Mhaisalkar, S.; Sum, T.C. Long-Range Balanced Electronand Hole-Transport Lengths in Organic-Inorganic CH3NH3PbI3. *Science* **2013**, *342*, 344–347. [CrossRef]
12. Nie, W.; Tsai, H.; Asadpour, R.; Blancon, J.-C.; Neukirch, A.J.; Gupta, G.; Crochet, J.J.; Chhowalla, M.; Tretiak, S.; Alam, M.A.; et al. High-efficiency solution-processed perovskite solar cells with millimeter-scale grains. *Science* **2015**, *347*, 522–525. [CrossRef] [PubMed]
13. Tan, H.; Jain, A.; Voznyy, O.; Lan, X.; García de Arquer, F.P.; Fan, J.Z.; Quintero-Bermudez, R.; Yuan, M.; Zhang, B.; Zhao, Y.; et al. Efficient and stable solution-processed planar perovskite solar cells via contact passivation. *Science* **2017**, *355*, 722–726. [CrossRef] [PubMed]
14. Yella, A.; Lee, H.W.; Tsao, H.N.; Yi, C.Y.; Chandiran, A.K.; Nazeeruddin, M.K.; Diau, E.W.G.; Yeh, C.Y.; Zakeeruddin, S.M.; Gratzel, M. Porphyrin-Sensitized Solar Cells with Cobalt (II/III)-Based Redox Electrolyte Exceed 12 Percent Efficiency. *Science* **2011**, *334*, 629–634. [CrossRef] [PubMed]
15. Wu, Y.; Yan, D.; Peng, J.; Duong, T.; Wan, Y.; Phang, P.; Shen, H.; Wu, N.; Barugkin, C.; Fu, X.; et al. Monolithic perovskite/silicon-homojunction tandem solar cell with over 22% efficiency. *Energy Environ. Sci.* **2017**, *10*, 2472–2479. [CrossRef]
16. Tidhar, Y.; Edri, E.; Weissman, H.; Zohar, D.; Hodes, G.; Cahen, D.; Rybtchinski, B.; Kirmayer, S. Crystallization of Methyl Ammonium Lead Halide Perovskites: Implications for Photovoltaic Applications. *J. Am. Chem. Soc.* **2014**, *136*, 13249–13256. [CrossRef]
17. Jiang, Y.; Yu, B.-B.; Liu, J.; Li, Z.-H.; Sun, J.-K.; Zhong, X.-H.; Hu, J.-S.; Song, W.-G.; Wan, L.-J. Boosting the Open Circuit Voltage and Fill Factor of QDSSCs Using Hierarchically Assembled ITO@Cu2S Nanowire Array Counter Electrodes. *Nano Lett.* **2015**, *15*, 3088–3095. [CrossRef] [PubMed]
18. Xiao, M.; Huang, F.; Huang, W.; Dkhissi, Y.; Zhu, Y.; Etheridge, J.; Gray-Weale, A.; Bach, U.; Cheng, Y.B.; Spiccia, L. A fast deposition-crystallization procedure for highly efficient lead iodide perovskite thin-film solar cells. *Angew. Chem.* **2014**, *126*, 10056–10061. [CrossRef]
19. Dai, Q.; Patel, K.; Donatelli, G.; Ren, S. Magnetic Cobalt Ferrite Nanocrystals For an Energy Storage Concentration Cell. *Angew. Chem. Int. Ed.* **2016**, *55*, 10439–10443. [CrossRef]
20. Seo, J.; Park, S.; Kim, Y.C.; Jeon, N.J.; Noh, J.H.; Yoon, S.C.; Seok, S.I. Benefits of very thin PCBM and LiF layers for solution-processed p–i–n perovskite solar cells. *Energy Environ. Sci.* **2014**, *7*, 2642–2646. [CrossRef]
21. Unger, E.L.; Hoke, E.T.; Bailie, C.D.; Nguyen, W.H.; Bowring, A.R.; Heumuller, T.; Christoforo, M.G.; McGehee, M.D. Hysteresis and transient behavior in current-voltage measurements of hybrid-perovskite absorber solar cells. *Energy Environ. Sci.* **2014**, *7*, 3690–3698. [CrossRef]
22. Wang, Q.; Shao, Y.; Dong, Q.; Xiao, Z.; Yuan, Y.; Huang, J. Large fill-factor bilayer iodine perovskite solar cells fabricated by a low-temperature solution-process. *Energy Environ. Sci.* **2014**, *7*, 2359–2365. [CrossRef]
23. Jiang, Q.; Zhao, Y.; Zhang, X.; Yang, X.; Chen, Y.; Chu, Z.; Ye, Q.; Li, X.; Yin, Z.; You, J. Surface passivation of perovskite film for efficient solar cells. *Nat. Photonics* **2019**, *13*, 460–466. [CrossRef]
24. Zhang, W.; Liu, X.; He, B.; Gong, Z.; Zhu, J.; Ding, Y.; Chen, H.; Tang, Q. Interface Engineering of Imidazolium Ionic Liquids toward Efficient and Stable CsPbBr3 Perovskite Solar Cells. *ACS Appl. Mater. Interfaces* **2020**, *12*, 4540–4548. [CrossRef]
25. Guo, M.; Lin, C.-Y.; Liou, S.-J.; Chang, Y.J.; Li, Y.; Li, J.; Wei, M. D–A–π–A organic sensitizer surface passivation for efficient and stable perovskite solar cells. *J. Mater. Chem. A* **2021**, *9*, 25086–25093. [CrossRef]
26. Zhu, X.; Du, M.; Feng, J.; Wang, H.; Xu, Z.; Wang, L.; Zuo, S.; Wang, C.; Wang, Z.; Zhang, C.; et al. High-Efficiency Perovskite Solar Cells with Imidazolium-Based Ionic Liquid for Surface Passivation and Charge Transport. *Angew. Chem. Int. Ed.* **2021**, *60*, 4238–4244. [CrossRef]
27. Kadam, K.D.; Rehman, M.A.; Kim, H.; Rehman, S.; Khan, M.A.; Patil, H.; Aziz, J.; Park, S.; Abdul Basit, M.; Khan, K.; et al. Enhanced and Passivated Co-doping Effect of Organic Molecule and Bromine on Graphene/HfO2/Silicon Metal–Insulator–Semiconductor (MIS) Schottky Junction Solar Cells. *ACS Appl. Energy Mater.* **2022**, *5*, 10509–10517. [CrossRef]
28. Rehman, M.A.; Park, S.; Khan, M.F.; Bhopal, M.F.; Nazir, G.; Kim, M.; Farooq, A.; Ha, J.; Rehman, S.; Jun, S.C.; et al. Development of directly grown-graphene–silicon Schottky barrier solar cell using co-doping technique. *Int. J. Energy Res.* **2022**, *46*, 11510–11522. [CrossRef]

Disclaimer/Publisher's Note: The statements, opinions and data contained in all publications are solely those of the individual author(s) and contributor(s) and not of MDPI and/or the editor(s). MDPI and/or the editor(s) disclaim responsibility for any injury to people or property resulting from any ideas, methods, instructions or products referred to in the content.

Article

Identification of Elastoplastic Constitutive Model of GaN Thin Films Using Instrumented Nanoindentation and Machine Learning Technique

Ali Khalfallah [1,2,3,*], Amine Khalfallah [4] and Zohra Benzarti [1,5]

1. CEMMPRE, Department of Mechanical Engineering, University of Coimbra, Rua Luís Reis Santos, 3030-788 Coimbra, Portugal; zohra.benzarti@uc.pt
2. Laboratoire de Génie Mécanique, École Nationale d'Ingénieurs de Monastir, Université de Monastir, Av. Ibn El-Jazzar, Monastir 5019, Tunisia
3. DGM, Institut Supérieur des Sciences Appliquées et de Technologie de Sousse, Université de Sousse, Cité Ibn Khaldoun, Sousse 4003, Tunisia
4. Departamento de Engenharia Informática, Faculdade de Ciências e Tecnologia, Universidade de Coimbra, Pólo II—Pinhal de Marrocos, 3030-290 Coimbra, Portugal; uc2022215115@student.uc.pt
5. Laboratory of Multifunctional Materials and Applications (LaMMA), Department of Physics, Faculty of Sciences of Sfax, University of Sfax, Soukra Road km 3.5, B.P. 1171, Sfax 3000, Tunisia
* Correspondence: ali.khalfallah@dem.uc.pt; Tel.: +351-933-411-080

Abstract: This study presents a novel inverse identification approach to determine the elastoplastic parameters of a 2 µm thick GaN semiconductor thin film deposited on a sapphire substrate. This approach combines instrumented nanoindentation with finite element (FE) simulations and an artificial neural network (ANN) model. Experimental load–depth curves were obtained using a Berkovich indenter. To generate a comprehensive database for the inverse analysis, FE models were constructed to simulate load–depth responses across a wide range of GaN thin film properties. The accuracy of both 2D and 3D simulations was compared to select the optimal model for database generation. The Box–Behnken design-based data sampling method was used to define the number of simulations and input variables for the FE models. The ANN technique was then employed to establish the complex mapping between the simulated load–depth curves (input) and the corresponding stress–strain curve (output). The generated database was used to train and test the ANN model. Then, the learned ANN model was used to achieve high accuracy in identifying the stress–strain curve of the GaN thin film from the experimental load–depth data. This work demonstrates the successful application of an inverse analysis framework, combining experimental nanoindentation tests, FE modeling, and an ANN model, for the characterization of the elastoplastic behavior of GaN thin films.

Keywords: nanoindentation; GaN thin films; elastoplastic; load–depth curves; machine learning; finite element method

1. Introduction

Gallium nitride (GaN) semiconductors including their doped forms using various doping elements, e.g., Si, Al, In, Mg, Zn, C, etc., which are grown using the metalorganic vapor deposition method, have emerged as a pivotal class of materials in modern electronics due to their exceptional electronic, optical, and thermal properties that have revolutionized various technological domains [1,2]. With their wide bandgap and exceptional physical properties, GaN thin films have found applications ranging from light-emitting diodes (LEDs) [3,4] and power electronics [5] to radio frequency (RF) devices [6], microelectromechanical systems (MEMSs) [7], optoelectronic sensors [8], and they have even been considered for applications in severe radiative environments, as in the spatial or nuclear industry [9]. While much attention has been directed towards GaN thin films' electrical and optical features [10,11], the mechanical properties of GaN thin films have gained substantial

interest in recent years as well [12–14]. The mechanical characteristics of GaN thin films hold immense significance due to their impact on the reliability, performance, and overall functionality of devices, which are essential for various electronic applications, enhancing the mechanical integrity of devices, improving efficiency, and increasing the lifespan of electronic products. Determining the mechanical behavior of GaN thin films is essential not only for optimizing their design and performance but also for unlocking new avenues for their integration into cutting-edge technologies [15]. Traditional experiments such as tensile and compression tests are commonly used for the mechanical characterization of the elastoplastic behavior of bulk materials. However, these macro techniques are not suitable to measure the stress–strain curve, reflecting the elastoplastic behavior of materials at the nanoscale and mesoscale because of the size effect phenomenon issue [16,17]. Fortunately, the rapid development of devices for load and depth measurement has led to the emergence of nanoindentation techniques, which are more effective than traditional tensile or compression experiments to determine the stress–strain curves [18]. Nowadays, nanoindentation techniques are widely used to measure the elastoplastic properties of materials at the nanoscale [12,19–21]. Consequently, these methods are mostly adopted in various fields of materials science, such as thin film ceramics, metals, semiconductors, polymers, and biomaterials [22–25].

In recent years, artificial neural networks (ANNs) and other machine learning (ML) tools have been used in many different investigations including the prediction of precipitated secondary phase volume fraction, indentation curves, and the forecast of the microstructure, key mechanical properties, fatigue, creep, and ductile fracture properties among other applications [14,26–28]. Therefore, these methods have become a standard way to study and interpret complex experimental data and to predict how the mechanical properties of materials, as a function of a wide range of parameters, will change depending on their microstructure and other factors. Nanoindentation-based numerical methods for identifying material-related elastoplastic properties can make outstanding progress in the determination of accurate material properties, particularly the stress–strain relationship of the thin film's material. However, it remains a challenging task to conclusively determine whether the elastoplastic characteristics of a material can be uniquely identified [29,30]. More research has revealed that the intricate connections between substrates and thin film elastoplastic characteristics impede the use of classical computational techniques [31].

Recently, Park et al. [32] and Lu et al. [33] developed an ANN-based inverse analysis methodology to identify the mechanical properties of bulk metallic glass materials and metals through ML from instrumented nanoindentation using specific parameters such as total indentation energy, material stiffness, and maximum load. Meanwhile, Jeong et al. [34] developed an ANN-based inverse method to determine the stress–strain curves for various bulk metallic materials using finite element simulation datasets of load–depth (P-h) curves. Wang et al. [35] developed the nanoindentation method to identify the elastic parameters of a ZnO thin film as a transversely isotropic material by combining the nanoindentation test with finite element method (FEM) simulation.

In this study, a novel inverse identification-based instrumented nanoindentation test is developed to identify, for the first time, the elastoplastic parameters of a 2 µm thickness GaN semiconductor thin film deposited on a sapphire substrate. Experimental (P-h) curves are obtained using a Berkovich indenter. Then, 2D and 3D finite element (FE) models are developed to generate a database of simulated (P-h) curves for various ranges of film properties. Based on this range of properties, the design of an experiment using Box–Behnken design-based data sampling is established to define the number of simulations and the used input variables to run the numerical simulations. The accuracy of the simulation results using either 2D or 3D FE models are compared to finally use the target FE model for the generation of the database. The complex mapping between the input and output variables in the inverse analysis is achieved using the ANN model. The generated database of (P-h) curves is used for the ANN model training and testing. Once the ANN model is optimized (i.e., adequately trained and tested), the experimental (P-h) curve of the GaN film is introduced to determine the stress–strain curve, representing the elastoplastic properties of GaN thin film.

This study demonstrates the efficiency of the inverse method using the ANN model for the accurate identification of the elastoplastic properties of GaN thin films from experimental nanoindentation data.

2. Materials and Nanoindentation Experiments

2.1. Growth Process

The gallium nitride (GaN) sample was grown on (0001) c-plane sapphire substrates using an atmospheric pressure metalorganic chemical vapor deposition (MOCVD) technique. Principally, 10 standard liters per minute (slm) of trimethylgallium (TMG) and 3 slm of ammonia (NH_3) were used as precursors for gallium and nitrogen, respectively. The carrier gas was 3 slm of nitrogen (N_2). First, the sapphire substrate was nitridated under ammonia and nitrogen for 10 min at 1110 °C. Then, a 30 nm GaN buffer layer was deposited on the substrate at a temperature of 600 °C. Thereafter, the temperature rose from 600 °C to 1110 °C for the subsequent growth of a GaN layer with a thickness of approximately 2 µm.

2.2. Nanoindentation Experiments

Nanoindentation experiments were conducted using a NanoTest™ NT1 nanoindentation instrument (NanoMaterials, Ltd., Wrexham, UK), which was equipped with a diamond pyramid-shaped Berkovich-type indenter. The Berkovich indenter has a three-sided pyramidal shape with a half angle of 65.3°. Prior to the measurements, the nanoindentation system was fully calibrated using fused silica as a standard sample. The tip-end curvature radius of the Berkovich indenter was determined using Hertzian analysis, yielding a value of 500 nm. All nanoindentation tests were performed at room temperature, using the load rate-control mode. The maximum applied loads (P_{max}) reached up to 15 mN. To ensure the reproducibility of the results, a minimum of twenty independent tests were conducted on the GaN sample. During the nanoindentation tests, the maximum load was held constant for a dwell period of 30 s to account for the thermal drift. Additionally, to prevent interference between neighboring indents, each indentation was spaced apart by 30 µm. Notably, all indentation depths remained below 10% of the film thickness, guaranteeing that the influence of the sapphire substrate on the properties of the GaN film was negligible.

3. Identification Methodology

3.1. Finite Element Modeling

Finite element modeling (FEM) has emerged as the leading and most prevalent technique for simulating the nanoindentation process [36,37]. Finite element analysis is used to simulate the elastoplastic behavior of the GaN thin film. The numerical model is built and employed to produce datasets of nanoindentation load–depth curves. The mechanical responses are obtained based on various ranges of the elastoplastic material properties of the GaN thin film. To better account for the realistic contact between the indenter and the indentation imprint, three FE models are built. First, a 2D axisymmetric model is considered with a 2D equivalent conical-shaped Berkovich indenter. Second, a 3D model is developed with a 3D equivalent conical-shaped Berkovich indenter. Finally, a 3D model is constructed using a 3D Berkovich indenter by considering the indenter tip radius. The accuracy of the models' response is recorded along with the computational time cost, for the purpose of comparison. FE models are constructed using ABAQUS Standard. In these FE models, thin films are assumed as deformable semi-infinite media, and the maximum indenter depth is kept shallow regarding the total film thickness. However, seeking simplicity, the indenter is assumed as a rigid body [38,39]. This assumption simplifies the computational model by negating the need to account for indenter deformation. The rationale behind this assumption lies in the significant stiffness disparity between the indenter, such as a diamond indenter in this study, and the GaN sample material. However, there are some examples of materials (as a general rule, we refer here to materials with an elastic stiffness more than 10% of that of the indenter and a yield strain higher than 1%, as well as to penetration depths less than ~5 times the characteristic tip defect length of the indenter) where accounting for the indenter deformation is important to obtain accurate hardness and modulus values [40]. The appropriate boundary conditions are applied depending on the used FE model among the three built models. For all models, the nodes on the bottom are subjected to fixed boundary conditions. Notably, mesh refinement is particularly applied in the region underneath the indenter, where very small elements of around 5 nanometers of size are used. But, as being far away from the indentation zone, the element size is gradually augmented to reduce time computation without losing the accuracy of the simulation results.

The 2D axisymmetric model is constructed using a total of 8353, 4-node bilinear elements with reduced integration (CAX4R) and 7922 nodes (see Figure 1a, top). The indenter in the axisymmetric assumption is an equivalent conical-shaped Berkovich indenter with a half angle of 70.3°, as the projected area is identical to that of the original pyramidal Berkovich indenter [41]. The conical-shaped Berkovich indenter is featured by a tip radius of 500 nm equivalent to the indenter tip radius used in experimental nanoindentation tests. As illustrated in Figure 1b, specific boundary conditions are applied to this type of FE model. The bottom surface of the deformed media is fixed for all degree of freedom (i.e., fixed boundary condition is implemented), while axisymmetric boundary conditions are applied to the left side in the z-direction. A reference point (RF) is defined to represent the rigid body indenter, and a displacement of 170 nm is associated with this reference point. Furthermore, the contact between the indenter and the film imprint is modeled using the implemented *surface to surface* contact condition in Abaqus. The friction is taken into consideration using the penalty method [42]. The 3D FE model with a 3D equivalent conical-shaped Berkovich indenter is built using a 10-node quadratic tetrahedron (the C3D10 element in Abaqus) that offers high accuracy for complex geometries. This makes it a preferred choice for 3D FE nanoindentation simulations where capturing the precise contact mechanics between the indenter and the material is vital. For symmetry reasons, one-quarter of the model was considered and defined by 119,005 nodes and a total of 83,443 elements (see Figure 1a, bottom). The 3D equivalent conical-shaped Berkovich indenter is a 3D analytic rigid body defined by a half angle of 70.3° and a tip radius of 500 nm. The third FE model is a 3D FE model with a Berkovich indenter featured by a tip radius of 500 nm (see Figure 1c). The deformable part of the FE model is similar to the second model; however, the indenter is a discrete rigid body and meshed using C3D4, a 4-node linear tetrahedral element. Figure 1d shows the complete 3D FE model used to run nanoindentation simulations. The thin film behavior is assumed as elastoplastic with isotropic linear elastic behavior and obeys von Mises yield criterion and the isotropic strain hardening law of Swift-type.

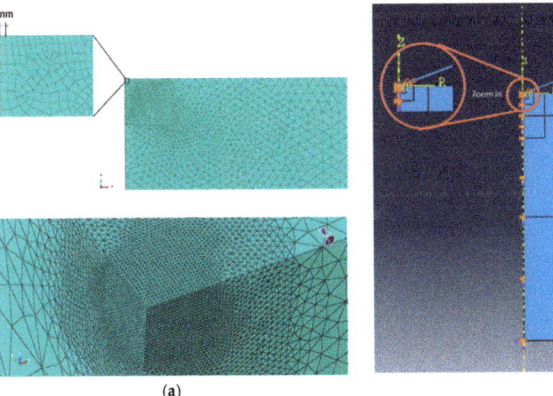
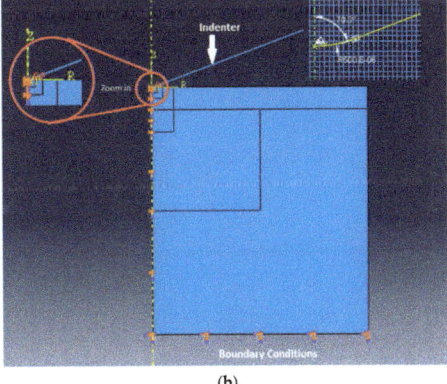

(a) (b)

Figure 1. *Cont.*

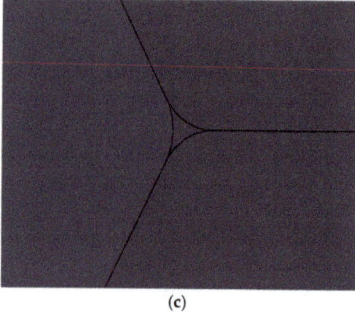
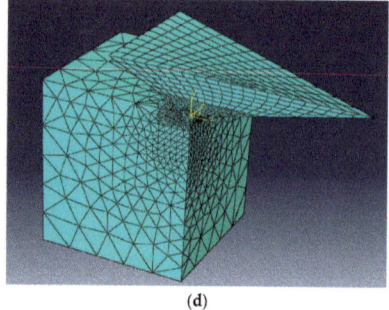

(c)　　　　　　　　　　　　　　(d)

Figure 1. Finite element models used in the numerical simulation of the nanoindentation (deformable parts): (**a**) the 2D axisymmetric FE model (top) and the 3D FE model (bottom) (deformable part); (**b**) the applied boundary conditions for the 2D FE model and definition of the 500 nm indenter tip radius in this case; (**c**) the definition of the 500 nm Berkovich indenter's tip radius; (**d**) the 3D FE model using the Berkovich indenter, where the tip radius is 500 nm, as shown in Figure 1c.

3.2. Elastoplastic Constitutive Model

Commonly, the uniaxial tensile test carried out on a dog-bone sample (at millimetric scale) is used to characterize the mechanical behavior of bulk materials and accurately provide the stress–strain curve. However, for thin films, this conventional method is not suitable due to limitations in their applicability at micro/nanometric scales. Alternative techniques, such as nanoindentation, are necessary to obtain the appropriate mechanical properties.

In this study, we assume that the GaN material behavior obeys the class of elastoplastic materials, for which the total strain is decomposed into elastic and plastic parts, as given in Equation (1).

$$\varepsilon = \varepsilon_y + \varepsilon_p \quad (1)$$

where ε, ε_y, and ε_p represent the total, elastic, and plastic strains.

Commonly, to analyze the elastoplastic behavior of various materials, by investigating the inverse analysis of materials using nanoindentation, the assumption for isotropic materials is employed. This assumption relates the stress and strain through a piecewise function, as given in Equation (2).

$$\sigma = \begin{cases} E\varepsilon & (\sigma \leq \sigma_y) \\ K\varepsilon^n & (\sigma \geq \sigma_y) \end{cases} \quad (2)$$

where E, σ_y, K, and n are the Young's modulus, the yield stress, the hardening modulus, and the hardening exponent, respectively. In the plastic regime, the flow stress, representing the elastoplastic behavior, could be expressed by the following equation.

$$\sigma = \sigma_y \left(1 + \frac{E}{\sigma_y}\varepsilon_p\right)^n \quad (3)$$

3.3. Data Sampling and Database Generation

Data sampling for generating a database is an essential step in the development of ANN models. The goal of data sampling is to select a representative set of data that can be used to train, test, and validate ANN models. The Box–Behnken design (BBD) is a type of experimental design that can be used for efficient data sampling from a high-dimensional space. The BBD is a fractional factorial design that is rotatable, meaning that the effects of the factors are not confounded. This makes the BBD a good choice for sampling data for ANN training, testing, and validation, as it ensures that the effects of the factors are accurately estimated [43]. The BBD is used to obtain the design matrix of the FE simulations. Table 1 lists the design matrix used to generate the FE results of the nanoindentation of the GaN thin film. In this work, the generated FE dataset contains 81 load–depth curves corresponding to the maximum applied load of 15 mN. Then, 25 data points are used for each curve, among them, the total work to the plastic work ratio (W_p/W_t) and the stiffness (dP/dh) are included along with (P-h) data to consider with accuracy the indentation information of the unloading curve (see Figure 2). Therefore, a matrix of [25 × 81] is introduced as inputs in the ANN model.

Table 1. Design matrix using DoE and BBD for finite element simulation.

Test No.	σ_y (GPa)	E (GPa)	n (-)
1	5	150	0.02
2	5	150	0.05
3	5	150	0.09
4	5	300	0.02
5	5	300	0.05
...	...	...	...
80	6	450	0.09
81	7	450	0.09

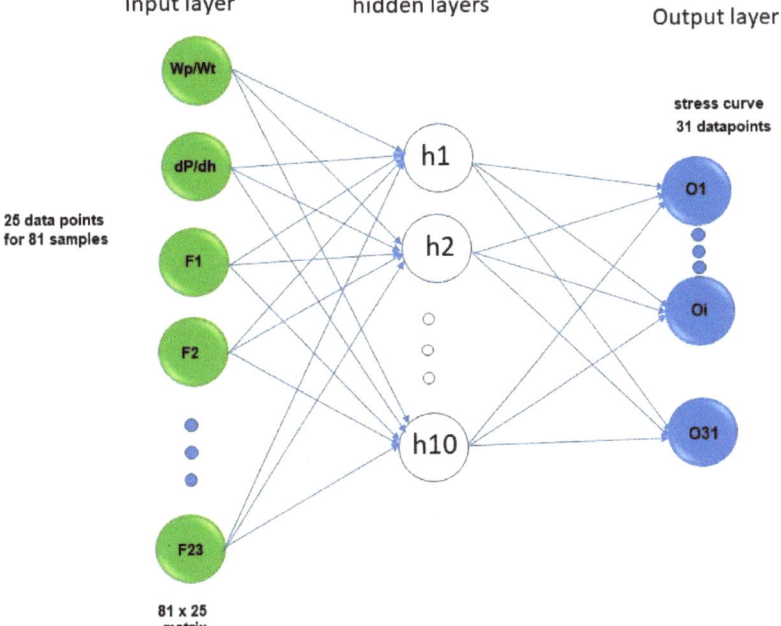

Figure 2. The feed-forward architecture of the ANN model, which is used to determine the stress–strain curve of the GaN thin film. The ANN model receives 81 sets of training data consisting of nanoindentation load–depth curves, (W_p/W_t), and (dP/dh) in order to consider the indentation information of the unloading curve; each set contains 25 input data points, forming then a [25 × 81] input matrix.

3.4. Artificial Neural Network (ANN)

Artificial neural networks (ANNs) are a type of supervised machine learning (ML) algorithms that are inspired by the human brain. They are composed of interconnected nodes, called neurons, that learn to process information and make predictions by adjusting their weights and biases during the training process. The performance of an ANN is based on the quality of the training data and the number of neurons and layers in the network. This allows for the construction of a functional dependency between inputs and outputs. In this work, we developed an inverse analysis method for identifying the elastoplastic parameters of GaN thin films using an ANN model which learns from simulated nanoindentation load–depth curves.

Figure 2 shows the feed-forward ANN structure with one hidden layer connecting the input and output layers. It is worth noting that the performance of the ANN model substantially depends on the number of neurons in the hidden layer and the learning rate. Therefore, a trial-and-error method was used to tune these parameters for the better accuracy of the ANN model. The training sets uses backpropagation with the Levenberg-Marquardt (LM) algorithm to update and determine the weights and biases of the neurons of the ANN model. The activation functions used in the hidden and output layer are the Logarithmic sigmoid (Logsig) and linear (Purelin) function, respectively. The input and output datasets were normalized using a linear transformation. This normalization was achieved to improve the training speed of the ANN model and to reduce the chances of the model becoming trapped in local minima. Additionally, normalization helps to prevent the ANN model from finding multiple solutions to the inverse problem [44,45].

It is important to note that the use of a limited number of training datasets for an ANN model can make it difficult to obtain an effective network predictivity. However, it is possible to train ANN models with limited datasets using regularization techniques, such as either Bayesian regularization [46] or the early stopping method to avoid the problem of ANN overfitting. Overfitting is a critical issue that can affect an ANN model's performance, when the ANN model prioritizes fitting the training data too precisely, at the expense of its ability to generalize well to new, unseen data. Bayesian regularization does not require a validation subset, but it is more computationally expensive. In this work, the so-called early stopping method is used to avoid the overfitting of the network during the training process. Therefore, this method is used to improve the generalization performance of the network. The input datasets are randomly divided into three subsets: the training, validation, and testing subsets. The training subset is used to learn the ANN model, the validation subset is used to assess the generalization capacity of the model, and the testing subset is used to assess the performance of the model on unknown data. The input dataset is randomly partitioned into 70% for training, 15% for validation, and 15% for testing. The error on the validation set is monitored during the training process. The training stops automatically when the validation error starts to increase instead of continuously decreasing the validation data's mean square error (MSE). To verify the performance and effectiveness of the network learning, additional virtual responses were randomly generated using a finite element model. The parameters of these virtual responses were known and within the range of the considered parameters. These virtual responses were then used as extra virtual experimental inputs in the trained ANN model, assuming that the corresponding parameters were unknown. This is an effective way to re-evaluate the performance capacity of the trained network using additional input datasets that are unknown to the network, thus avoiding multiple solutions to the inverse problem. More details about the feature of the used ANN model are summarized in Table 2.

Table 2. The characteristics of the used ANN model.

Feature	Feed-Forward Network
Learning method	Supervised learning
Training method	Levenberg–Marquart backpropagation
Activation function	Log sigmoid (Logsig) Linear (Purelin)
Number of layers	3 layers: (input, hidden, output) ([81 × 25], 10, [1 × 31])
Number of hidden neurons	10

Table 2. *Cont.*

Feature	Feed-Forward Network
Learning rate	0.01
Performance metric	Mean squared error: (MSE)

Figure 3 shows the flowchart of the inverse identification approach used in this work to identify the stress–strain curve of the elastoplastic behavior. The identification procedure can be described in the following steps: firstly, the experimental data (load–depth nanoindentation curve) are obtained through the averaging of twenty experimental nanoindentation tests; secondly, a highly accurate finite element model is built for the numerical simulation of the nanoindentation; thirdly, data sampling was performed based on the BBD to determine and prepare the sets of the data inputs for the FE model to generate the numerical datasets; and fourthly, the dataset of the numerical load–depth curves is applied for the training, validation, and test procedures.

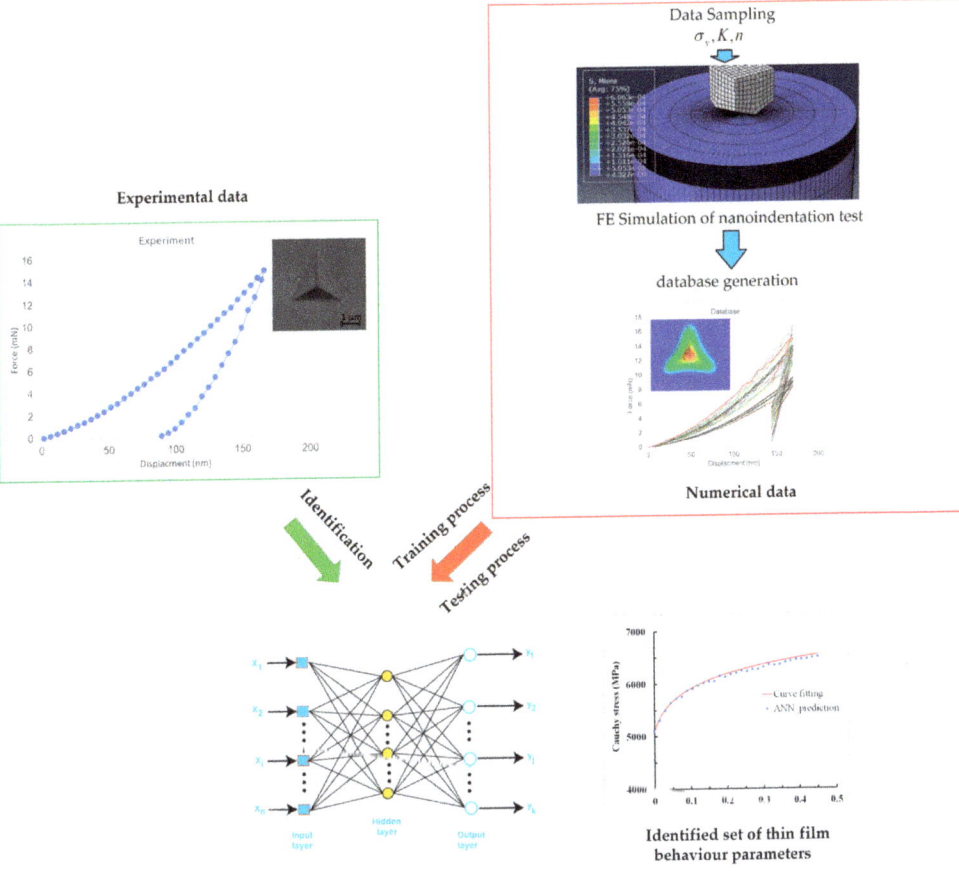

Figure 3. A flowchart of the inverse identification approach used to uniquely identify the stress–strain curve of the thin film.

4. Results and Discussion

Figure 4a shows the averaged curve over the twenty experimental load–depth curves. Furthermore, the scanning electron microscopy (SEM) image depicts the indentation imprint on the GaN sample obtained by applying a load of 500 mN on the Berkovich indenter, as displayed in Figure 4b. Table 3 lists, for the GaN thin film and the sapphire substrate, the hardness and the Young's modulus. The mechanical properties of the GaN thin film are obtained from analyzing the experimental load–depth curves using the Oliver–Pharr method. This table also presents the elastic recovery (i.e., which refers to the amount of deformation that the material recovers elastically after the indenter has been removed) along with the maximum applied load and the resulting maximal depth recorded at this load. In comparison, the sapphire substrate exhibits higher mechanical properties compared to GaN. The substrate's hardness is reported as 27.5 ± 2 GPa, with an elastic modulus of 420.6 ± 20 GPa. The specific data regarding the maximum load, maximum depth, and elastic recovery for the sapphire substrate are not available.

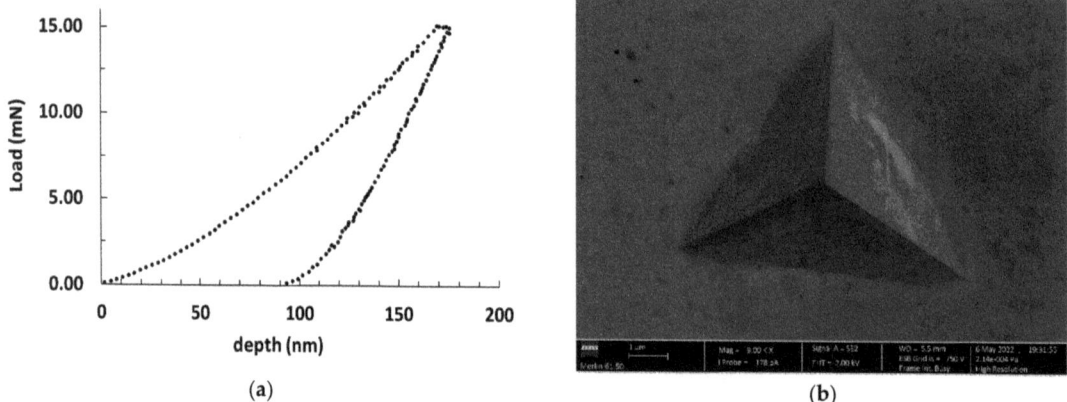

Figure 4. (a) The experimental load–depth curve averaged over the twenty experimental nanoindentation test results; (b) an SEM image of the imprint of the Berkovich indenter at the maximum load of 500 mN on the top surface of the GaN sample.

Table 3. Mechanical properties obtained from analyzing the twenty load–unload curves using the Oliver–Pharr method. The mechanical properties of the sapphire substrate are provided from the literature.

Material	Hardness (GPa)	E (GPa)	Max Load (mN)	Max Depth (nm)	Degree of the Elastic Recovery %
GaN	19.5 ± 0.5	356 ± 8	15.09 ± 0.02	170.06 ± 1.65	28 ± 9
Sapphire substrate [47]	27.5 ± 2	420.6 ± 20	-	-	-

Figure 5 shows the numerical load–depth curves obtained using the three finite element models: (i) 2D axisymmetric, (ii) 3D equivalent conical-shaped Berkovich indenter, and (iii) 3D Berkovich indenter (see Figure 1). The three models exhibit comparable predictive capabilities regarding the nanoindentation load curve. The unload curve demonstrates a high degree of similarity. Nevertheless, at the end of the unloading phase, the 3D Berkovich model displays minor deviations from the prediction of both the 2D axisymmetric and 3D equivalent conical-shaped Berkovich indenter models. Notably, the use of the 3D Berkovich model demands significant computational resources and time. In pursuit of an optimal balance between computational accuracy and efficiency, it becomes evident that the 2D axisymmetric model emerges as the preferred choice. Therefore, this model is adopted to conduct the finite element simulations of nanoindentation tests in the subsequent analyses.

To decouple material properties from the influence of factors like the friction coefficient and indenter tip radius in the case of nanoindentation, Figure 6 presents an analysis of the numerical model's response to variations in these factors. By assessing their effects using nanoindentation simulations, we can gain valuable insights into the actual mechanical behavior of the thin films characterized by nanoindentation. This allows us to understand the sensitivity of the FE model to changes in these critical variables, providing a more nuanced interpretation of the results and their implications on experimental nanoindentation data. Figure 6a depicts that the friction coefficient has no visible effect on the nanoindentation load–depth curve, even for high values. This is because friction primarily affects lateral forces and frictional interactions between the indenter and the film's contact surface. While friction may influence the lateral frictional force during loading and unloading phases, it typically has a negligible impact on the overall vertical force–displacement relationship captured by the load–depth curve, as also reported by Wang [48]. In contrast, Figure 6b clearly shows the visual influence of the indenter tip radius on the nanoindentation curve. The curves depict the results for several distinct tip radii, ranging from 50 nm to 500 nm. As the curves reach the same indentation depth (e.g., the simulations are run using imposed displacement), the load values differ significantly. Notably, the curve representing the smaller tip radius (50 nm) exhibits a lower load compared to the larger tip radius (500 nm) at the same depth. This behavior arises from the varying contact area created by different tip geometries. The larger tip with a 500 nm radius distributes the applied force over a broader area, necessitating a higher overall load to achieve the same level of indentation compared to the sharper tip that concentrates the force on a smaller area. Wang [48] reported that the change in the tip radius results in a significant effect on the load displacement data. The authors found the load to increase with a larger tip radius at the same maximum indentation depth. The sensitivity of the load–depth response to the indenter tip radius underscores a limitation of the rigid indenter assumption in nanoindentation simulations. While this assumption simplifies calculations, it neglects tip deformation, which becomes more pronounced as indenter radii decrease and loads increase. This can result in inaccurate load–depth relationships, errors in extracted mechanical properties, and misinterpretations of material behavior. However, when the indenter tip radius is sufficiently large, the assumption of using a rigid indenter in finite element analysis is valid, as shown in this work.

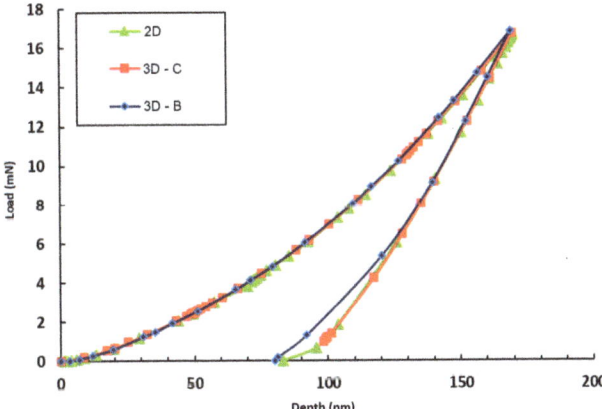

Figure 5. Load–depth curves obtained using three FE models: 2D equivalent conical-shaped Berkovich indenter model, 3D equivalent conical-shaped Berkovich indenter, and 3D with Berkovich indenter.

Figure 7a shows the training performance of the ANN model over epochs. It includes three lines, one representing the training set error (in blue), the validation set error (green line), and the test set error (red line). The mean squared error (MSE) of the training set decreases steadily, indicating improvement in the ANN model's performance on the training set. The MSE of the validation and the test processes are also decreasing but with more fluctuations, suggesting improvement in the model's performance on the validation set and test set with some variability. The best performance of the ANN model is achieved with the lowest mean squared error (MSE) for the validation and test sets at approximately 1.022×10^{-3}. The increase trend after the 14th epoch of the MSE of the validation set indicates an overfitting issue of the ANN model. Therefore, the early stopping regularization method is activated to avoid overfitting, which reduces the ANN model's performance. This MSE plot is a metric used to monitor the model's learning progress, aiming to minimize errors on the training, validation, and test sets. Figure 7b displays the R-plots, which illustrate the regression analysis, showcasing the correlation between the target and the actual outputs of the ANN model. The linear regression coefficients of the training and validation and test processes are depicted during the network's learning process. The correlation coefficient (R) value of 0.999 (approaching 1) is attained, signifying strong correlation. Additionally, the minimum MSE value is achieved, while the overfitting of the ANN model is mitigated by applying the early stopping regularization method, as shown in Figure 7a, indicating the construction of the high-performance ANN model, ready to be used for prediction for input experimental data.

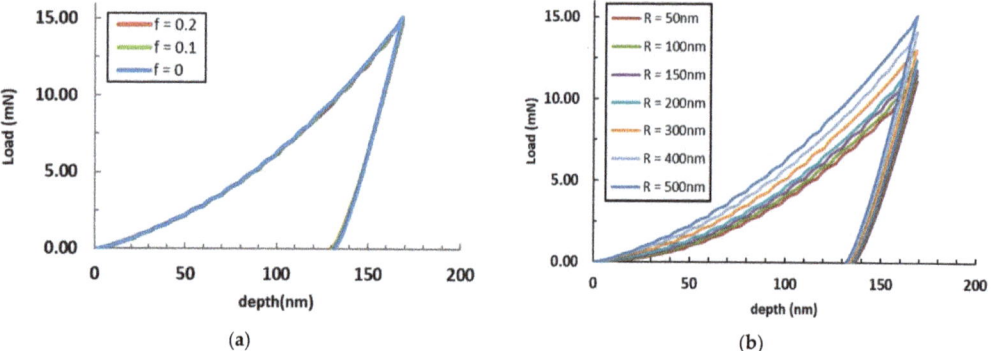

Figure 6. The simulation effect of the (**a**) friction coefficient and (**b**) indenter tip radius on the load–depth curve.

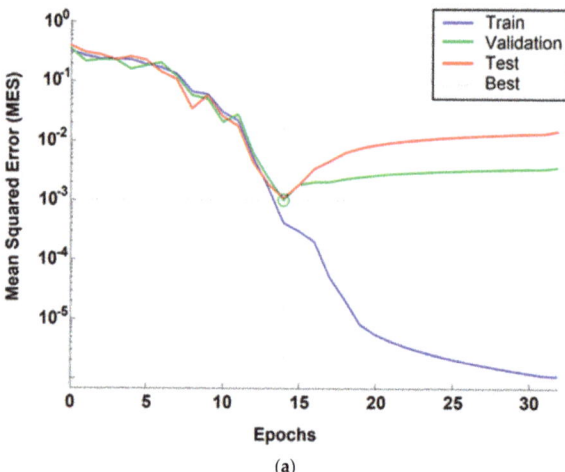

Figure 7. *Cont.*

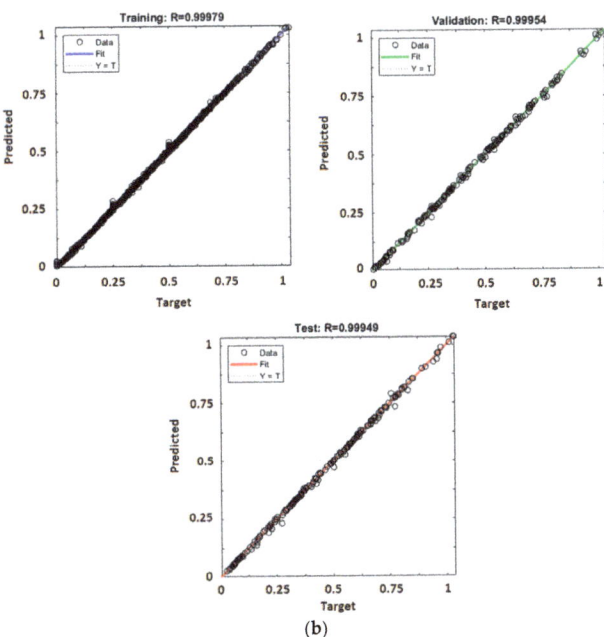

Figure 7. (**a**) The MSE evolution of the training validation and test set performances of the ANN model over epochs; (**b**) the correlation between predicted data by the ANN model and the target values.

Figure 8 exhibits the stress–strain curve determined using the developed ANN model and the verification of the capacity of the obtained stress–strain curve to produce the experimental responses using finite element simulation. Figure 8a showcases the stress–strain data (blue solid circles "•") predicted by the ANN model and using the experimental load–depth data as input in the trained ANN model. The red line ("−") represents the curve fitting of the data points using the elastoplastic constitutive model (power law model) given in Equation (3). It is noted that this curve fitting is in quite good agreement with the data points. The material parameters obtained based on the stress–strain curve fitting are listed in Table 4.

Table 4. The best fit elastoplastic parameters for Equation (3) of the GaN thin film obtained by the fitting of the stress–strain curve using the power law expression.

Yield Stress σ_y (MPa)	Young's Modulus E (GPa)	Power Law Exponent n (-)
5105.7	365.4	0.073

Figure 8b displays the comparison between the experimental load–depth curve and the curve obtained by numerical simulation using the stress–strain curve identified by the ANN model (data points). This result ascertains the performance of the used identification procedure to determine the elastoplastic behavior of the GaN thin film. It is clearly observed that the curve calculated by the finite element method is in very good agreement with the experimental curve. This represents a good verification of the obtained solution, which reflects the accuracy and performance of the developed inverse identification-based ANN model for deducing the constitutive equation of thin film using nanoindentation experiments. Furthermore, it could be mentioned that the ratio between the measured hardness H and the identified yield stress σ_y approximately verifies the empirical well-known Tabor formula ($H \approx 3 \times \sigma_y$), which is the most applicable for relatively hard materials like metals and ceramics, where plastic deformation dominates the indentation response. It is worth noting that the identified Young's modulus (365.4 GPa) is relatively close to the measured value (356 ± 8 GPa), determined from nanoindentation experiments using the Oliver–Pharr approach. On the other hand, the power law exponent with a value of 0.073 is very small. This low value is likely attributed to the dominant plasticity mechanisms occurring during the nanoindentation process. This behavior reflects the weak work-hardening exhibited by the GaN thin film material, which is commonly observed, at nanoscale, in stress–strain curves for ceramics and metals. In molecular dynamic (MD) simulations of metallic behavior at the atomistic scale, near-perfect plasticity (plastic flow with shallow strain hardening) is often found [49,50]. Subsequently, it is noteworthy that the identified behavior model (Equation (3)) is an empirical power law equation typically used to describe the elastoplastic behavior of metallic materials at the macroscale. In this study, we applied this equation to assess its capability in describing the mechanical behavior of a thin film at the nanoscale. Further investigation using MD simulations could provide valuable insights into the plastic behavior and potentially reveal correlations with the analysis based on macroscale relations. Furthermore, it is suggested to couple the finite element model with an MD model to thoroughly study material behaviors at different length scales.

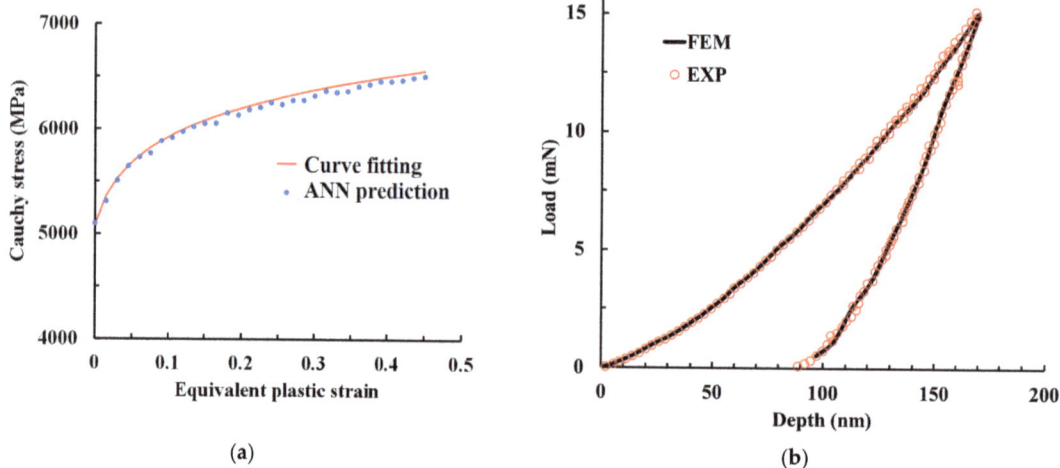

Figure 8. The identification of the stress–strain curve and numerical verification (validation) of the ANN model prediction: (**a**) The stress–strain curve (blue solid circles) predicted by the ANN model, determining the elastoplastic behavior of the GaN thin film. The red line represents the fitting plot of the data using Equation (3); (**b**) the verification of the inverse identification-based ANN model outputs by simulating the P-h curve using the stress–strain curve predicted by the ANN model. The comparison between the experimentally measured and simulated load–depth curves shows good agreement, indicating an accurate and efficient inverse identification approach.

5. Conclusions

This study demonstrated the effectiveness of an inverse identification approach using artificial neural networks (ANNs) for extracting the elastoplastic parameters of GaN thin films from load–depth nanoindentation data. Three finite element models were built and evaluated, revealing that a 2D axisymmetric model could accurately replicate the computationally expensive 3D model's outputs. However, the load–depth response remained sensitive to the indenter tip radius, highlighting the importance of considering this factor in the simulation of nanoindentation for enhanced result accuracy. The trained/validated ANN model, built upon the 2D model-generated database, successfully predicted the GaN thin film's stress–strain curve. The verification capacity of this stress–strain curve in describing the elastoplastic behavior of the GaN thin film was achieved through a finite element simulation of the nanoindentation test. The excellent performance of the developed ANN model underscores its capability to predict the actual stress–strain curve of the GaN thin film based on experimental nanoindentation data. This approach offers a computationally efficient and accurate method for identifying elastoplastic parameters of thin films, leveraging the combined power of nanoindentation testing, finite element simulations, and ANNs.

Author Contributions: Conceptualization, methodology, and software, A.K. (Ali Khalfallah), A.K. (Amine Khalfallah), and Z.B.; validation, A.K. (Ali Khalfallah) and Z.B.; formal analysis, A.K. (Ali Khalfallah); investigation, A.K. (Amine Khalfallah) and Z.B.; data curation, A.K. (Amine Khalfallah) and Z.B.; writing—original draft preparation, A.K. (Ali Khalfallah); writing—review and editing, A.K. (Ali Khalfallah) and Z.B.; visualization. All authors have read and agreed to the published version of the manuscript.

Funding: This research received no external funding.

Institutional Review Board Statement: Not applicable.

Informed Consent Statement: Not applicable.

Data Availability Statement: Data are available on request.

Conflicts of Interest: The authors declare no conflicts of interest.

References

1. Touré, A.; Halidou, I.; Benzarti, Z.; Fouzri, A.; Bchetnia, A.; El Jani, B. Characterization of Low Al Content AlxGa1−xN Epitaxial Films Grown by Atmospheric-Pressure MOVPE. *Phys. Status Solidi* **2012**, *209*, 977–983. [CrossRef]
2. Benzarti, Z.; Khalfallah, A.; Bougrioua, Z.; Evaristo, M.; Cavaleiro, A. Understanding the Influence of Physical Properties on the Mechanical Characteristics of Mg-Doped GaN Thin Films. *Mater. Chem. Phys.* **2023**, *307*, 128182. [CrossRef]
3. Benzarti, Z.; Sekrafi, T.; Bougrioua, Z.; Khalfallah, A.; El Jani, B. Effect of SiN Treatment on Optical Properties of In$_x$Ga$_{1-x}$N/GaN MQW Blue LEDs. *J. Electron. Mater.* **2017**, *46*, 4312–4320. [CrossRef]
4. Bayram, C.; Shiu, K.T.; Zhu, Y.; Cheng, C.W.; Sadana, D.K.; Teherani, F.H.; Rogers, D.J.; Sandana, V.E.; Bove, P.; Zhang, Y.; et al. Engineering Future Light Emitting Diodes and Photovoltaics with Inexpensive Materials: Integrating ZnO and Si into GaN-Based Devices. *Oxide-Based Mater. Devices IV* **2013**, *8626*, 86260L. [CrossRef]
5. Prado, E.O.; Bolsi, P.C.; Sartori, H.C.; Pinheiro, J.R. An Overview about Si, Superjunction, SiC and GaN Power MOSFET Technologies in Power Electronics Applications. *Energies* **2022**, *15*, 5244. [CrossRef]
6. Runton, D.W.; Trabert, B.; Shealy, J.B.; Vetury, R. History of GaN: High-Power RF Gallium Nitride (GaN) from Infancy to Manufacturable Process and Beyond. *IEEE Microw. Mag.* **2013**, *14*, 82–93. [CrossRef]
7. Lin, J.T.; Wang, P.; Shuvra, P.; McNamara, S.; McCurdy, M.; Davidson, J.; Walsh, K.; Alles, M.; Alphenaar, B. Impact of X-ray Radiation on GaN/AlN MEMS Structure and GaN HEMT Gauge Factor Response. In Proceedings of the 2020 IEEE 33rd International Conference on Micro Electro Mechanical Systems (MEMS), Vancouver, BC, Canada, 18–22 January 2020; pp. 968–971. [CrossRef]
8. Ambacher, O.; Eickhoff, M.; Link, A.; Hermann, M.; Stutzmann, M.; Bernardini, F.; Fiorentini, V.; Smorchkova, Y.; Speck, J.; Mishra, U.; et al. Electronics and Sensors Based on Pyroelectric AlGaN/GaN Heterostructures: Part A: Polarization and Pyroelectronics. *Phys. Status Solidi C Conf.* **2003**, 1878–1907. [CrossRef]
9. Lv, L.; Ma, J.G.; Cao, Y.R.; Zhang, J.C.; Zhang, W.; Li, L.; Xu, S.R.; Ma, X.H.; Ren, X.T.; Hao, Y. Study of Proton Irradiation Effects on AlGaN/GaN High Electron Mobility Transistors. *Microelectron. Reliab.* **2011**, *51*, 2168–2172. [CrossRef]
10. Rocco, E.; Licata, O.; Mahaboob, I.; Hogan, K.; Tozier, S.; Meyers, V.; McEwen, B.; Novak, S.; Mazumder, B.; Reshchikov, M.; et al. Hillock Assisted P-Type Enhancement in N-Polar GaN:Mg Films Grown by MOCVD. *Sci. Rep.* **2020**, *10*, 1426. [CrossRef]
11. Azimah, E.; Zainal, N.; Hassan, Z.; Shuhaimi, A.; Bahrin, A. Electrical and Optical Characterization of Mg Doping in GaN. *Adv. Mater. Res.* **2013**, *620*, 453–457.
12. Boughrara, N.; Benzarti, Z.; Khalfallah, A.; Evaristo, M.; Cavaleiro, A. Comparative Study on the Nanomechanical Behavior and Physical Properties Influenced by the Epitaxial Growth Mechanisms of GaN Thin Films. *Appl. Surf. Sci.* **2022**, *579*, 152188. [CrossRef]
13. Tsai, C.H.; Jian, S.R.; Juang, J.Y. Berkovich Nanoindentation and Deformation Mechanisms in GaN Thin Films. *Appl. Surf. Sci.* **2008**, *254*, 1997–2002. [CrossRef]
14. Khalfallah, A.; Benzarti, Z. Mechanical Properties and Creep Behavior of Undoped and Mg-Doped GaN Thin Films Grown by Metal–Organic Chemical Vapor Deposition. *Coatings* **2023**, *13*, 1111. [CrossRef]
15. Laxmikant Vajire, S.; Prashant Singh, A.; Kumar Saini, D.; Kumar Mukhopadhyay, A.; Singh, K.; Mishra, D. Novel Machine Learning-Based Prediction Approach for Nanoindentation Load-Deformation in a Thin Film: Applications to Electronic Industries. *Comput. Ind. Eng.* **2022**, *174*, 108824. [CrossRef]

16. Peng, L.; Liu, F.; Ni, J.; Lai, X. Size Effects in Thin Sheet Metal Forming and Its Elastic–Plastic Constitutive Model. *Mater. Des.* **2007**, *28*, 1731–1736. [CrossRef]
17. Raulea, L.V.; Goijaerts, A.M.; Govaert, L.E.; Baaijens, F.P.T. Size Effects in the Processing of Thin Metal Sheets. *J. Mater. Process. Technol.* **2001**, *115*, 44–48. [CrossRef]
18. Pathak, S.; Kalidindi, S.R. Spherical Nanoindentation Stress-Strain Curves. *Mater. Sci. Eng. R. Rep.* **2015**, *91*, 1–36. [CrossRef]
19. Golovin, Y.I. Nanoindentation and Mechanical Properties of Solids in Submicrovolumes, Thin near-Surface Layers, and Films: A Review. *Phys. Solid State* **2008**, *50*, 2205–2236. [CrossRef]
20. Liu, K.; Ostadhassan, M.; Bubach, B. Applications of Nano-Indentation Methods to Estimate Nanoscale Mechanical Properties of Shale Reservoir Rocks. *J. Nat. Gas Sci. Eng.* **2016**, *35*, 1310–1319. [CrossRef]
21. Boughrara, N.; Benzarti, Z.; Khalfallah, A.; Oliveira, J.C.; Evaristo, M.; Cavaleiro, A. Thickness-Dependent Physical and Nanomechanical Properties of $Al_xGa_{1-x}N$ Thin Films. *Mater. Sci. Semicond. Process.* **2022**, *151*, 107023. [CrossRef]
22. Jian, S.R.; Tasi, C.H.; Huang, S.Y.; Luo, C.W. Nanoindentation Pop-in Effects of Bi_2Te_3 Thermoelectric Thin Films. *J. Alloys Compd.* **2015**, *622*, 601–605. [CrossRef]
23. Zhou, G.; Guo, J.; Zhao, J.; Tang, Q.; Hu, Z. Nanoindentation Properties of 18CrNiMo7-6 Steel after Carburizing and Quenching Determined by Continuous Stiffness Measurement Method. *Metals* **2020**, *10*, 125. [CrossRef]
24. Ohmura, T.; Wakeda, M. Pop-In Phenomenon as a Fundamental Plasticity Probed by Nanoindentation Technique. *Materials* **2021**, *14*, 1879. [CrossRef]
25. Qian, L.; Zhao, H. Nanoindentation of Soft Biological Materials. *Micromachines* **2018**, *9*, 654. [CrossRef]
26. Li, Y.; Li, S. Deep Learning Based Phase Transformation Model for the Prediction of Microstructure and Mechanical Properties of Hot-Stamped Parts. *Int. J. Mech. Sci.* **2022**, *220*, 107134. [CrossRef]
27. Sun, X.; Liu, Z.; Wang, X.; Chen, X. Determination of Ductile Fracture Properties of 16MND5 Steels under Varying Constraint Levels Using Machine Learning Methods. *Int. J. Mech. Sci.* **2022**, *224*, 107331. [CrossRef]
28. Merayo, D.; Rodríguez-Prieto, A.; Camacho, A.M. Prediction of Mechanical Properties by Artificial Neural Networks to Characterize the Plastic Behavior of Aluminum Alloys. *Materials* **2020**, *13*, 5227. [CrossRef]
29. Long, X.; Dong, R.; Su, Y.; Chang, C. Critical Review of Nanoindentation-Based Numerical Methods for Evaluating Elastoplastic Material Properties. *Coatings* **2023**, *13*, 1334. [CrossRef]
30. Li, Z.; Ye, Y.; Zhang, G.; Guan, F.; Luo, J.; Wang, P.; Zhao, J.; Zhao, L. Research on Determining Elastic–Plastic Constitutive Parameters of Materials from Load Depth Curves Based on Nanoindentation Technology. *Micromachines* **2023**, *14*, 1051. [CrossRef]
31. Smith, J.L. Advances in Neural Networks and Potential for Their Application to Steel Metallurgy. *Mater. Sci. Technol.* **2020**, *36*, 1805–1819. [CrossRef]
32. Park, S.; Fonseca, J.H.; Marimuthu, K.P.; Jeong, C.; Lee, S.; Lee, H. Determination of Material Properties of Bulk Metallic Glass Using Nanoindentation and Artificial Neural Network. *Intermetallics* **2022**, *144*, 107492. [CrossRef]
33. Lu, L.; Dao, M.; Kumar, P.; Ramamurty, U.; Karniadakis, G.E.; Suresh, S. Extraction of Mechanical Properties of Materials through Deep Learning from Instrumented Indentation. *Proc. Natl. Acad. Sci. USA* **2020**, *117*, 7052–7062. [CrossRef]
34. Jeong, K.; Lee, H.; Kwon, O.M.; Jung, J.; Kwon, D.; Han, H.N. Prediction of Uniaxial Tensile Flow Using Finite Element-Based Indentation and Optimized Artificial Neural Networks. *Mater. Des.* **2020**, *196*, 109104. [CrossRef]
35. Wang, J.S.; Zheng, X.J.; Zheng, H.; Song, S.T.; Zhu, Z. Identification of Elastic Parameters of Transversely Isotropic Thin Films by Combining Nanoindentation and FEM Analysis. *Comput. Mater. Sci.* **2010**, *49*, 378–385. [CrossRef]
36. Cheng, S.W.; Chen, B.S.; Jian, S.R.; Hu, Y.M.; Le, P.H.; Tuyen, L.T.C.; Lee, J.W.; Juang, J.Y. Finite Element Analysis of Nanoindentation Responses in Bi_2Se_3 Thin Films. *Coatings* **2022**, *12*, 1554. [CrossRef]
37. Shui, S. Progress and Challenges in Finite Element Simulation of Nanoindentation of Ion-Irradiated Materials. *J. Phys. Conf. Ser.* **2021**, *1885*, 032039. [CrossRef]
38. Bressan, J.D.; Tramontin, A.; Rosa, C. Modeling of Nanoindentation of Bulk and Thin Film by Finite Element Method. *Wear* **2005**, *258*, 115–122. [CrossRef]
39. Barkachary, B.M.; Joshi, S.N. Numerical Simulation and Experimental Validation of Nanoindentation of Silicon Using Finite Element Method. In *Advances in Computational Methods in Manufacturing: Select Papers from ICCMM 2019*; Springer: Singapore, 2019; pp. 861–875. [CrossRef]
40. Keryvin, V.; Charleux, L.; Bernard, C.; Nivard, M. The Influence of Indenter Tip Imperfection and Deformability on Analysing Instrumented Indentation Tests at Shallow Depths of Penetration on Stiff and Hard Materials. *Exp. Mech.* **2017**, *57*, 1107–1113. [CrossRef]
41. Shi, Z.; Feng, X.; Huang, Y.; Xiao, J.; Hwang, K.C. The Equivalent Axisymmetric Model for Berkovich Indenters in Power-Law Hardening Materials. *Int. J. Plast.* **2010**, *26*, 141–148. [CrossRef]
42. Lee, H.; Huen, W.Y.; Vimonsatit, V.; Mendis, P. An Investigation of Nanomechanical Properties of Materials Using Nanoindentation and Artificial Neural Network. *Sci. Rep.* **2019**, *9*, 13189. [CrossRef]
43. Ktari, Z.; Leitão, C.; Prates, P.A.; Khalfallah, A. Mechanical Design of Ring Tensile Specimen via Surrogate Modelling for Inverse Material Parameter Identification. *Mech. Mater.* **2021**, *153*, 103673. [CrossRef]
44. Nino-Adan, I.; Portillo, E.; Landa-Torres, I.; Manjarres, D. Normalization Influence on ANN-Based Models Performance: A New Proposal for Features' Contribution Analysis. *IEEE Access* **2021**, *9*, 125462–125477. [CrossRef]

45. Huang, L.; Qin, J.; Zhou, Y.; Zhu, F.; Liu, L.; Shao, L. Normalization Techniques in Training DNNs: Methodology, Analysis and Application. *IEEE Trans. Pattern Anal. Mach. Intell.* **2023**, *45*, 10173–10196. [CrossRef] [PubMed]
46. Burden, F.; Winkler, D. Bayesian Regularization of Neural Networks. In *Methods in Molecular Biology*; Livingstone, D.J., Ed.; Humana Press: Totowa, NJ, USA, 2008; Volume 458, pp. 23–42. ISBN 9781588297181.
47. Mao, W.G.; Shen, Y.G.; Lu, C. Nanoscale Elastic-Plastic Deformation and Stress Distributions of the C Plane of Sapphire Single Crystal during Nanoindentation. *J. Eur. Ceram. Soc.* **2011**, *31*, 1865–1871. [CrossRef]
48. Wang, T.H.; Fang, T.H.; Lin, Y.C. A Numerical Study of Factors Affecting the Characterization of Nanoindentation on Silicon. *Mater. Sci. Eng. A* **2007**, *447*, 244–253. [CrossRef]
49. Ivashchenko, V.I.; Turchi, P.E.A.; Shevchenko, V.I. Simulations of the Mechanical Properties of Crystalline, Nanocrystalline, and Amorphous SiC and Si. *Phys. Rev. B—Condens. Matter Mater. Phys.* **2007**, *75*, 085209. [CrossRef]
50. Schiøtz, J.; Vegge, T.; Di Tolla, F.D.; Jacobsen, K.W. Atomic-Scale Simulations of the Mechanical Deformation of Nanocrystalline Metals. *Phys. Rev. B—Condens. Matter Mater. Phys.* **1999**, *60*, 11971–11983. [CrossRef]

Disclaimer/Publisher's Note: The statements, opinions and data contained in all publications are solely those of the individual author(s) and contributor(s) and not of MDPI and/or the editor(s). MDPI and/or the editor(s) disclaim responsibility for any injury to people or property resulting from any ideas, methods, instructions or products referred to in the content.

Article

The Influence of Lyophobicity and Lyophilicity of Film-Forming Systems on the Properties of Tin Oxide Films

Elena Dmitriyeva [1], Igor Lebedev [1,*], Ekaterina Bondar [1], Anastasia Fedosimova [1], Abzal Temiraliev [1], Danatbek Murzalinov [1], Sayora Ibraimova [1], Bedebek Nurbaev [1], Kasym Elemesov [2] and Bagila Baitimbetova [1]

[1] Institute of Physics and Technology, Satbayev University, Ibragimov 11, Almaty 050013, Kazakhstan; dmitriyeva@sci.kz (E.D.); bondar@sci.kz (E.B.); a.fedossimova@sci.kz (A.F.); a.temiraliyev@sci.kz (A.T.); d.murzalinov@sci.kz (D.M.); ibraimova@sci.kz (S.I.); b.nurbayev@sci.kz (B.N.); b.baitimbetova@sci.kz (B.B.)
[2] Institute of Energy and Mechanical Engineering Named after A. Burkitbaeva, Satbayev University, Satpayev 22, Almaty 050000, Kazakhstan; info@sci.kz
* Correspondence: lebedev@sci.kz; Tel.: +7-707-6015-013

Abstract: In this work, the effects of lyophobicity and lyophilicity of film-forming systems on the properties of thin nanostructured films was studied. Systematic series of experiments were carried out with lyophilic film-forming systems: $SnCl_4$/EtOH, $SnCl_4$/EtOH/NH_4F, $SnCl_4$/EtOH/NH_4OH and lyophobic systems: SnO_2/EtOH and SnO_2/EtOH/NH_4F. Film growth mechanisms are determined depending on the type of film-forming system. The surface of the films was studied using a scanning electron microscope and an optical microscope. The spectrophotometric method is used to study the transmission spectra and the extinction coefficient. The surface resistance of the films was determined using the four-probe method. The quality factor and specific conductivity of the films are calculated. It was found that the addition of a fluorinating agent (NH_4F) to a film-forming system containing SnO_2 in the form of a dispersed phase does not lead to an increase in the specific conductivity of the films. X-ray diffraction analysis proved the incorporation of fluorine ions into the structure of the film obtained from the $SnCl_4$/EtOH/NH_4F system by the presence of $SnOF_2$ peaks. In films obtained from SnO_2/EtOH/NH_4F systems, there are no $SnOF_2$ peaks. In this case, ammonium fluoride crystallizes as a separate phase and decomposes into volatile compounds.

Keywords: sol-gel; thin films; SnO_2; dendritic structures; specific conductivity; sensitivity to ethanol

Citation: Dmitriyeva, E.; Lebedev, I.; Bondar, E.; Fedosimova, A.; Temiraliev, A.; Murzalinov, D.; Ibraimova, S.; Nurbaev, B.; Elemesov, K.; Baitimbetova, B. The Influence of Lyophobicity and Lyophilicity of Film-Forming Systems on the Properties of Tin Oxide Films. *Coatings* **2023**, *13*, 1990. https://doi.org/10.3390/coatings13121990

Academic Editors: Ali Khalfallah and Zohra Benzarti

Received: 13 October 2023
Revised: 14 November 2023
Accepted: 20 November 2023
Published: 23 November 2023

Copyright: © 2023 by the authors. Licensee MDPI, Basel, Switzerland. This article is an open access article distributed under the terms and conditions of the Creative Commons Attribution (CC BY) license (https://creativecommons.org/licenses/by/4.0/).

1. Introduction

SnO_2 thin films possess exceptional properties, including a wide bandgap (Eg = 3.6 eV), n-type conductivity, transparency of more than 80% in the visible wavelength range, chemical stability, and others [1–3]. As a result, SnO_2 has been widely used in various applications such as solar cells [4–6], gas sensors [7–9], transistors [10–12], optoelectronic devices, and transparent conductive electrodes [13–15]. SnO_2 films are doped and modified in composition to meet the increasing demand for functional coatings. Both metallic particles [16,17] and non-metallic ions are used as doping additives. Of the non-metals, fluorine ions are prevalent [18–23]. The bandgap is reduced and the oxidation potential of holes is improved by doping the material with fluorine atoms [24,25]. Tin oxide films doped with fluorine and indium [26] are used in photocatalysis [27–29], perovskite solar cells [30–32], electrochemistry [33–35], and biosensors [36–38]. Great success in obtaining films of tin oxide doped with fluorine has been achieved using the sol-gel method. This method is based on the transition of film-forming systems into a gel. Gel formation can occur from both a lyophilic system and a lyophobic system. Considerable attention has been directed toward the investigation of lyophilic and lyophobic systems. In particular, thermohydrodynamics exhibit differences in lyophilic and lyophobic systems [39,40], and diverse interactions of a nanodroplet with a nanoprotrusion are observed [41]. The introduction of additives often leads to a change in the characteristics of film-forming systems, which affects the size of the particles obtained, the density and porosity of the films, the specific surface area, and the surface-to-volume ratio change. Positron annihilation spectroscopy is the most informative method for the study of porosity and free volume [42].

However, several questions persist when opting to use either a lyophilic or lyophobic system for acquiring sol-gel films. Specifically, what is the mechanism of film growth in a particular system? Can the growth of films be described using mathematical equations? How do additives affect one or both systems? Is the alloying element consistently incorporated into the film structure? Do the properties of the resulting films change equivalently upon the addition of additives to a lyophobic or lyophilic system?

In this work, it is shown that the film structure and its functional properties can differ significantly for lyophilic and lyophobic systems. This is because lyophilic systems incorporate an element from a dopant into the film composition, unlike lyophobic systems.

2. Materials and Methods

Film-forming systems were obtained from: tin tetrachloride crystalline hydrate $SnCl_4 \cdot 5H_2O$ (>98% pure grade), rectified alcohol, corresponding to State Standard [43], aqueous ammonia solution (>98% pure grade), and an NH_4F dopant (>98% pure grade). The experimental procedure is shown in Figure 1.

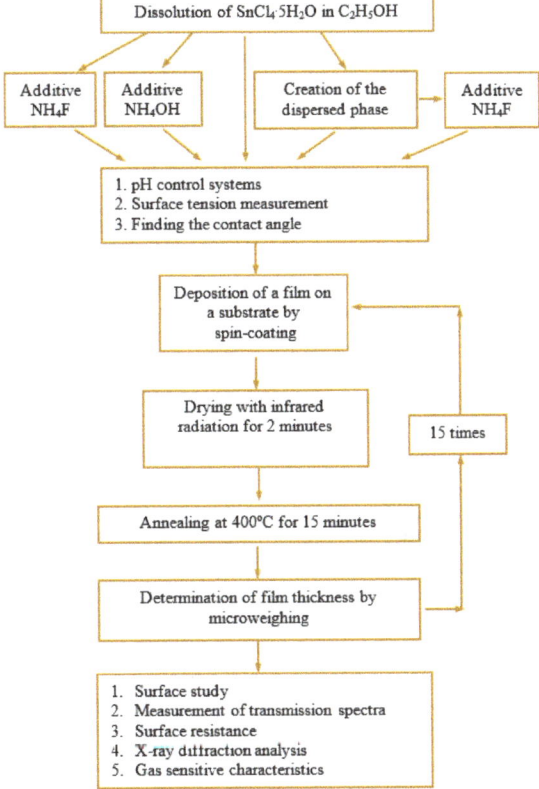

Figure 1. Scheme of the experiment.

Five film-forming systems were obtained:
1. $SnCl_4$/EtOH: $SnCl_4 \cdot 5H_2O$ was dissolved in ethanol. The concentration of tin ions in the system was 0.13 mol/L.

2. SnCl$_4$/EtOH/NH$_4$F: The fluorinating agent, NH$_4$F, was added to the SnCl$_4$/EtOH system. The ratio of tin ions to fluorine ions was 10/4. The NH$_4$F crystals were dissolved after being stirred for 2 h at 140 rpm and concurrently heated to 35 °C.
3. SnCl$_4$/EtOH/NH$_4$OH: 20% alcohol solution of ammonia was added to the SnCl$_4$/EtOH system until the pH level reached the same as that of the SnCl$_4$/EtOH/NH$_4$F system (pH = 1.80).
4. SnO$_2$/EtOH: A concentrated aqueous ammonia solution was added to the SnCl$_4$/EtOH system until the complete precipitation of tin hydroxide. The solvent and water associated with tin oxide was removed by heating to 100 °C and stirring at a speed of 160 rpm. After drying, a white powder was obtained. Ethanol was used to fill the tin oxide powder to obtain a solution with a tin oxide concentration of 0.13 mol/L. The vessel's contents were stirred at a speed of 100 rpm without any heating until the precipitate transitioned into solution. The procedure lasted for 4 h.
5. Film-forming system containing SnO$_2$ in the form of a dispersed phase with the addition of NH$_4$F (SnO$_2$/EtOH/NH$_4$F): crystals of ammonium fluoride were added to the system containing SnO$_2$ in the form of a dispersed phase. The ratio of tin ions to fluorine ions was 10/4.

The pH level of the systems was assessed using a pH meter known as "pH-150 M". The surface tension was determined through the stalagmometric method, and the spreading drop method was applied to study the wetting angle of film-forming systems with the substrate surface.

Film-forming systems were applied onto glass slides using spin-coating. The rotation speed was 3000 rpm for 4 s. Infrared radiation was used for 2 min at 80 °C to dry the samples. Subsequently, the sample was annealed in a muffle furnace for 15 min at 400 °C. After cooling the sample, the deposition process was repeated. A total of 15 layers were applied.

The thickness of the film was determined through the microbalance approach. The JSM-6490LA, JEOL analytical scanning electron microscope, and MPE-11 optical microscope were utilized to study the surface structure of the obtained SnO$_2$ films. The transmission and absorption spectra were measured on a UNICO spectrophotometer, while the surface resistance was quantified using the four-probe method. X-ray diffraction spectra were recorded on a DRON-6 diffractometer to complete the analysis. The study investigated the sensitivity to ethanol vapour using an experimental apparatus capable of maintaining the working chamber's temperature with an accuracy of at least ± 2 °C, ranging from room temperature to 450 °C.

3. Results and Discussion

3.1. Formation of Films

Control of the acidity of the film-forming systems indicated that the SnCl$_4$/EtOH system has a pH of 0.18. The addition of NH$_4$F significantly changes this value to pH 1.80. The resulting SnCl$_4$/EtOH/NH$_4$F system has an acidity close to that of Sn(OH)$_4$ formation (pH 2.0), which can lead to a change in the film structure. In order to accurately interpret subsequent measurements and detect the effects of fluorine ions on the properties of the obtained films, the SnCl$_4$/EtOH/NH$_4$OH system with pH = 1.80 was made. The pH level of film-forming systems that contained SnO$_2$ in a dispersed phase was 3.62. With the addition of NH$_4$F, the pH level was 3.58. The surface tension of all colloidal systems was $21.55 \times 10^3 \pm 0.05 \times 10^3$ N/m. The contact angle of wetting on the glass substrate was found to be close to zero for all systems, demonstrating that the surface was sufficiently wettable to produce homogeneous layers via spin-coating.

The change in sample mass as the film was deposited on the substrate was measured after each layer. From this data the variation in film thickness was calculated. The results obtained are shown in Figure 2.

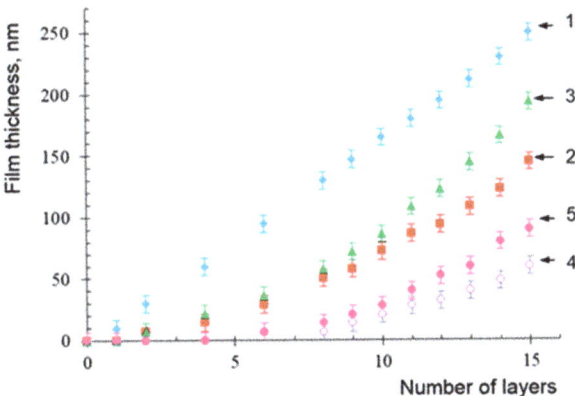

Figure 2. Film thickness depending on the number of applied layers for the film-forming system: 1—SnCl$_4$/EtOH, 2—SnCl$_4$/EtOH/NH$_4$F, 3—SnCl$_4$/EtOH/NH$_4$OH, 4—SnO$_2$/EtOH, 5—SnO$_2$/EtOH/NH$_4$F.

Figure 2 illustrates the linear correlation between the number of deposited layers and the film thickness, as determined by the micro-weighting technique, in the SnCl$_4$/EtOH film-forming system.

To obtain a more accurate assessment of how the estimated film thickness varies with the number of layers applied, we will use the least squares method of mathematical analysis. The fundamental principle of which can be expressed by the following equation:

$$\sum_i e_i^2 = \sum_i (y_i - f_i(x))^2 \rightarrow min \qquad (1)$$

where e_i is the deviation of the i-th experimental value from the corresponding value of the approximating function, y_i is the experimental value, and $f_i(x)$ is the approximating function. The value of the approximation reliability is determined by the formula:

$$R^2 = 1 - \frac{\sum_{i=1}^n (y_i - f(x_i))^2}{\left(\sum_{i=1}^n y_i^2\right) - \frac{\left(\sum_{i=1}^n y_i\right)^2}{n}} \qquad (2)$$

where R^2 is the value of approximation confidence. Approximation functions describing the change in film thickness on the number of applied layers, and the approximation confidence value are shown in Table 1.

Table 1. Approximating functions.

The Composition of the Film-Forming System	Approximation Function	Approximation Confidence Level (R^2)
SnCl$_4$/EtOH	$y = 16.755x - 4.2498$	0.9994
SnCl$_4$/EtOH/NH$_4$F	$y = 0.7716x^2 + 1.0978x + 0.7338$	0.9989
SnCl$_4$/EtOH/NH$_4$OH	$y = 0.476x^2 + 2.4008x - 1.1223$	0.9978
SnO$_2$/EtOH	$y = 0.6135x^2 - 3.3234x + 2.3959$	0.9958
SnO$_2$/EtOH/NH$_4$F	$y = 0.3788x^2 - 1.8943x + 1.3543$	0.9926

All functions have an approximation confidence value greater than 0.9; therefore, the change in film thickness depending on the number of deposited layers for the chosen film-forming systems was described accurately. Table 1 illustrates that for the SnCl$_4$/EtOH film-forming system, the linear function provides a high confidence description of the change in film thickness as the number of layers deposited increases. The thickness of the films is quadratically dependent on the number of layers in other film-forming systems.

This indicates that with each subsequent layer, the mass change is greater than the previous one. This is possible in the case of the formation of "uneven" porous layers, when each subsequent layer fills the pores of the previous one and forms its own. This is shown schematically in Figure 3.

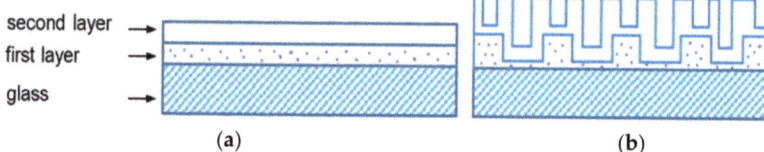

Figure 3. Layer formation scheme: (**a**) linear, (**b**) non-linear.

After applying 15 layers (Figure 1), the thickness of the films obtained from film-forming systems was: 250 ± 7 nm for $SnCl_4$/EtOH; 152 ± 7 nm for $SnCl_4$/EtOH/NH_4F; 193 ± 7 nm for $SnCl_4$/EtOH/NH_4OH; 60 ± 7 nm for SnO_2/EtOH; and 90 ± 7 nm for SnO_2/EtOH/NH_4F.

The films obtained from the $SnCl_4$/EtOH lyophilic film-forming system exhibit superior adhesion to both the substrate and other layers due to their greater thickness. Films obtained from lyophobic film-forming systems containing SnO_2 in the form of a dispersed phase have the smallest thickness. This highlights their lower adhesion to the glass substrate and other layers.

Thus, the films grown from the $SnCl_4$/EtOH film-forming system proceed according to the Frank–van der Merwe layer-by-layer growth mechanism. The growth of films obtained from a film-forming system containing SnO_2 in the form of a dispersed phase was carried out according to the Volmer–Weber island growth mechanism. And the growth of films with additives is carried out using the Stransky–Krystanov layer-by-layer-plus-island growth mechanism.

3.2. Film Structure

SEM images of the film surfaces investigated are illustrated in Figure 4. Observing Figure 4a, it is evident that films produced from the $SnCl_4$/EtOH lyophilic system exhibit a smoother surface compared to other films. The crystallization process in films transpires on the substrate's heated surface as HCl molecules evaporate. In doing so, HCl molecules hinder the formation of $Sn(OH)_4$, as expressed in the chemical reaction:

$$SnCl_4 + 4H_2O \rightarrow Sn(OH)_4 + 4HCl. \tag{3}$$

The SEM images in Figure 4b,c reveal that the films possess diverse dendritic structures [44,45]. The development of dendritic structure is caused by a change in the acidity of film-forming systems due to the introduction of NH_4F and NH_4OH. Accelerated crystallization occurs from nonequilibrium conditions [46–48]. The introduction of NH_4OH into the system not only leads to a change in acidity, but also adds functional OH^- ions. This leads to the growth of dendritic axes of the first order (see Figure 4c). SEM images in Figure 4d,f depict films prepared from lyophobic systems. We observe dendritic structures up to 10 μm in size on the surface, which are grains formed under accelerated crystallization conditions (see Figure 4d). Figure 4e shows agglomerates formed from SnO_2 grains combined with an NH_4F additive.

Figure 4. SEM images of the surface of films: (**a**) $SnCl_4$/EtOH; (**b**) $SnCl_4$/EtOH/NH_4F; (**c**) $SnCl_4$/EtOH/NH_4OH; (**d**) SnO_2/EtOH; (**e**) SnO_2/EtOH/NH_4F.

3.3. Optical Spectra and Surface Resistance

Figure 5 shows the transmission spectra of the films under study. The transmittance of the glass substrate is 88%–91% in the wavelength range from 340 nm to 2500 nm (curve 1 in Figure 5a).

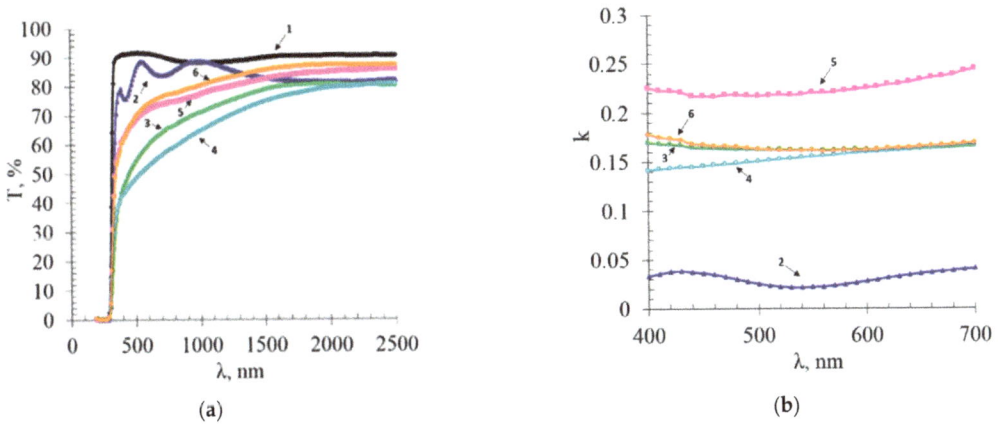

Figure 5. Spectra of tin dioxide films: (**a**) transmission; (**b**) extinction coefficient. 1—the glass substrate, 2—$SnCl_4$/EtOH, 3—$SnCl_4$/EtOH/NH_4F, 4—$SnCl_4$/EtOH/NH_4OH, 5—SnO_2/EtOH, 6—SnO_2/EtOH/NH_4F.

For films with dendritic structures, the transmittance increases with increasing wavelength. For the purpose of considering the variation in film thicknesses, we use the "extinction coefficient" parameter which is computed as follows [49,50]:

$$\alpha = \frac{1}{d} \cdot \ln\left(\frac{I_0}{I}\right). \tag{4}$$

Using the equation

$$\alpha = 4\pi \cdot \frac{k}{\lambda} \tag{5}$$

we obtained an expression for the extinction coefficient:

$$k = \alpha \cdot \frac{\lambda}{4\pi} = \left(\frac{1}{d}\right) \cdot \ln\left(\frac{I_0}{I}\right) \cdot \frac{\lambda}{4\pi}, \tag{6}$$

where I_0 is the intensity of incident light, I is the intensity of light transmitted through a layer with thickness d, α is the absorption index of the medium, and k is the extinction coefficient.

The extinction coefficient of the glass substrate is about 10^{-5} and aligns with the abscissa axis. In the selected wavelength range (curve 2 in Figure 5b), the film obtained from the $SnCl_4$/EtOH film-forming system displays an extinction coefficient of 0.021–0.038. The extinction coefficients of films formed from the film-forming systems $SnCl_4$/EtOH/NH_4F, $SnCl_4$/EtOH/NH_4OH, and SnO_2 dispersed in NH_4F coincide with one another when wavelengths greater than 600 nm. The film obtained from the system containing SnO_2 in a dispersed phase displays the highest extinction coefficient.

The transmission spectra of the films from the $SnCl_4$/EtOH film-forming system illustrate interference peaks, observed in the case of plane-parallel films as a consequence of light interference reflected by two film surfaces [51,52]. The interference pattern relies on the phase difference of the propagating waves, expressed by the following conditions transmitted through light:

for minimum:

$$2dn_1\sqrt{n_{21}^2 - \sin^2\alpha} = m\lambda. \tag{7}$$

for maximum:

$$2dn_1\sqrt{n_{21}^2 - \sin^2\alpha} = (2m-1)\frac{\lambda}{2}. \tag{8}$$

As the beam was directed perpendicular to the surface, we can determine that the angle $\alpha = 0° \Rightarrow \sin^2\alpha = 0$. Assuming the refractive index of air as unity ($n_1 = 1$), we obtain the minimum value:

$$2dn_{21} = m\lambda. \tag{9}$$

for maximum:

$$2dn_{21} = (2m-1)\frac{\lambda}{2}. \tag{10}$$

We have three extremes, two of which were lows and one was a high. After substituting the experimental values of λ at the extremum point, we solved a system of three equations with two unknowns, and upon performing the calculations, we found that d is equal to 313 ± 18 nm, and $n = 2.03$. The variation between the film thickness obtained by microbalancing and the thickness calculated from the transmission spectra was attributed to the use of cassiterite density (7 g/cm^3) in the calculations. Based on the difference in thickness values, we find the actual density of the film, which is 5.59 g/cm^3.

The condition for interference peaks to appear is not met in other films.

One of the areas of application for films based on tin oxide is transparent conductive coatings. In this case, the considered transmittance and surface resistance are important.

To determine the quality of the obtained films, a parameter known as the quality factor is used. It is determined using the Haacke relation [53,54]:

$$\Phi = (T/100)^{10}/R_{sh}, \quad (11)$$

where Φ is the quality factor, and T and R_{sh} are the transmission coefficients and surface resistance of the film, respectively. The average transmittance for the visible wavelength range was taken when calculating the quality factor.

For a correct assessment of the contribution of doping additives to the conductivity of the films, calculations of resistivity and conductivity were made. Resistivity is related to surface resistance and film thickness by the relationship:

$$\rho = R_{sh} * d \quad (12)$$

where ρ is the resistivity, R_{sh} is the surface resistance, and d is the film thickness. Specific conductivity is the reciprocal of resistivity and is equal to $1/\rho$. Surface resistance measurement data and calculation results are shown in Table 2.

Table 2. Surface resistance, resistivity, conductivity and quality factor of the studied films.

The Composition of the Film-Forming System	R_{sh}, kOm/sq.	ρ, Om·cm	$1/\rho$, Om^{-1}·cm^{-1}	Φ, 10^{-7} Om^{-1}
SnCl$_4$/EtOH	15.6 ± 1.4	0.390 ± 0.035	2.6 ± 0.2	146.8 ± 3.4
SnCl$_4$/EtOH/NH$_4$F	6.7 ± 0.6	0.097 ± 0.008	10.3 ± 0.8	6.3 ± 0.6
SnCl$_4$/EtOH/NH$_4$OH	15.4 ± 1.6	0.255 ± 0.026	3.9 ± 0.4	0.9 ± 0.1
SnO$_2$/EtOH	78.9 ± 6.9	0.512 ± 0.044	1.9 ± 0.2	4.5 ± 0.4
SnO$_2$/EtOH/NH$_4$F	69.4 ± 8.3	0.590 ± 0.070	1.7 ± 0.2	6.2 ± 0.5

Specific conductivity depends on the number and mobility of charge carriers:

$$1/\rho = en_e\mu_e + en_h\mu_h, \quad (13)$$

where e is the electron charge, n_e is the number of electrons, μ_e is the electron mobility, n_h is the number of holes, and μ_h is the hole mobility.

Since thin films of tin dioxide are n-type semiconductors and have a higher concentration of electrons than holes, the second term on the right side of Equation (13) is usually neglected. This equation takes the form [53] at a good approximation:

$$1/\rho = en_e\mu_e. \quad (14)$$

According to the equation above, the conductivity ($1/\rho$) is expected to be influenced by n_e and μ_e. Table 2 shows that the addition of NH$_4$F to the SnCl$_4$/EtOH/NH$_4$F film-forming system leads to a several-fold increase in specific conductivity. This confirms the presence of fluorine ions as additional sources of free charge carriers in the films [55]. The addition of an aqueous ammonia solution to the SnCl$_4$/EtOH/NH$_4$OH film-forming system also led to an increase in the specific conductivity. Apparently, due to an unshared electron pair in the nitrogen atom. The addition of NH$_4$F to the film-forming system containing SnO$_2$ in the form of a dispersed phase did not lead to an increase in the specific electrical conductivity, but even slightly decreased it.

The answer to this phenomenon was found in the study of films in an optical microscope. Figure 6 shows the surface of films obtained from a film-forming system containing SnO$_2$/EtOH and SnO$_2$/EtOH/NH$_4$F.

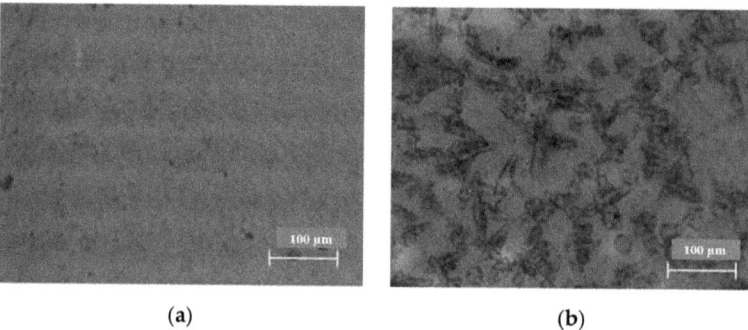

Figure 6. The surface of films obtained from a film-forming system containing: (a) SnO_2/EtOH; (b) SnO_2/EtOH/NH_4F.

Figure 6a shows the dense structural connection between particles in the suspension. The calculated small thickness of the films led to the assumption that the adhesion of the films to the glass substrate reduced in comparison with those obtained from the $SnCl_4$/EtOH, $SnCl_4$/EtOH/NH_4F, and $SnCl_4$/EtOH/NH_4OH film-forming systems. However, as can be seen from Figure 6a the film is uniform over the entire surface and there are no areas of the substrate not covered by the film. The minimal thickness of the film can be attributed to the low adhesion forces between the particles within the dispersed phase of the film-forming system. Figure 6b shows the surface of films obtained from a film-forming system containing SnO_2 in the form of a dispersed phase with the addition of ammonium fluoride. In this area, one can observe depressions in the form of large dendritic structures, the dimensions of which reach 100 μm along the first-order axes.

Based on the analysis of Figure 6b, it can be assumed that the addition of ammonium fluoride to a film-forming system containing SnO_2 in the form of a dispersed phase does not lead to a uniform distribution of the fluorinating agent in the film volume. This leads to the formation of separate SnO_2 and NH_4F phases.

Indeed, NH_4F forms colorless crystals, which, upon heating, decompose in two stages [56]:

1-st stage. At 167 °C, ammonium fluoride decomposes into gaseous ammonia (NH_3) and ammonium hydrofluoride (NH_4HF_2) according to the reaction:

$$\text{>167 °C} \\ 2NH_4F \rightarrow NH_4(HF_2) + NH_3\uparrow. \tag{15}$$

2-nd stage. At 238 °C, ammonium hydrofluoride decomposes into gaseous ammonia (NH_3) and gaseous hydrogen fluoride (HF) according to the reaction:

$$\text{>238 °C} \\ NH_4(HF_2) \rightarrow NH_3\uparrow + 2HF\uparrow. \tag{16}$$

Considering that the films were annealed at 400 °C, the NH_4F dendritic structures formed during drying decomposed into gaseous compounds. Instead of ammonium fluoride, voids formed after annealing. That is, defects that reduce the mean free path and, consequently, the mobility of charge carriers. The mobility of charge carriers is related to the mean free path by the expression:

$$\mu = \frac{q}{m} \cdot t_{av} = \frac{q}{m} \cdot \frac{l_{av}}{V_T}, \tag{17}$$

where q is the electron charge, m is the effective electron mass, t_{av} is the mean free path of an electron between two successive collisions with defects in the crystal lattice, l_{av} is the electron mean free path, and V_T is the thermal velocity.

Thus, the addition of NH$_4$F into a film-forming system containing SnO$_2$ in the form of a dispersed phase leads to a decrease in the specific conductivity. Due to the formation of a separate NH$_4$F phase, which decomposes into volatile compounds, forming voids in the film, which reduce the mean free path of charge carriers. X-ray diffraction analysis confirmed the presence of fluorine ions in the films obtained from the SnCl$_4$/EtOH/NH$_4$F film-forming systems. The X-ray patterns are shown in Figure 7.

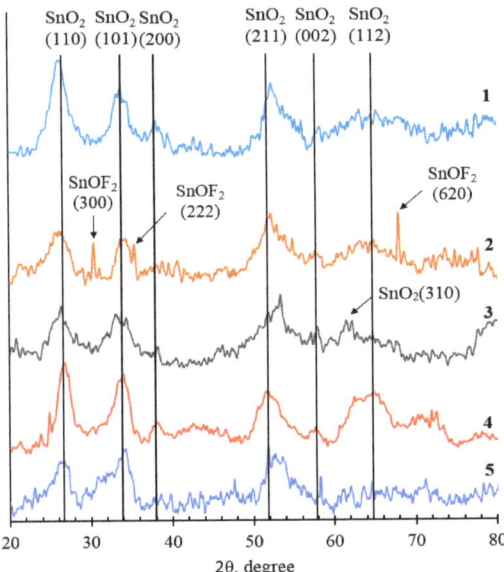

Figure 7. X-ray patterns of films. The composition of the film-forming systems: 1—SnCl$_4$/EtOH, 2—SnCl$_4$/EtOH/NH$_4$F, 3—SnCl$_4$/EtOH/NH$_4$OH, 4—SnO$_2$/EtOH, 5—SnO$_2$/EtOH/NH$_4$F.

The X-ray patterns (Figure 7) demonstrate that the films comprised SnO$_2$ crystallites. On the X-ray diffraction pattern of the film obtained from the film-forming system SnCl$_4$/EtOH/NH$_4$F, SnOF$_2$ peaks are observed, which indicates the successful incorporation of tin ions into the film structure. The absence of NH$_4$F peaks in the X-ray diffraction pattern of films obtained from a film-forming system containing SnO$_2$ in the form of a dispersed phase with the addition of ammonium fluoride confirms the assumption that ammonium fluoride crystallites decompose.

The average crystallite size (D) was determined from the broadening of X-ray lines using the Scherrer formula [57]:

$$D = (k\lambda)/(\beta\cos\theta) \tag{18}$$

where k is the Scherrer constant, usually taken as 0.9, but its value strongly depends on the shape of the crystallites [58], λ is the X-ray wavelength, θ is the Bragg diffraction angle, and β is the diffraction line broadening measured at half its maximum intensity (rad). Table 3 shows the calculation results.

As can be seen from Table 3, the size of crystallites in films ranges from 3.5 nm to 6.3 nm. Thus, the formation of films can be represented as a diagram shown in Figure 8. Lyophilic film-forming systems (Figure 8a–c), after application to substrates, form a layer of tin hydroxide, in the form of α-form.

Table 3. The sizes of the crystallitis SnO$_2$.

The Composition of the Film-Forming System	SnO$_2$ Crystallitis Sizes on Planes		
	(110)	(101)	(211)
SnCl$_4$/EtOH	5.1 ± 0.3 nm	6.3 ± 0.3 nm	5.1 ± 0.3 nm
SnCl$_4$/EtOH/NH$_4$F	4.0 ± 0.2 nm	6.1 ± 0.3 nm	5.2 ± 0.3 nm
SnCl$_4$/EtOH/NH$_4$OH	3.6 ± 0.2 nm	3.7 ± 0.2 nm	3.5 ± 0.2 nm
SnO$_2$/EtOH	4.0 ± 0.2 nm	4.6 ± 0.2 nm	3.8 ± 0.2 nm
SnO$_2$/EtOH/NH$_4$F	4.1 ± 0.2 nm	4.5 ± 0.2 nm	3.8 ± 0.2 nm

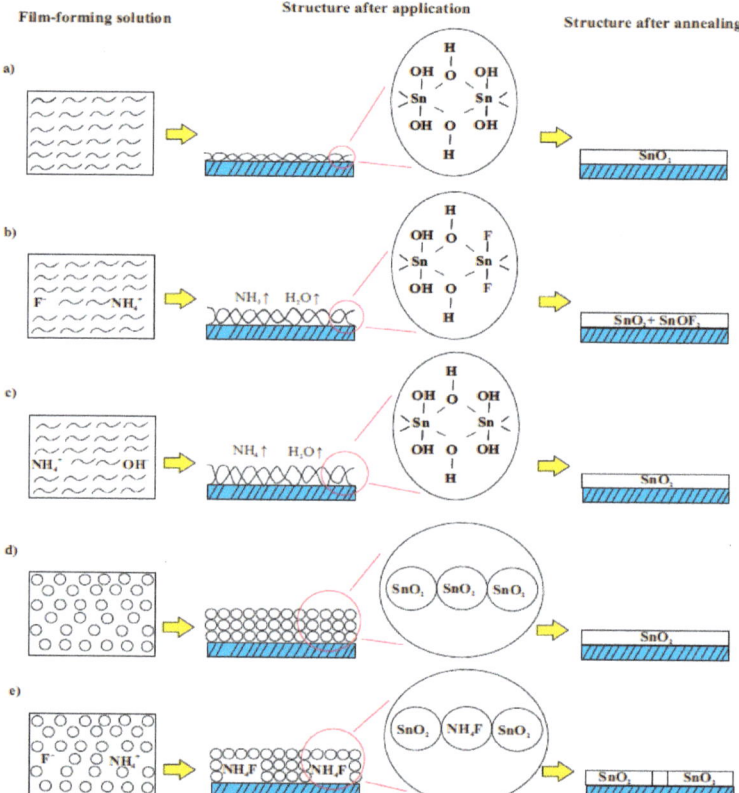

Figure 8. Layer formation scheme: (a) SnCl$_4$/EtOH; (b) SnCl$_4$/EtOH/NH$_4$F; (c) SnCl$_4$/EtOH/NH$_4$OH; (d) SnO$_2$/EtOH; (e) SnO$_2$/EtOH/NH$_4$F.

When NH$_4$F is added into the lyophilic system, SnOF$_2$ is formed. Film annealing leads to the formation of nanocrystalline SnO$_2$. Lyophobic film-forming systems (Figure 8d,e) form films from already formed SnO$_2$ particles. The addition of NH$_4$F to the lyophobic system does not lead to the formation of SnOF$_2$. Crystallization of NH$_4$F occurs, which decomposes into volatile compounds, leaving voids in the film.

3.4. Sensitivity to Ethanol

In terms of electronic structure, SnO$_2$ is a Lewis acid because it can accept a lone pair of electrons in its unfilled $5s^25p^2$ orbital, while ethanol vapors (C$_2$H$_5$OH) can be Lewis bases because they include an OH$^-$ group. Thus, the acid–base interaction can be responsible for the strong adsorption of ethanol vapor on SnO$_2$. Upon adsorption, alcohol molecules tend to dissociate into an H atom to form a surface alkoxide, a surface hydroxyl, and the former tends to further convert to an aldehyde or ketone [59]. It is generally accepted that adsorbed oxygen (presumably O^-_{ads}) or even surface lattice oxygen (O^{2-}_{late}) takes part in the oxidation of organic molecules. Electrons are returned to SnO_2, and resistance decreases when oxidation occurs [60]. The interaction between the analyzed gas and the oxygen present on the surface of the tin oxide film largely determines its response to gas. Sensitivity properties are consequently significantly influenced by morphology, surface area, and any defective states. Additionally, the operational temperature has a strong impact on the characteristics of metal oxide-based gas sensors [61].

Figure 9a shows the temperature dependence of the sensitivity of thin SnO_2 films to ethanol vapor (1 mg/L). Among all studied films, the highest sensitivity is obtained at a substrate temperature of 230 °C.

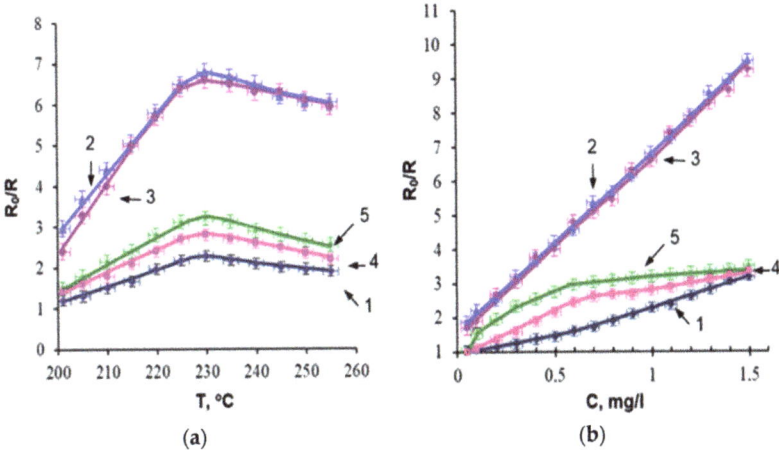

Figure 9. Sensitivity to ethanol vapors. Composition of film-forming systems: 1—$SnCl_4$/EtOH, 2—$SnCl_4$/EtOH/NH_4F, 3—$SnCl_4$/EtOH/NH_4OH, 4—SnO_2/EtOH, 5—SnO_2/EtOH/NH_4F. (**a**) From the temperature of the films with a concentration of 1 mg/L; (**b**) From the concentration of ethanol at 230 °C.

Since, at temperatures below 100 °C, the main adsorbed forms of oxygen are O^{2-}, and at temperatures < 300 °C—O^-, then, in our case, the interaction of ethanol vapor with the film, leading to a decrease in resistance, occurs according to the reaction [62]:

$$C_2H_5OH_{(gas)} + 6O_{(ads)} \rightarrow 3H_2O + 2CO_2 + 6e^- \quad (T < 300 \ °C). \tag{19}$$

The experiment investigated film sensitivity to varying concentrations of ethanol vapours at 230 °C and is illustrated in Figure 9b. The dendritic structured films demonstrated higher ethanol sensitivity than films produced from the $SnCl_4$/EtOH film-forming system. The increase in sensitivity correlates with the increase in surface area available for adsorption-desorption reactions. The sensitivity of films obtained from film-forming systems containing SnO_2 in the form of a dispersed phase changes little at an ethanol vapor concentration of more than 0.6 mg/L. This is attributed to the low mass of the film and, as a result, a restricted amount of reactive adsorbed oxygen ions. There is a "saturation" of active centers.

The sensitivities of films obtained from the $SnCl_4$/EtOH/NH_4F and $SnCl_4$/EtOH/NH_4OH film-forming systems differ within the measurement accuracy and are linear in the selected concentration range. These films exhibit sensitivity to low-concentration ethanol vapor (0.05 mg/L (26 ppm)).

The response time is an important characteristics of gas sensors, and Figure 10 illustrates the response time values of the studied films when exposed to ethanol vapor at a concentration of 1 mg/L.

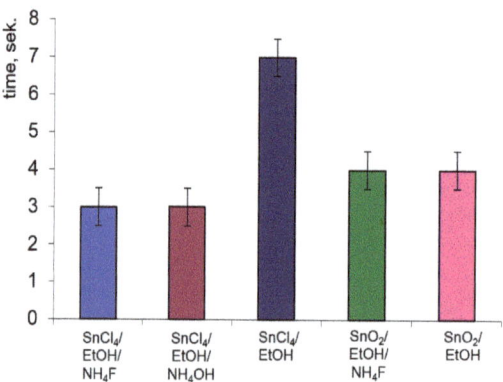

Figure 10. SnO_2 films response to the presence of ethanol vapor (1 mg/L).

Figure 10 shows that the films synthesized from the $SnCl_4$/EtOH/NH_4F and $SnCl_4$/EtOH/NH_4OH film-forming systems have the shortest response time. Reducing the response time by a factor of 2 when NH_4F and NH_4OH are added to the film-forming system is associated with the formation of a developed specific surface area of the film.

Consequently, the films produced from the $SnCl_4$/EtOH/NH_4F and $SnCl_4$/EtOH/NH_4OH film-forming systems possess high sensitivity to ethanol vapour and a short response time.

4. Conclusions

The film growth from the lyophilic system ($SnCl_4$/EtOH) proceeded according to the layer-by-layer Frank–van der Merwe growth mechanism. The growth of films from the lyophobic system (SnO_2/EtOH) was carried out according to the Volmer–Weber island growth mechanism. And the growth of films with additives was carried out according to the Stranski–Krystanov layer-by-island growth mechanism.

Films obtained from the lyophilic system have smoother surfaces than those obtained from the lyophobic system. Foams from the lyophobic system have a dendritic structure.

Adding NH_4F to the lyophilic system increases film conductivity, while it has no effect on specific conductivity in a lyophobic system.

The addition of ammonium fluoride to the lyophilic system leads to the incorporation of fluorine ions into the film structure, resulting in the formation of $SnOF_2$ crystallites. Meanwhile, in the lyophobic mode, separate phases of SnO_2 and NH_4F are formed. Upon annealing, NH_4F decomposes into volatile compounds, leaving behind voids.

The films obtained from the $SnCl_4$/EtOH/NH_4F and $SnCl_4$/EtOH/NH_4OH film-forming systems show equivalent sensitivity within the limits of measurement accuracy and linearity within the selected concentration range. A sensitivity to low concentrations of ethanol vapour (0.05 mg/L (26 ppm)) is also evident.

Author Contributions: Methodology, E.D. and I.L.; formal analysis, E.D., A.F., E.B. and S.I.; investigation, E.D., I.L., A.F., E.B., A.T., D.M., S.I., B.N., K.E. and B.B.; writing—original draft preparation, E.D., I.L., A.F. and E.B.; writing—review and editing, E.D. and I.L.; project administration, E.D. and I.L.; funding acquisition, E.D. All authors have read and agreed to the published version of the manuscript.

Funding: This research is funded by the Science Committee of the Ministry of Education and Science of the Republic of Kazakhstan, grant numbers: BR18574141 and AP09058002.

Institutional Review Board Statement: Not applicable.

Informed Consent Statement: Not applicable.

Data Availability Statement: Data are contained within the article.

Conflicts of Interest: The authors declare no conflict of interest.

References

1. Ramanauskas, R.; Iljinas, A.; Marcinauskas, L.; Milieska, M.; Kavaliauskas, Z.; Gecevicius, G.; Capas, V. Deposition and application of Indium-Tin-Oxide films for defrosting windscreens. *Coatings* **2022**, *12*, 670. [CrossRef]
2. Chen, Z.N.; Chen, S.M. Efficient and stable quantum-dot light-emitting diodes enabled by Tin Oxide multifunctional electron transport layer. *Adv. Opt. Mater.* **2022**, *10*, 2102404. [CrossRef]
3. Donercark, E.; Guler, S.; Ciftpinar, E.H.; Kabacelik, I.; Koc, M.; Ercelebi, A.C.; Turan, R. Impact of oxygen partial pressure during Indium Tin Oxide sputtering on the performance of silicon heterojunction solar cells. *Mater. Sci. Eng. B Adv. Funct. Solid State Mater.* **2022**, *281*, 115750. [CrossRef]
4. Kuznetsova, S.; Knalipova, O.; Chen, Y.W.; Kozik, V. The joint effect of doping with tin(IV) and heat treatment on the transparency and conductivity of films based on titanium dioxide as photoelectrodes of sensitized solar cells. *Nanosyst. Phys. Chem. Math.* **2022**, *13*, 193–205. [CrossRef]
5. Villarreal, C.C.; Sandoval, J.I.; Ramnani, P.; Terse-Thakoor, T.; Vi, D.; Mulchandani, A. Graphene compared to fluorine-doped tin oxide as transparent conductor in ZnO dye-sensitized solar cells. *J. Environ. Chem. Eng.* **2022**, *10*, 107551. [CrossRef]
6. Pozov, S.M.; Andritsos, K.; Theodorakos, I.; Georgiou, E.; Ioakeimidis, A.; Kabla, A.; Melamed, S.; de la Vega, F.; Zergioti, I.; Choulis, S.A. Indium Tin Oxide-Free inverted organic photo-voltaics using laser-induced forward transfer silver nanoparticle embedded metal grids. *ACS Appl. Electron. Mater.* **2022**, *4*, 2689–2698. [CrossRef]
7. Ambardekar, V.; Bhowmick, T.; Bandyopadhyay, P.P. Understanding on the hydrogen detection of plasma sprayed tin oxide/tungsten oxide (SnO_2/WO_3) sensor. *Int. J. Hydrogen Energy* **2022**, *47*, 15120–15131. [CrossRef]
8. Grushevskaya, E.A.; Ibraimova, S.A.; Dmitriyeva, E.A.; Lebedev, I.A.; Mukhamedshina, D.M.; Fedosimova, A.I.; Serikkanov, A.S.; Temiraliev, A.T. Sensitivity to ethanol vapour of thin films SnO_2 doped with fluorine. *Eurasian Chem. Technol. J.* **2019**, *21*, 13–17. [CrossRef]
9. Mukhamedshina, D.M.; Fedosimova, A.I.; Dmitriyeva, E.A.; Lebedev, I.A.; Grushevskaya, E.A.; Ibraimova, S.A.; Serikkanov, A.S. Influence of plasma treatment on physical properties of thin SnO_2 films obtained from $SnCl_4$ solutions with additions of NH_4F and NH_4OH. *Eurasian Chem. Technol. J.* **2019**, *21*, 57–61. [CrossRef]
10. Lee, Y.G.; Choi, W.S. Effect of PGMEA addition on Zinc-Tin-Oxide thin-film transistor fabricated by inkjet-printing process. *Adv. Eng. Mater.* **2022**, *24*, 2200128. [CrossRef]
11. Jin, R.S.; Shi, K.L.; Qiu, B.B.; Huang, S.H. Photoinduced-reset and multilevel storage transistor memories based on antimony-doped tin oxide nanoparticles floating gate. *Nanotechnology* **2022**, *33*, 025201. [CrossRef] [PubMed]
12. Tang, Y.L.; Huang, C.H.; Nomura, K. Vacuum-free liquid-metal-printed 2D Indium-Tin Oxide thin-film transistor for oxide inverters. *ACS Nano* **2022**, *16*, 3280–3289. [CrossRef]
13. Huang, S.Y.; Liu, Y.N.; Yang, F.; Wang, Y.; Yu, T.; Ma, D.L. Metal nanowires for transparent conductive electrodes in flexible chromatic devices: A review. *Environ. Chem. Lett.* **2022**, *20*, 3005–3037. [CrossRef]
14. Gilshtein, E.; Bolat, S.; Sevilla, G.T.; Cabas-Vidani, A.; Clemens, F.; Graule, T.; Tiwari, A.N.; Romanyuk, Y.E. Inkjet-printed conductive ITO patterns for transparent security systems. *Adv. Mater. Technol.* **2020**, *5*, 2000369. [CrossRef]
15. Kim, J.; Murdoch, B.J.; Partridge, J.G.; Xing, K.; Qi, D.C.; Lipton-Duffin, J.; McConville, C.F.; van Embden, J.; Gaspera, E.D. Ultrasonic spray pyrolysis of antimony-doped tin oxide transparent conductive coatings. *Adv. Mater. Interfaces* **2020**, *7*, 2000655. [CrossRef]
16. Shilova, O.A.; Gubanova, N.N.; Matveev, V.A.; Ivanova, A.G.; Arsentiev, M.Y.; Pugachev, K.E.; Ivankova, E.M.; Kruchinina, I.Y. Processes of film-formation and crystallization in catalytically active 'spin-on glass' silica films containing Pt and Pd nanoparticles. *J. Mol. Liq.* **2019**, *288*, 110996. [CrossRef]
17. Gubanova, N.N.; Matveev, V.A.; Shilova, O.A. Bimetallic Pt/Pd nanoparticles in sol–gel-derived silica films and xerogels. *J. Sol Gel Sci. Technol.* **2019**, *92*, 367–375. [CrossRef]
18. Muhammad, A.; Hassan, Z.; Mohammad, S.M.; Rajamanickam, S.; Abed, S.M.; Ashiq, M.G.B. Realization of UV-C absorption in ZnO nanostructures using fluorine and silver co-doping. *Colloid Interface Sci. Commun.* **2022**, *47*, 100588. [CrossRef]
19. Wu, K.D.; Feng, Y.F.; Xie, Y.D.; Zhang, J.M.; Xiong, D.P.; Chen, L.; Feng, Z.Y.; Wen, K.H.; He, M. Ti_3C_2/fluorine-doped carbon as anode material for high performance potassium-ion batteries. *J. Alloys Compd.* **2023**, *938*, 168430. [CrossRef]
20. Guo, Y.; Tang, Z.H.; Liu, Y.; Lu, Y.; Wu, W.W.; Yin, Z.D.; Wu, X.H. Boosting the multi-electron reaction capability of $Na_4MnCr(PO_4)_3$ by a fluorine doping strategy for sodium-ion batteries. *J. Alloys Compd.* **2023**, *937*, 168429. [CrossRef]
21. Chen, Y.L.; Huang, Y.P.; Tian, H.L.; Ye, L.Q.; Li, R.P.; Chen, C.C.; Dai, Z.X.; Huang, D. Fluorine-doped $BiVO_4$ photocatalyst: Preferential cleavage of C−N bond for green degradation of glyphosate. *J. Environ. Sci.* **2023**, *127*, 60–68. [CrossRef]
22. Nusupov, K.K.; Beisenkhanov, N.B.; Valitova, I.V.; Dmitrieva, E.A.; Zhumagaliuly, D.; Shilenko, E.A. Structural studies of thin silicon layers repeatedly implanted by carbon ions. *Phys. Solid State* **2006**, *48*, 1255–1267. [CrossRef]
23. Lu, K.; Wang, Z.Y.; Wu, Y.X.; Zhai, X.W.; Wang, C.X.; Li, J.; Wang, Z.M.; Li, X.Y.; He, Y.X.; An, T. Synergistic effect of F doping and WO_3 loading on electrocatalytic oxygen evolution. *Chem. Eng. J.* **2023**, *451*, 138590. [CrossRef]

24. Ye, F.; Zeng, J.J.; Cai, X.M.; Su, X.Q.; Wang, B.; Wang, H.; Roy, V.A.L.; Tian, X.Q.; Li, J.W.; Zhang, D.P. Doping cuprous oxide with fluorine and its band gap narrowing. *J. Alloys Compd.* **2017**, *721*, 64–69. [CrossRef]
25. Ding, Y.F.; Yin, S.F.; Cai, M.Q. Enhanced photocatalytic toluene oxidation performance induced by two types of cooperative fluorine doping in polymeric carbon nitride with the first-principles calculations. *J. Colloid Interface Sci.* **2023**, *630*, 452–459. [CrossRef] [PubMed]
26. Almaev, A.V.; Kopyev, V.V.; Novikov, V.A.; Chikiryaka, A.V.; Yakovlev, N.N.; Usseinov, A.B.; Karipbayev, Z.T.; Akilbekov, A.T.; Koishybayeva, Z.K.; Popov, A.I. ITO Thin Films for Low-Resistance Gas Sensors. *Materials* **2023**, *16*, 342. [CrossRef] [PubMed]
27. Liu, X.S.; Wang, G.; Zhi, H.; Dong, J.; Hao, J.; Zhang, X.; Wang, J.; Li, D.T.; Liu, B.S. Synthesis of the Porous ZnO Nanosheets and $TiO_2/ZnO/FTO$ Composite films by a low-temperature hydrothermal method and their applications in photocatalysis and electrochromism. *Coatings* **2022**, *12*, 695. [CrossRef]
28. Niu, F.J.; Wang, D.G.; Williams, L.J.; Nayak, A.; Li, F.; Chen, X.Y.; Troian-Gautier, L.; Huang, Q.; Liu, Y.M.; Brennaman, M.K.; et al. A Semiconductor-mediator-catalyst artificial photosynthetic system for photoelectrochemical water oxidation. *Chem. A Eur. J.* **2022**, *28*, e202102630. [CrossRef]
29. Yang, S.H.; Ke, X.; Zhang, M.L.; Luo, D.X. Decoration of PdAg dual-metallic alloy nanoparticles on Z-Scheme α-Fe_2O_3/CdS for manipulable products via photocatalytic reduction of carbon dioxide. *Front. Chem.* **2022**, *10*, 937543. [CrossRef]
30. Han, D.; Wu, C.; Zhao, Y.; Xiao, L.; Zhao, Z. Ion implantation-modified fluorine-doped tin oxide by zirconium with continuously tunable work function and its application in perovskite solar cells. *ACS Appl. Mater. Interfaces* **2017**, *9*, 42029–42034. [CrossRef]
31. Latif, H.; Liu, J.; Mo, D.; Wang, R.; Zeng, J.; Zhai, P.F.; Sattar, A. Effect of target morphology on morphological, optical and electrical properties of FTO thin film deposited by pulsed laser deposition for mapbbr3 perovskite solar cell. *Surf. Interfaces* **2021**, *24*, 101117. [CrossRef]
32. Afzaal, M.; Yates, H.M.; Walter, A.; Nicolay, S. Improved FTO/NiO_x interfaces for inverted planar triple-cation perovskite solar cells. *IEEE J. Photovolt.* **2019**, *9*, 1302–1308. [CrossRef]
33. Yang, X.; Liu, R.; Lei, Y.; Li, P.; Wang, K.; Zheng, Z.; Wang, D. Dual Influence of reduction annealing on diffused Hematite/FTO junction for enhanced photoelectrochemical water oxidation. *ACS Appl. Mater. Interfaces* **2016**, *8*, 16476–16485. [CrossRef]
34. Mu, S.; Shi, Q. Photoelectrochemical properties of bare fluorine doped tin oxide and its electrocatalysis and photoelectocatalysis toward cysteine oxidation. *Electrochim. Acta* **2016**, *195*, 59–67. [CrossRef]
35. Cirocka, A.; Zarzeczańska, D.; Wcisło, A. Good choice of electrode material as the key to creating electrochemical sensors—Characteristics of carbon materials and transparent conductive oxides (TCO). *Materials* **2021**, *14*, 4743. [CrossRef]
36. Burnat, D.; Sezemsky, P.; Lechowicz, K.; Koba, M.; Janczuk-Richter, M.; Janik, M.; Stranak, V.; Niedziółka-Jönsson, J.; Bogdanowicz, R.; Śmietana, M. Functional fluorine-doped tin oxide coating for opto-electrochemical label-free biosensors. *Sens. Actuators B Chem.* **2022**, *367*, 132145. [CrossRef]
37. Nascimento, R.A.S.; Mulato, M. Mechanisms of ion detection for fet-sensors using FTO: Role of cleaning process, ph sequence and electrical resistivity. *Mater. Res.* **2017**, *20*, 1369–1379. [CrossRef]
38. Huang, G.-K.; Gupta, S.; Lee, C.-Y.; Tai, N.-H. Acid-treated carbon nanotubes/polypyrrole/fluorine-doped tin oxide electrodes with high sensitivity for saliva glucose sensing. *Diam. Relat. Mater.* **2022**, *129*, 109385. [CrossRef]
39. Liu, J.W.; Minh, D.Q.; Amberg, G. Thermohydrodynamics of boiling in binary compressible fluids. *Phys. Rev. E* **2015**, *92*, 043017. [CrossRef]
40. Shchekin, A.K.; Gosteva, L.A.; Lebedeva, T.S. Thermodynamic properties of stable and unstable vapor shells around lyophobic nanoparticles. *Phys. A Stat. Mech. Its Appl.* **2020**, *560*, 125105. [CrossRef]
41. Liang, Y.E.; Weng, Y.H.; Hsieh, I.F.; Tsao, H.K.; Sheng, Y.J. Attractive encounter of a nanodrop toward a nanoprotrusion. *J. Phys. Chem. C* **2017**, *121*, 7923–7930. [CrossRef]
42. Klym, H.; Karbovnyk, I.; Piskunov, S.; Popov, A.I. Positron Annihilation Lifetime Spectroscopy Insight on Free Volume Conversion of Nanostructured $MgAl_2O_4$ Ceramics. *Nanomaterials* **2021**, *11*, 3373. [CrossRef]
43. GOST 5962-13; Rectified Ethyl Alcohol from Food Raw Materials. Russian GOST: Moscow, Russia, 2014.
44. Sui, R.; Charpentier, P.A.; Marriott, R.A. Metal oxide-related dendritic structures: Self-assembly and applications for sensor, catalysis, energy conversion and beyond. *Nanomaterials* **2021**, *11*, 1686. [CrossRef] [PubMed]
45. Małgorzata, Z.; Tadeusz, K.; Beata, S.; Paweł, S. Investigation of the dendritic structure influence on the electrical and mechanical properties diversification of the continuously casted copper strand. *Materials* **2020**, *13*, 5513. [CrossRef]
46. Murzalinov, D.; Dmitriyeva, E.; Lebedev, I.; Bondar, E.A.; Fedosimova, A.I.; Kemelbekova, A. The effect of pH solution in the sol–gel process on the structure and properties of thin SnO_2 films. *Processes* **2022**, *10*, 1116. [CrossRef]
47. Temiraliev, A.; Tompakova, N.; Fedosimova, A.; Dmitriyeva, E.; Lebedev, I.; Grushevskaya, E.; Mukashev, B.; Serikkanov, A. Birth and fusion in a sol-gel process with low diffusion. *Eurasian Phys. Tech. J.* **2020**, *17*, 132–137. [CrossRef]
48. Shilova, O.A. Fractals, morphogenesis and triply periodic minimal surfaces in sol-gel-derived thin films. *J. Sol. Gel. Sci. Technol.* **2020**, *95*, 599–608. [CrossRef]
49. Chen, X.; Bai, R.; Huang, M. Optical properties of amorphous Ta_2O_5 thin films deposited by RF magnetron sputtering. *Opt. Mater.* **2019**, *97*, 109404. [CrossRef]
50. Zuo, Y.; Guo, L.; Liu, W.; Ding, J. Measurement of the scattering matrix and extinction coefficient of the chaff corridor. *IEEE Access* **2020**, *8*, 206755–206769. [CrossRef]

51. Dmitrieva, E.A.; Mukhamedshina, D.M.; Beisenkhanov, N.B.; Mit', K.A. The effect of NH_4F and NH_4OH on the structure and physical properties of thin SnO_2 films synthesized by the sol-gel method. *Glass Phys. Chem.* **2014**, *40*, 31–36. [CrossRef]
52. Mukhamedshina, D.M.; Mit', K.A.; Beisenkhanov, N.B.; Dmitriyeva, E.A.; Valitova, I.V. Influence of plasma treatments on the microstructure and electrophysical properties of SnO_x thin films synthesized by magnetron sputtering and sol–gel technique. *J. Mater. Sci. Mater. Electron.* **2008**, *19*, 382–387. [CrossRef]
53. Tarighi, A.; Mashreghi, A. Dependence of photovoltaic properties of spray-pyrolyzed F-doped SnO_2 thin film on spray solution preparation method. *J. Electron. Mater.* **2019**, *48*, 7827–7835. [CrossRef]
54. Miranda, H.; Velumani, S.; Samudio Pérez, C.A.; Krause, J.C.; D'Souza, F.; De Obaldía, E.; Ching-Prado, E. Effects of changes on temperature and fluorine concentration in the structural, optical and electrical properties of SnO_2:F thin films. *J. Mater. Sci. Mater. Electron.* **2019**, *30*, 15563–15581. [CrossRef]
55. Dmitriyeva, E.A.; Mukhamedshina, D.M.; Mit, K.A.; Lebedev, I.A.; Girina, I.I.; Fedosimova, A.I.; Grushevskya, E.A. Doping of fluorine of tin dioxide films synthesized by sol-gel method. *News Natl. Acad. Sci. Repub. Kazakhstan* **2019**, *433*, 73–79. [CrossRef]
56. Smorokov, A.A.; Kantaev, A.S.; Borisov, V.A. Research of titanomagnetite concentrate decomposition by means of ammonium fluoride and ammonium hydrogen fluoride. *AIP Conf. Proc.* **2019**, *2143*, 020022. [CrossRef]
57. Konan, F.K.; Hartiti, B.; Batan, A.; Aka, B. X-ray diffraction, XPS, and Raman spectroscopy of coated ZnO:Al (1–7 at%) nanoparticles. *E J. Surf. Sci. Nanotechnol.* **2019**, *17*, 163–168. [CrossRef]
58. Konan, F.K.; Ncho, J.S.; Nkuissi, H.J.T.; Hartiti, B.; Boko, A. Influence of the precursor concentration on the morphological and structural properties of zinc oxide (ZnO). *Mater. Chem. Phys.* **2019**, *229*, 330–333. [CrossRef]
59. Kim, K.W.; Cho, P.S.; Kim, S.J.; Lee, J.H.; Kang, C.Y.; Kim, J.S.; Yoon, S.J. The selective detection of C_2H_5OH using SnO_2–ZnO thin film gas sensors prepared by combinatorial solution deposition. *Sens. Actuators B* **2007**, *123*, 318–324. [CrossRef]
60. Gong, H.; Wang, Y.J.; Teo, S.C.; Huang, L. Interaction between thin-film tin oxide gas sensor and five organic vapors. *Sens. Actuators B* **1999**, *54*, 232–235. [CrossRef]
61. Li, X.; Gua, Z.; Cho, J.; Sun, H.; Kurup, P. Tin–copper mixed metal oxide nanowires: Synthesis and sensor response to chemical vapors. *Sens. Actuators B* **2011**, *158*, 199–207. [CrossRef]
62. Motsoeneng, R.G.; Kortidis, I.; Ray, S.S.; Motaung, D.E. Designing SnO_2 nanostructure-based sensors with tailored selectivity toward propanol and ethanol vapors. *ACS Omega* **2019**, *4*, 13696–13709. [CrossRef] [PubMed]

Disclaimer/Publisher's Note: The statements, opinions and data contained in all publications are solely those of the individual author(s) and contributor(s) and not of MDPI and/or the editor(s). MDPI and/or the editor(s) disclaim responsibility for any injury to people or property resulting from any ideas, methods, instructions or products referred to in the content.

Communication

The Facile Construction of Anatase Titanium Dioxide Single Crystal Sheet-Connected Film with Observable Strong White Photoluminescence

Tao He [1,2,3], Dexin Wang [1,2,3], Yu Xu [1,2,3],* and Jing Zhang [1,2,3],*

1. New Energy Material and Device, College of Science, Donghua University, Shanghai 201620, China
2. Textile Key Laboratory for Advanced Plasma Technology and Application, China National Textile & Apparel Council, Shanghai 201620, China
3. Magnetic Confinement Fusion Research Center, Ministry of Education of the People's Republic of China, Shanghai 201620, China
* Correspondence: yuxu@dhu.edu.cn (Y.X.); jingzh@dhu.edu.cn (J.Z.)

Abstract: Deposited by a reactive atmospheric pressure non-thermal $TiCl_4/O_2/Ar$ plasma, anatase TiO_2 single crystal sheet-connected film exhibits two large exposed {001} facets and a high concentration of oxygen defects. Strong white photoluminescence centered at 542 nm has been observed with naked eyes, whose internal quantum efficiency is 0.62, and whose intensity is comparable to that of commercial fluorescent lamp interior coatings. Based on the simulation results of a hybrid global–analytical model developed on this atmospheric pressure non-equilibrium plasma system, the mechanism of a self-confined growth of single crystal sheets was proposed. A high concentration of oxygen defects is in situ incorporated into the anatase crystal lattice without damaging its crystallographic orientation. This method opens a new way to construct 3D porous metal-oxide single crystal sheet-connected films with two exposing high energy surfaces and a large concentration of oxygen defects.

Keywords: atmospheric pressure plasma; TiO_2; photoluminescence; crystal growth; thin film

Citation: He, T.; Wang, D.; Xu, Y.; Zhang, J. The Facile Construction of Anatase Titanium Dioxide Single Crystal Sheet-Connected Film with Observable Strong White Photoluminescence. *Coatings* 2024, 14, 292. https://doi.org/10.3390/coatings14030292

Academic Editor: Maria Vittoria Diamanti

Received: 22 January 2024
Revised: 21 February 2024
Accepted: 26 February 2024
Published: 28 February 2024

Copyright: © 2024 by the authors. Licensee MDPI, Basel, Switzerland. This article is an open access article distributed under the terms and conditions of the Creative Commons Attribution (CC BY) license (https://creativecommons.org/licenses/by/4.0/).

1. Introduction

Appropriately designed anatase TiO_2 single crystal sheet-connected films with two exposed {001} reactive facets and highly reduced characteristics are highly desirable for scientific research and practical applications [1–3]. Desirable characteristics include directly connected boundaries, high carrier mobility, low incident light reflection loss, large specific surface area, and large reactive surfaces. Constructing such architecture films is expected to have interesting photo, electromagnetic, cytokine, chemical, and catalyst effects and applications to photocatalysts, photonics crystals, spintronic devices, anticancer or gene therapy modalities, photo/electrochromic devices, gas sensors, chemical degradation, water splitting, solar cells, etc. [1,2,4–7]. Recently, anatase micro/nano powders composed of single crystal sheets or flowerlike clusters with a large percentage of exposed {001} faces have been synthesized through a high concentration of hydrofluoric acid (HF) as a crystal shape controller [3]. Anatase TiO_2 architecture films were constructed from its powders through post-binding and annealing processes. This could have problems of nanoparticle coagulation, weak boundary connection, or crystal phase transformation. Post-vacuum annealing, ion implantation, plasma treatment, or electron bombardment are often applied to reduce TiO_2 surface via the introduction of oxygen defects, and to improve its surface reactivity [8–10]. Plasma sputtering and plasma-enhanced chemical vapor deposition (PECVD) have been applied to fabricate anatase crystalline films, but with no obvious exposed {001} facets [11]. The growth of anatase single crystal sheet-connected architecture film with two exposed {001} facets and a high surface reactivity is an area that has had little success until now, to the best of our knowledge. Two challenges have to be taken to obtain such structured anatase films in one deposition step. One is that the highly reactive surface diminishes rapidly during crystal growth; this is due to the minimization of surface energy [12]. The other is that the regular crystal lattice growth is usually interrupted by incorporating oxygen defects into the crystal facets and interconnecting omnidirectional crystal sheets into the film architecture. This suggests that a well-designed crystal growth process has to balance the opposite aspects of high surface reactivity, crystal lattice regularity, defect introduction, and crystal sheet interconnection.

Hydrothermal [13] and sol-gel [14] methods are widely employed in the preparation of {001} facets TiO_2 films. The processes are usually complicated and can last for several hours. Fluorine ions (F^-) are always used as crystal shape controllers in the process [15]. Atmospheric pressure plasma deposition has been well studied in recent decades due to its low-cost equipment, simple steps, and its capability for continuous process [16,17]. Meanwhile, the plasma provides a unique (non-thermal equilibrium) deposition environment when compared to traditional chemical process.

In this paper, we provide a one-step, crystal-shape, controller-free method to deposit single crystal sheet-connected {001} facets TiO_2 film. The TiO_2 film exhibits unusually strong white photoluminescence due to the oxygen defects generated during the plasma deposition.

2. Experimental

The apparatus was a homemade atmospheric pressure plasma reactor, composed of two coaxial quartz glass tubes or parallel quartz plates with a gap of 1.5 mm (Figure 1a). A 13.56 MHz radio-frequency power supply (RF-10S/PWT, Advanced Energy Co., Ltd., Denver, CO, USA) was applied to ignite the plasma in the $5.4 \times 2.0 \times 0.15$ cm^3 space between the two electrodes. The power was set at 100 W. Argon (Ar, 99.99%, Shanghai Central Gases Co., Ltd., Shanghai, China) was used as the discharge gas, oxygen (O_2, 99.99%, Shanghai Central Gases Co., Ltd., China) was used as the oxidant, and titanium tetrachloride ($TiCl_4$, 98.0%, Sinopharm Chemical Reagent Co., Ltd., Shanghai, China) was selected as the titanium precursor. A bubbler containing $TiCl_4$ was immersed in a water bath which was heated and maintained at 40 °C. The connecting tube from the bubbler to the discharge chamber was kept at 70 °C using a heating band in ordwer to prevent the $TiCl_4$ vapor from condensing. $TiCl_4/O_2/Ar$ (0.17/30/203 sccm) was fed into the reactor through mass flow controllers. The deposition time was 1 h.

The morphology and crystal structure of the deposited films were examined using a field emission scanning electron microscope (FESEM, Hitachi S-4800, Tokyo, Japan) and a transmission electron microscope (TEM, JEM-2100, JEOL Ltd., Tokyo, Japan). The chemical states of the samples were investigated using X-ray photoelectron spectroscopy (XPS, ThermoFisher Scientific ESCALAB 250Xi, Waltham, MA, USA), equipped with Al-Kα X-ray radiation (1486 eV). For XPS analysis, Avantage software (version 5.934, from Thermo Fisher) was employed to deconvolve and analyze the detailed binding energy and the ratios of species. The photoluminescence (PL) of all the samples was measured using a 325 nm He-Cd laser (Renishaw) with a pump power of 30 mW.

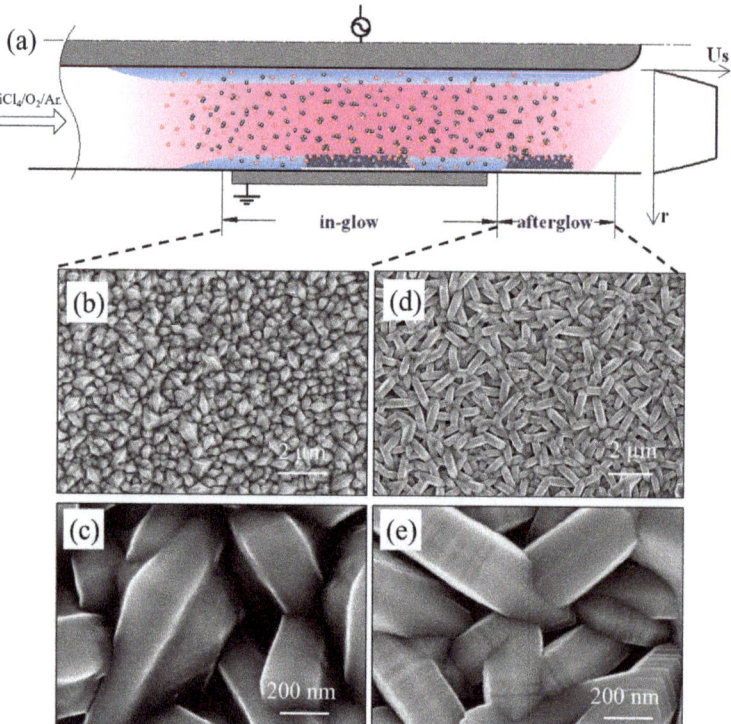

Figure 1. (**a**) Scheme of the plasma deposition equipment. The SEM images of the films, (**b**) in-glow, (**d**) afterglow. (**c**,**e**) are the relative magnified images corresponding to (**b**,**d**).

3. Results and Discussion

The FE-SEM images of the deposited films in the in-glow zone and in the after-glow zone are shown in Figure 1b,d, respectively. They show that both films are composed of single crystal sheets which connect with each other. According to the symmetries of anatase TiO_2, the two square surfaces and the eight isosceles trapezoidal surfaces in Figure 1c are, respectively, recognized to be {001} and {101} facets. Importantly, almost all the connected crystal sheets stand vertically on the substrate, especially in the after-glow zone, which are exposed high {001} facets, as shown in Figure 1e. Cl^- plays an important role in promoting the {001} facets of TiO_2 [18]. The plasma sheath will repel negatively charged ions or particles, causing them to be unable to reach the surface of the substrate and participate in chemical reactions [19]. In the "in-glow" region, the sheath is strong, but in the after-glow region, the sheath is weak. A weak plasma sheath leads to more Cl ions reaching the surface, which promotes the growth of the film 001 facets.

Figure 2a shows the XRD spectra of the films deposited in the in-glow and after-glow conditions. Both films exhibit the typical pattern for anatase TiO_2. Peaks are identified at 2θ values of 25.4°, 38.1°, 48.2°, 53.9°, 55.2°, 62.9°, 68.9°, 70.5°, and 75.5°, corresponding to the crystal planes of (101), (004), (200), (105) (211), (204), (116), (200), and (215), respectively, agreeing with the standard XRD pattern of anatase TiO_2 (JCPDS files No. 21-1272). Figure 2b shows the Raman spectra of the TiO_2 films. These are in agreement with the XRD results. The samples show the five Raman-active modes of E_g, E_g, A_{1g}, A_{1g}, and E_g, corresponding to 144, 197, 399, 515, and 639 cm^{-1} [20].

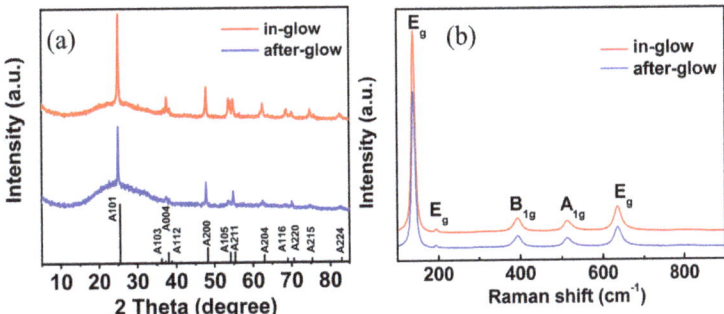

Figure 2. (a) XRD, and (b) Raman spectra of TiO$_2$ thin films deposited in in-glow and after-glow.

Figure 3a–d show the single crystal growth process in the 120 min deposition in the after-glow condition. After 1 min (Figure 3a), the TiO$_2$ seeds are observed on the substrate, and, after 10 min (Figure 3b), the film shows then step-growth of the crystals. The crystals keep growing and then presented both {001} and {101} facets (Figure 3c) after 30 min. Figure 3d shows the crystals after deposition for 2 h. Huge TiO$_2$ crystals with exposed high {101} facets are presented. High resolution transmission electron microscopy (HR-TEM) images and selected-area electron diffraction (SAED) patterns are given in Figure 3e,f, which clearly show the single crystal characteristics. As shown in Figure 3e and in its inset, the (200) and (020) atomic planes have a lattice spacing of 0.189 nm, and the SAED spot pattern is well-indexed into the zone, ascribing the two flat square surfaces to the {001} facets of the anatase single crystal [21]. The rotational Moiré fringes and the corresponding SAED spot pattern in Figure 3f and in its inset confirm the existence of the dislocation structure of the crystal sheets [22], which corresponds to the dislocation steps in Figure 1. The results indicate that grain sizes and shapes grow with the deposited time, showing the 001 facet TiO$_2$ crystalline growth processes. The density of these films, alongside the presence of visible voids in the composition, has not yet been studied.

According to the Wulff construction, the lower reactive {101} facet of anatase is more thermodynamically stable (surface energy = 0.44 J m^{-2}) than the higher reactive {001} facet (surface energy = 0.90 J m^{-2}), and dominates more than 90% of the single-crystalline particles of anatase under equilibrium conditions [15]. In this atmospheric pressure non-equilibrium reactive plasma system, high densities of reactive species, such as Ar$^+$, Ar$_2^+$, e, TiOCl2, O, Cl$^-$, Cl$_2$, ClO, etc., can be generated in the plasma gas phase [23]. It is demonstrated that a self-confined growth and the connections of the anatase TiO$_2$ single crystal sheets with the two exposed {001} reactive crystal facets, as well as Cl ions all play a role in the control of surface morphology in TiO$_2$ crystal growth [18]. Because the binding energy of Cl-Cl (D_0^{Cl-Cl} = 242.6 kJ mol^{-1}) is much lower than that of Cl-Ti (D_0^{Cl-Ti} = 494 kJ mol^{-1}), the bonding of Cl to the TiO$_2$ crystal surface will lower the surface energy of the (001) surface to almost the same extent as that of the (101) surface, and will also form anatase TiO$_2$ crystals with large {001} facets. Therefore, under the circumstances of a high density of Cl$^-$ containing precursor species, the newly formed nanosized anatase TiO$_2$ nuclei come with exposed {001} facets. Subsequently, these loosely and randomly packed anatase TiO$_2$ nuclei redevelop into branched or cross linked crystal sheets with large {001} facets. When the branched or cross-linked crystal sheets grow large enough via a self-assembly extension recrystallization process (a self-confined Ostwald process in the reactive plasma) and combine together, a porous 3D film architecture, composed of micro/nano anatase TiO$_2$ single crystal sheets with two exposed {001} facets, can be constructed directly on the substrate surface without any catalysts or templates after a certain time of deposition.

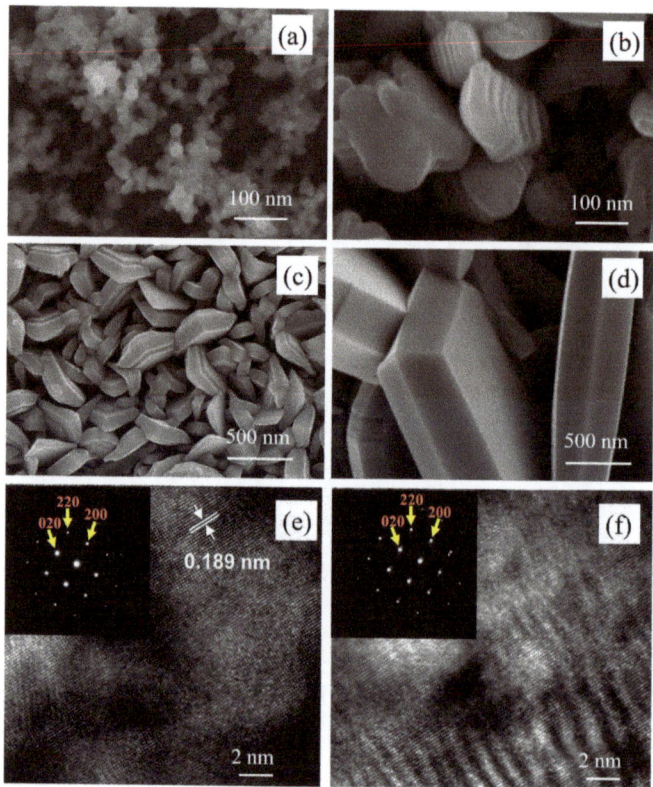

Figure 3. The SEM images of the TiO$_2$ film deposited in after-glow conditions for (**a**) 1 min, (**b**) 10 min, (**c**) 30 min, and (**d**) 120 min. HRTEM images and SAED (insets) of (**e**) the {001} facet, and (**f**) rotational Moiré fringes. The insets in (**e**,**f**) are the selected-area electron diffraction pattern, and arrows indicate the corresponding lattice parameters.

A high concentration of oxygen defects is in situ introduced into the anatase TiO$_2$ single crystal sheets. The O 1s band is obviously different from that of pure unreduced anatase TiO$_2$, and has a shoulder peak at about 531.5 eV, as shown in Figure 4a,b. The O 1s band can be resolved and assigned to three valence states at 530.1 ± 0.1 eV, 531.5 ± 0.1 eV, and 533.0 ± 0.1 eV. The peak with a binding energy at 531.6 ± 0.1 eV, which is 1.6 eV higher than the common Ti-O bond, is caused by the lower electron density of oxygen ions when compared to that of the O^{2-}. This can be attributed to the oxygen ion O$^-$ with specific coordination, particularly integrated into the bulk structure near the surface and associated with oxygen defects in the subsurface [24]. The resolved band at 531.5 ± 0.1 eV covers, respectively, 41.0% and 25.2% of the O 1s spectrum for the TiO$_2$ film deposited in the in-glow zone and in the after-glow zone. The peak of 533.1 ± 0.1 eV is related to the Ti-OH bond [25]. According to the references, O 532 eV is often related to defects or oxygen vacancy [26]. And the oxygen vacancy in TiO$_2$ has a great influence on photoluminescence, as they will induce a higher number of photoexcited carriers to recombine cause more photoexcited carriers to be recombined [27]. The thermal effect of plasma treatment can accelerate the formation of the above-mentioned defects in TiO$_2$. And, because of the destruction of O-Ti-O and Ti-O-Ti bonds via the plasma bombard, the residual electrons can migrate from the lattice Ti and O atom, and then create the defective lattice structure of TiO$_2$ [28]. The thin film in the "in-glow" period will be subjected to more plasma bombarding and heat than the "after-glow" film. The high-resolution XPS spectra of Ti 2p are shown in

Figure 4c,d. The peaks with a binding energy at 458 eV and 464 eV are Ti^{4+}. The thin film deposited in the in-glow period shows Ti^{3+} at the binding energy of 457.2 eV [29]. However, the thin film deposited in the after-glow period does not show any Ti^{3+} peak.

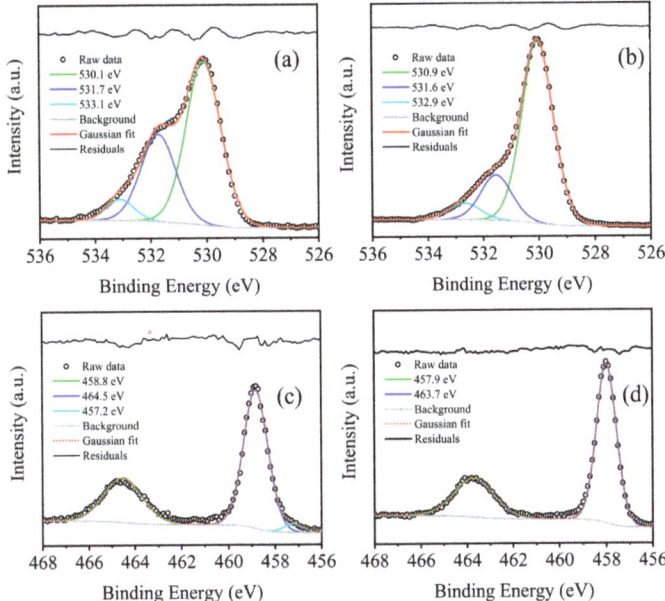

Figure 4. (**a**,**b**) represent the resolved peaks of O 1s spectrum, and (**c**,**d**) represent the Ti 2p spectrum of the films obtained in the in-glow zone and in the afterglow zone, respectively.

The large percentage of oxygen defects in the film is confirmed by the PL spectrum of the film. As seen from Figure 5, the deposited films exhibit light emissions from 390 nm to 700 nm, centered at 542 nm, which is strong enough to be viewed easily with the naked eye. The strong PL, centered at 542 nm, are attributed to oxygen vacancies (OVs) [30,31]. Furthermore, the PL intensity increases with the percentage of the oxygen defects, and is higher for the film deposited in the in-glow zone than in the after-glow zone. TiO_2 powder (Degussa, P25) and commercial fluorescent coating (get from Opple T8 fluorescent lamp) were also tested at the same condition for use as a contrast.

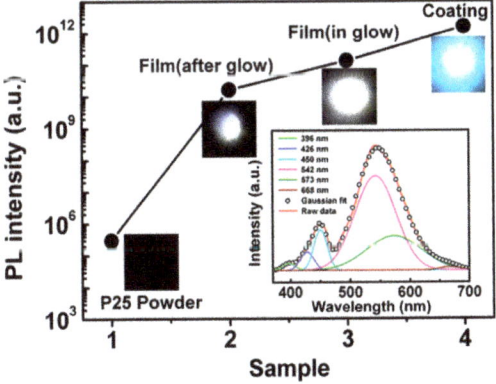

Figure 5. The PL intensity of (1) the P25 powder, (2) the after-glow film, (3) the in-glow film, and (4) the commercial fluorescent coating, as well as the PL spectrum (inset), of the in-glow film.

4. Conclusions

In summary, 3D porous anatase TiO$_2$ single crystal sheet-connected films with two large exposed {001} reactive facets and a high percentage of oxygen defects has been successfully deposited onto the substrate in one step, without any catalysts or templates. This provides a feasible way for the fast synthesis of 3D porous metal-oxide films that are composed of single crystal sheets connected with each other, and vertically aligned on the substrate. A self-confined crystal sheet growth mechanism has been proposed for the architecture formation, based on the hybrid analytical global model simulation and the experimental results. Unusually, strong white photoluminescence centered at 542 nm has been observed for the first time, and is mainly caused by the high percentage of oxygen defects introduced into the crystal in situ.

Author Contributions: Conceptualization, T.H.; methodology, T.H.; validation, T.H. and D.W.; investigation, Y.X.; writing—original draft preparation, T.H.; writing—review and editing, Y.X. and J.Z. All authors have read and agreed to the published version of the manuscript.

Funding: The National Natural Science Foundation of China NSFC (Nos. 12205040, 12075054).

Institutional Review Board Statement: Not applicable.

Informed Consent Statement: Not applicable.

Data Availability Statement: Data are contained within the article.

Conflicts of Interest: The authors declare no conflicts of interest.

References

1. O'Regan, B.; Grätzel, M. A low-cost, high-efficiency solar cell based on dye-sensitized colloidal TiO$_2$ films. *Nature* **1991**, *353*, 737–740. [CrossRef]
2. Diebold, U.; Ruzycki, N.; Herman, G.S.; Selloni, A. One step towards bridging the materials gap: Surface studies of TiO$_2$ anatase. *Catal. Today* **2003**, *85*, 93–100. [CrossRef]
3. Yang, H.G.; Sun, C.H.; Qiao, S.Z.; Zou, J.; Liu, G.; Smith, S.C.; Cheng, H.M.; Lu, G.Q. Anatase TiO$_2$ single crystals with a large percentage of reactive facets. *Nature* **2008**, *453*, 638–641. [CrossRef] [PubMed]
4. Fujishima, A.; Honda, K. Electrochemical Photolysis of Water at a Semiconductor Electrode. *Nature* **1972**, *238*, 37–38. [CrossRef] [PubMed]
5. Zhang, W.; Xue, J.; Shen, Q.; Jia, S.; Gao, J.; Liu, X.; Jia, H. Black single-crystal TiO$_2$ nanosheet array films with oxygen vacancy on {001} facets for boosting photocatalytic CO$_2$ reduction. *J. Alloys Compd.* **2021**, *870*, 159400. [CrossRef]
6. Tian, X.; Cui, X.; Lai, T.; Ren, J.; Yang, Z.; Xiao, M.; Wang, B.; Xiao, X.; Wang, Y. Gas sensors based on TiO$_2$ nanostructured materials for the detection of hazardous gases: A review. *Nano Mater. Sci.* **2021**, *3*, 390–403. [CrossRef]
7. Zhang, J.; Lei, Y.; Cao, S.; Hu, W.; Piao, L.; Chen, X. Photocatalytic hydrogen production from seawater under full solar spectrum without sacrificial reagents using TiO$_2$ nanoparticles. *Nano Res.* **2022**, *15*, 2013–2022. [CrossRef]
8. Aschauer, U.; He, Y.; Cheng, H.; Li, S.-C.; Diebold, U.; Selloni, A. Influence of Subsurface Defects on the Surface Reactivity of TiO$_2$: Water on Anatase (101). *J. Phys. Chem. C* **2010**, *114*, 1278–1284. [CrossRef]
9. Grätzel, M. Photoelectrochemical cells. *Nature* **2001**, *414*, 338–344. [CrossRef]
10. Cai, R.; Kubota, Y.; Shuin, T.; Sakai, H.; Hashimoto, K.; Fujishima, A. Induction of Cytotoxicity by Photoexcited TiO$_2$ Particles1. *Cancer Res.* **1992**, *52*, 2346–2348.
11. Lou, B.-S.; Chen, W.-T.; Diyatmika, W.; Lu, J.-H.; Chang, C.-T.; Chen, P.-W.; Lee, J.-W. High power impulse magnetron sputtering (HiPIMS) for the fabrication of antimicrobial and transparent TiO$_2$ thin films. *Curr. Opin. Chem. Eng.* **2022**, *36*, 100782. [CrossRef]
12. Han, X.; Kuang, Q.; Jin, M.; Xie, Z.; Zheng, L. Synthesis of Titania Nanosheets with a High Percentage of Exposed (001) Facets and Related Photocatalytic Properties. *J. Am. Chem. Soc.* **2009**, *131*, 3152–3153. [CrossRef]
13. Shahvardanfard, F.; Cha, G.; Denisov, N.; Osuagwu, B.; Schmuki, P. Photoelectrochemical performance of facet-controlled TiO$_2$ nanosheets grown hydrothermally on FTO. *Nanoscale Adv.* **2021**, *3*, 747–754. [CrossRef]
14. Yan, Y.; Keller, V.; Keller, N. On the role of BmimPF6 and P/F- containing additives in the sol-gel synthesis of TiO$_2$ photocatalysts with enhanced activity in the gas phase degradation of methyl ethyl ketone. *Appl. Catal. B Environ.* **2018**, *234*, 56–69. [CrossRef]
15. Butburee, T.; Kotchasarn, P.; Hirunsit, P.; Sun, Z.; Tang, Q.; Khemthong, P.; Sangkhun, W.; Thongsuwan, W.; Kumnorkaew, P.; Wang, H.; et al. New understanding of crystal control and facet selectivity of titanium dioxide ruling photocatalytic performance. *J. Mater. Chem. A* **2019**, *7*, 8156–8166. [CrossRef]
16. Xu, Y.; Zhang, Y.; He, T.; Ding, K.; Huang, X.; Li, H.; Shi, J.; Guo, Y.; Zhang, J. The Effects of Thermal and Atmospheric Pressure Radio Frequency Plasma Annealing in the Crystallization of TiO$_2$ Thin Films. *Coatings* **2019**, *9*, 357. [CrossRef]

17. Xu, Y.; Zhang, Y.; Li, L.; Ding, K.; Guo, Y.; Shi, J.; Huang, X.; Zhang, J. Synergistic Effect of Plasma Discharge and Substrate Temperature in Improving the Crystallization of TiO_2 Film by Atmospheric Pressure Plasma Enhanced Chemical Vapor Deposition. *Plasma Chem. Plasma Process.* **2019**, *39*, 937–947. [CrossRef]
18. Liu, G.; Yang, H.G.; Pan, J.; Yang, Y.Q.; Lu, G.Q.; Cheng, H.-M. Titanium Dioxide Crystals with Tailored Facets. *Chem. Rev.* **2014**, *114*, 9559–9612. [CrossRef]
19. Xu, Y.; He, T.; Zhang, Y.; Wang, H.; Guo, Y.; Shi, J.; Du, C.; Zhang, J. Insights into the low-temperature deposition of a dense anatase TiO_2 film via an atmospheric pressure pulse-modulated plasma. *Plasma Process. Polym.* **2021**, *18*, 2100050. [CrossRef]
20. Zhang, Y.; Wang, H.; He, T.; Li, Y.; Guo, Y.; Shi, J.; Xu, Y.; Zhang, J. The effects of radio frequency atmospheric pressure plasma and thermal treatment on the hydrogenation of TiO_2 thin film. *Plasma Sci. Technol.* **2023**, *25*, 065504. [CrossRef]
21. Ni, J.; Fu, S.; Wu, C.; Maier, J.; Yu, Y.; Li, L. Self-Supported Nanotube Arrays of Sulfur-Doped TiO_2 Enabling Ultrastable and Robust Sodium Storage. *Adv. Mater.* **2016**, *28*, 2259–2265. [CrossRef]
22. Zhu, M.; Chikyow, T.; Ahmet, P.; Naruke, T.; Murakami, M.; Matsumoto, Y.; Koinuma, H. A high-resolution transmission electron microscopy investigation of the microstructure of TiO_2 anatase film deposited on $LaAlO_3$ and $SrTiO_3$ substrates by laser ablation. *Thin Solid Film.* **2003**, *441*, 140–144. [CrossRef]
23. Leblanc, A.; Ding, K.; Lieberman, M.A.; Wang, D.X.; Zhang, J.; Jun Shi, J. Hybrid model of atmospheric pressure $Ar/O_2/TiCl_4$ radio-frequency capacitive discharge for TiO_2 deposition. *J. Appl. Phys.* **2014**, *115*, 183302. [CrossRef]
24. Sanjinés, R.; Tang, H.; Berger, H.; Gozzo, F.; Margaritondo, G.; Lévy, F. Electronic structure of anatase TiO_2 oxide. *J. Appl. Phys.* **1994**, *75*, 2945–2951. [CrossRef]
25. He, Z.; Que, W.; Chen, J.; He, Y.; Wang, G. Surface chemical analysis on the carbon-doped mesoporous TiO_2 photocatalysts after post-thermal treatment: XPS and FTIR characterization. *J. Phys. Chem. Solids* **2013**, *74*, 924–928. [CrossRef]
26. Lu, Y.; Liu, Y.-X.; He, L.; Wang, L.-Y.; Liu, X.-L.; Liu, J.-W.; Li, Y.-Z.; Tian, G.; Zhao, H.; Yang, X.-H.; et al. Interfacial co-existence of oxygen and titanium vacancies in nanostructured TiO_2 for enhancement of carrier transport. *Nanoscale* **2020**, *12*, 8364–8370. [CrossRef] [PubMed]
27. Shi, J.; Chen, J.; Feng, Z.; Chen, T.; Lian, Y.; Wang, X.; Li, C. Photoluminescence Characteristics of TiO_2 and Their Relationship to the Photoassisted Reaction of Water/Methanol Mixture. *J. Phys. Chem. C* **2007**, *111*, 693–699. [CrossRef]
28. Kong, X.; Xu, Y.; Cui, Z.; Li, Z.; Liang, Y.; Gao, Z.; Zhu, S.; Yang, X. Defect enhances photocatalytic activity of ultrathin TiO_2 (B) nanosheets for hydrogen production by plasma engraving method. *Appl. Catal. B Environ.* **2018**, *230*, 11–17. [CrossRef]
29. Bharti, B.; Kumar, S.; Lee, H.-N.; Kumar, R. Formation of oxygen vacancies and Ti^{3+} state in TiO_2 thin film and enhanced optical properties by air plasma treatment. *Sci. Rep.* **2016**, *6*, 32355. [CrossRef] [PubMed]
30. Lei, Y.; Zhang, L.D.; Meng, G.W.; Li, G.H.; Zhang, X.Y.; Liang, C.H.; Chen, W.; Wang, S.X. Preparation and photoluminescence of highly ordered TiO_2 nanowire arrays. *Appl. Phys. Lett.* **2001**, *78*, 1125–1127. [CrossRef]
31. Serpone, N.; Lawless, D.; Khairutdinov, R. Size Effects on the Photophysical Properties of Colloidal Anatase TiO_2 Particles: Size Quantization versus Direct Transitions in This Indirect Semiconductor? *J. Phys. Chem.* **1995**, *99*, 16646–16654. [CrossRef]

Disclaimer/Publisher's Note: The statements, opinions and data contained in all publications are solely those of the individual author(s) and contributor(s) and not of MDPI and/or the editor(s). MDPI and/or the editor(s) disclaim responsibility for any injury to people or property resulting from any ideas, methods, instructions or products referred to in the content.

Article

Retrieving the Intrinsic Microwave Permittivity and Permeability of Ni-Zn Ferrites

Artem Shiryaev [1,*], Konstantin Rozanov [1], Vladimir Kostishin [2], Dmitry Petrov [1], Sergey Maklakov [1], Arthur Dolmatov [1] and Igor Isaev [2]

[1] Institute for Theoretical and Applied Electromagnetics, Izhorskaya 13, Moscow 125412, Russia; k.rozanov@yandex.ru (K.R.); dpetrov-itae@yandex.ru (D.P.); squirrel498@gmail.com (S.M.); dolmatov.av@phystech.edu (A.D.)
[2] College of New Materials and Nanotechnologies, The National University of Science and Technology MISIS, Moscow 119991, Russia; drvgkostishyn@mail.ru (V.K.); isa@misis.ru (I.I.)
* Correspondence: artemshiryaev@mail.ru

Abstract: Mixing rules may be extremely useful for predicting the properties of composite materials and coatings. The paper is devoted to the study of the applicability of the mixing rules to permittivity and permeability and the possibility of retrieving the intrinsic properties of inclusions. Magnetically soft Ni-Zn ferrites are chosen as the object of the study due to their low permittivity and the negligible influence of the skin effect. Due to this, the microwave properties of bulk ferrites may be measured by standard techniques. It is suggested to perform the analysis of the microwave properties of composites filled with Ni-Zn ferrite powder in terms of the normalized inverse susceptibility defined as the volume fraction of inclusions divided by the effective dielectric or magnetic susceptibility of the composite. The measured properties of the bulk ferrite are compared with those obtained by mixing rules from composite materials. The experimental evidence for difference between the mixing rules for permittivity and permeability of a composite, which was previously predicted only theoretically, is obtained. The reason for the difference is considered to be the effect of non-ideal electrical contacts between neighboring inclusions. It is also experimentally shown that the measured permeability of the bulk material may differ from the retrieved one. The measured static permeability is 1400 and the retrieved one is 12. The reason for the discrepancy is the difference between the domain structures and demagnetizing fields of particles and bulk ferrite.

Keywords: composite materials; microwave permeability; ferrites; mixing rules

1. Introduction

Microwave technology is extensively used for telecommunications and radar engineering due to the ease of focusing into narrow beams, wide bandwidth, high data transmissions rates and small antenna sizes. Bulk ferrites and composite materials filled with ferrite powders are often used for developing novel materials and coatings for microwave applications [1–3]. Ferrite materials are predominantly applied for electromagnetic compatibility [4], anechoic chamber coatings [5] and radar absorbers [6] due to their high magnetic properties and low conductivity. The evaluation of their microwave permittivity ε and permeability μ is important for assessing the potential for practical use.

An important problem is the description of the effective properties of composites as a function of the properties of inclusions and the host matrix. Mixing rules [7,8] are often used to solve this task. A huge number of different mixing rules have been proposed, see, e.g., review [9]. The most commonly used of them are the Maxwell Garnet (MG) model, the effective medium theory (EMT), the asymmetric Bruggeman theory and the Landau–Lifshitz–Looyenga (LLL) formula. The comparison of the mixing rules with experimental data can be found in [10–17].

The standard approach to the analysis of the effective properties of composites assumes that magnetic inclusions may be characterized in terms of intrinsic permeability. It is usually believed that both the permittivity ε_{eff} and permeability μ_{eff} of composites should be described by the same mixing rule [18–21] at all frequencies. Thus, the mixing rules may be extremely useful for predicting the microwave properties of materials and coatings.

In contrast to the LLL formula, EMT and MG include the form factor of inclusions, N. Strictly speaking, EMT and MG formulas are valid only for $N = 1/3$, because, only in this case, they are consistent with the LLL theory at $\mu_i \to 1$. In practice, the form factor N differs from 1/3 even for spherical inclusions [22].

Composite materials filled with ferromagnetic inclusions are considered in most of the works devoted to the comparison of mixing rules with the experimental data. In contrast to bulk ferromagnets, the microwave properties of bulk ferrites may be measured by standard techniques due to their low permittivity and the negligible influence of the skin effect. This makes it possible to compare the measured intrinsic permeability with that found by mixing rules. It is also important that the permeability of ferrites has a strong frequency dispersion in the microwave range. This makes it possible to evaluate the applicability of mixing rules over a wide range of frequencies. An investigation of the properties of bulk and powdered ferrites was carried out in [23–25]. The variation in static permeability and the frequency dispersion parameters with the concentration of inclusions was evaluated using the coherent model approximation and the MG mixing rule. However, the effective permeability was poorly described by mixing rules, and the difference between theory and the experiment was not explained.

The difference between mixing rules for the permittivity and permeability was theoretically predicted in [26]. The magneto-dipole interaction in single-domain particles leads to the concentration dependence of the demagnetizing factor [27], while the depolarization factor is not affected by this interaction. The experimental evidence for difference between dielectric and magnetic mixing rules for composites has not been found in the literature. This paper is devoted to the experimental study of these issues.

Thus, the aims of this study are analysis of the possibility of retrieving the intrinsic properties of inclusions, the comparison of the measured properties of bulk ferrite and properties retrieved by mixing rules from the composite materials and the study of the applicability of the same mixing rule to permittivity and permeability. The investigation of the microwave magnetic and dielectric properties of Ni-Zn ferrites [28–30] is carried out using the inverse susceptibility approach described in the next section.

For the first time, it is experimentally shown that the permittivity and permeability of composite materials may obey different mixing rules and the measured permeability of the bulk material may differ from the retrieved one.

2. Materials and Methods

2.1. Research Method

The formulas relating the permittivities of composite ε_{eff}, host matrix ε_m, and inclusions ε_i for various mixing rules are given in Table 1, where p is the volume fraction of inclusions, and N is the form factor of inclusions. For the permeabilities of composite μ_{eff} and inclusions μ_i, the formulas are written identically to the permittivities. The mixing rules differ only slightly from each other at low fractions of inclusions, and it is often difficult to choose a model that correctly describes the properties of composites.

Table 1. Various mixing rules.

Mixing Rule	Formula	Number
The Maxwell Garnet (MG)	$\varepsilon_{eff} = \varepsilon_m \left(1 + \dfrac{p}{N(1-p) + \frac{1}{\varepsilon_i/\varepsilon_m - 1}}\right)$	(1)
The effective medium theory (EMT)	$p\dfrac{\varepsilon_i - \varepsilon_{eff}}{\varepsilon_{eff} + N(\varepsilon_i - \varepsilon_{eff})} + (1-p)\dfrac{\varepsilon_m - \varepsilon_{eff}}{\varepsilon_{eff} + N(\varepsilon_m - \varepsilon_{eff})} = 0$	(2)
The asymmetric Bruggeman theory	$\dfrac{\varepsilon_i - \varepsilon_{eff}}{\varepsilon_i - \varepsilon_m} = (1-p)\left(\dfrac{\varepsilon_{eff}}{\varepsilon_m}\right)^{1/3}$	(3)
The Landau–Lifshitz–Looyenga (LLL)	$\varepsilon_{eff} = \left[\left(\varepsilon_i^{1/3} - \varepsilon_m^{1/3}\right)p + \varepsilon_m^{1/3}\right]^3$	(4)

The approach proposed in [20] can facilitate this task. It is based on the analysis of the inverse normalized susceptibility η of the composite, which is defined for the permittivity and permeability, respectively, as

$$\eta = \dfrac{p}{\varepsilon_{eff}/\varepsilon_m - 1} \text{ or } \eta = \dfrac{p}{\mu_{eff} - 1}, \quad (5)$$

A consideration of the microwave properties of a composite in the form of normalized inverse susceptibility was shown as a useful tool for studying the effects determining the permeability of composites, such as a magnetic interaction between inclusions and a distribution of the particles in shape. It was shown in [20] that the concentration dependence of the inverse susceptibility may indicate the mixing rule that determines the properties of the composite. The calculated dependence of the susceptibility η on the volume fraction p for various mixing models is shown in Figure 1. The Figure shows the dependencies for the mixing rules given in Table 1. For example, the inverse susceptibility for the MG mixing model, see Equation (1), is written as

$$\eta = -Np + N + 1/\chi_i, \quad (6)$$

where $\chi_i = \mu_i - 1$ or $\varepsilon_i/\varepsilon_m - 1$ is the normalized susceptibility of inclusions.

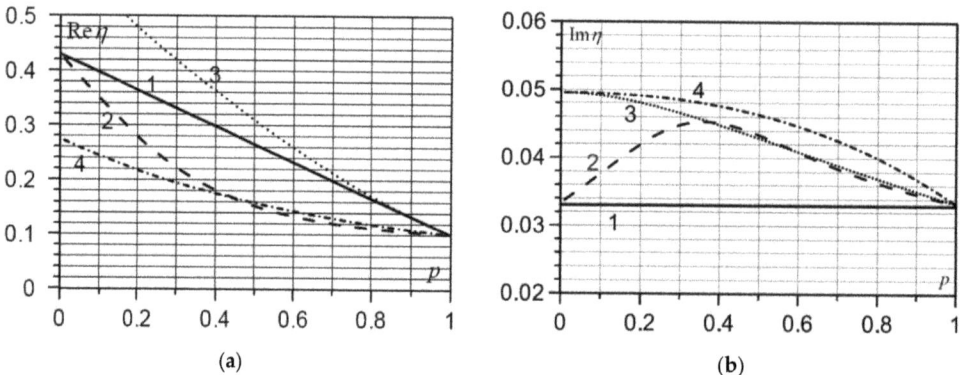

Figure 1. The calculated real (a) and imaginary (b) parts of the inverse susceptibility, η, plotted against the volume concentration of the inclusions in the composite. The numbers indicate the dependencies for the various mixing rules: 1—the MG model, 2—the EMT, 3—the asymmetric Bruggeman theory, 4—the LLL formula.

It is seen that the dependence of the real part of η on the volume fraction is linear, and the imaginary part does not depend on the fraction. The inclination of the line is determined by the effective form factor of the inclusions. The concentration dependence of the permittivity or permeability of the composite can be presented in a form that is more convenient for the analysis. Due to this, the distinctive features of the dependence become more pronounced and useful for understanding the phenomena that determine the properties of the composite.

The inverse susceptibility approach is used to describe the microwave properties of composites filled with ferrite powders for the first time. This approach makes it possible to clearly determine which mixing rule is applicable to the measured permittivity and permeability. A schematic illustration of the research approach is shown in Figure 2.

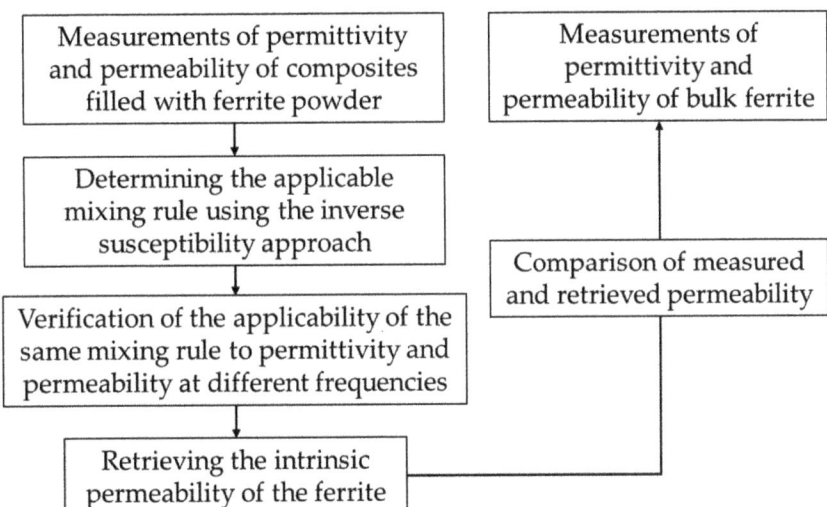

Figure 2. Schematic illustration of the research approach.

2.2. Materials under Study

The object of the study is magnetically soft Ni-Zn ferrite of 2000 NN brand. The chemical composition is $Ni_{0.32}Zn_{0.68}Fe_2O_4$. This ferrite was chosen for the research because of the high static permeability and the strong difference in microwave properties between the powder and bulk materials. In general, any material with low permittivity and conductivity can be suitable for this study. However, a small difference in the properties of the powder and bulk material could be attributed to measurement inaccuracy.

The ferrite is made at the National University of Science and Technology MISIS (Moscow, Russia) according to standard ceramic technology [31]. The mixture of initial oxides Fe_2O_3, ZnO and NiO is used as the starting material. The initial oxides are mixed and crushed in a rotary mill, the mixture is calcined, and the synthesized powders are comminuted in the vibration mill. Then, the obtained powder is sintered and pressed into the coaxial sample with an inner diameter of 7 mm and an outer one of 16 mm. The thickness of the sample is 8.9 mm. The density of the ferrite bulk is found from the volume and mass of the sample and is 5.0 g/cm^3. The found density coincides with the typical values for Ni-Zn ferrites.

Composites filled with both spherical and stone-like powders of ferrite are investigated. The spherical powder for the composite is obtained during the manufacturing process of the bulk ferrite before pressing and annealing. However, the color of the powder differs from that of the bulk ferrite, which may be due to a change in the material during annealing. The stone-like powder is obtained by grinding bulk ferrite with a diamond disk and sieving with a mesh size of 20 µm. The stone-like powder is obtained directly from the measured sample of the bulk ferrite, so the similarity between the powder and bulk materials is beyond doubt.

The photos of the studied powders are obtained using a scanning electron microscope. The SEM images of spherical and stone-like ferrite powders are shown in Figures 3a and 3b, respectively. It is seen that the size of the spherical particles is less than 2 µm, and the stone-like particles are strongly distributed in size.

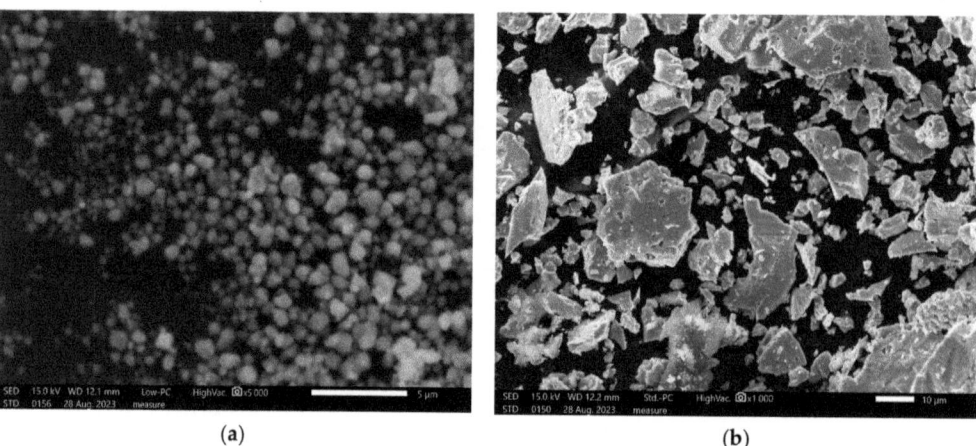

Figure 3. Scanning electron microscope (SEM) images of spherical (**a**) and stone-like (**b**) ferrite powders.

The samples of composites are manufactured by mixing calculated amounts of ferrite powder and paraffin wax. The components are heated until the paraffin melts; then, the powder is blended with the melted paraffin until it cools and solidifies. The resulting mixture is pressed into a sample to fit a 7/3 mm measuring coaxial cell. The volume fraction of ferrite powders is calculated from the thickness of the sample and the masses of the paraffin, filler, and the obtained sample. The concentration of inclusions in the samples is approximately 15, 27, 35, 48 and 61 vol.%. The permittivity of the paraffin host matrix is 2.25.

The elemental analysis of the materials under study is carried out by energy-dispersive X-ray spectroscopy (EDX) using a JEOL JCM-7000 scanning electron microscope. The penetration depth of the beam during analysis is less than 5 µm. The ferrite powders are dispersed on a conductive carbon adhesive tape, placed on an aluminum substrate and mounted in a microscope chamber. The powders covered the adhesive tape in a thick layer to neglect the errors during the EDX analysis. A sample of the bulk ferrite is split, and the analysis is carried out at the split site. The elemental analysis data for bulk ferrite are given in Table 2. The chemical compositions are the same for powder and bulk ferrite and coincide with the chemical formula.

Table 2. The elemental analysis of the studied Ni-Zn ferrite.

Materials	O, atom%	Fe, atom%	Ni, atom%	Zn, atom%
bulk ferrite	54	32	4.9	8.4

2.3. Measurement Techniques

The frequency dependencies of microwave permittivity $\varepsilon(f)$ and permeability $\mu(f)$ are measured with a vector network analyzer. The parameters of the bulk ferrites are measured by the coaxial technique [32,33] in a 16/7 mm coaxial line in the frequency range of 10 kHz to 1 GHz. The parameters of the composite materials filled with ferrite powders are measured by the transmission-reflection (Nicolson–Ross–Weir [34,35]) technique in a 7/3 mm coaxial line in the frequency range of 0.1 to 20 GHz. Additionally, the permeability of the bulk ferrite is determined from the quasi-static measurements of the inductance [32] of the coil with a ferrite ring in the frequency range of 1 kHz to 10 MHz.

3. Results

The measured frequency dependencies of permittivity and permeability of the composite materials filled with stone-like ferrite particles are shown in Figure 4. The color indicates the volume fraction of the ferrite powder in the sample. The frequency dependence of the imaginary permittivity is smoother than that characteristic of a conductive material.

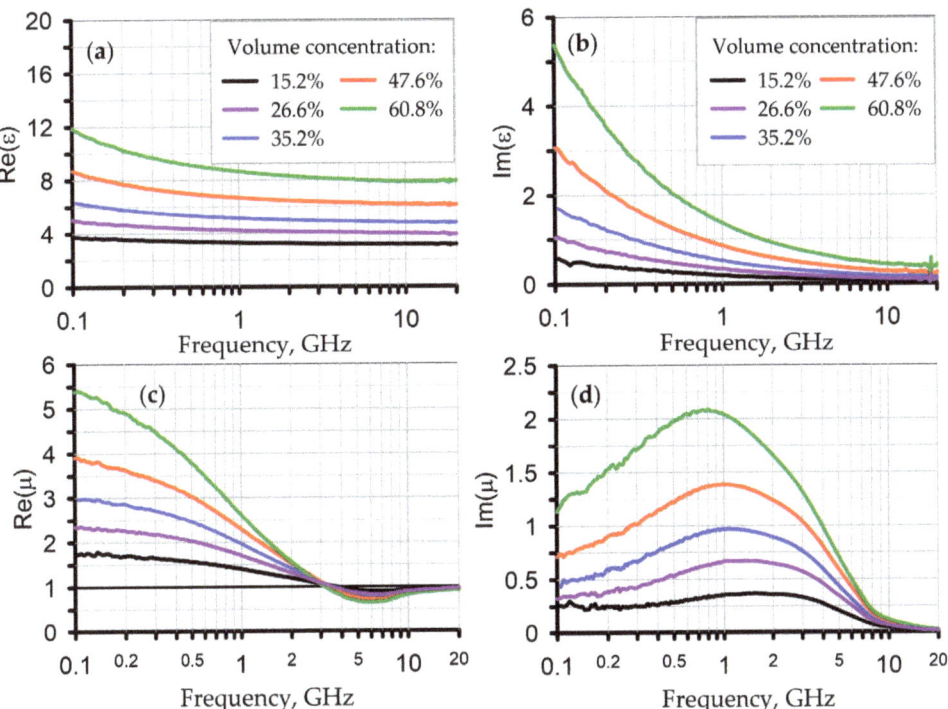

Figure 4. The measured frequency dependencies of microwave complex permittivity and permeability of composite materials filled with stone-like ferrite powder. (**a**) is the real permittivity; (**b**) is the imaginary permittivity; (**c**) is the real permeability; (**d**) is the imaginary permeability.

The frequency dependencies of permittivity and permeability of the bulk ferrite are shown in Figure 5. The permeability measured by the inductance technique is shown as gray lines. It is seen that measurements by two different techniques give the same result. It is worth noting that the measurement accuracy of microwave permittivity of the bulk ferrite is low due to gaps between the sample and the walls of the coaxial cell. The obtained frequency dependencies of permeability for the composite and for the bulk ferrite are quite different, even for the volume fraction of 60%. Both the magnitude and position of the magnetic loss differ greatly.

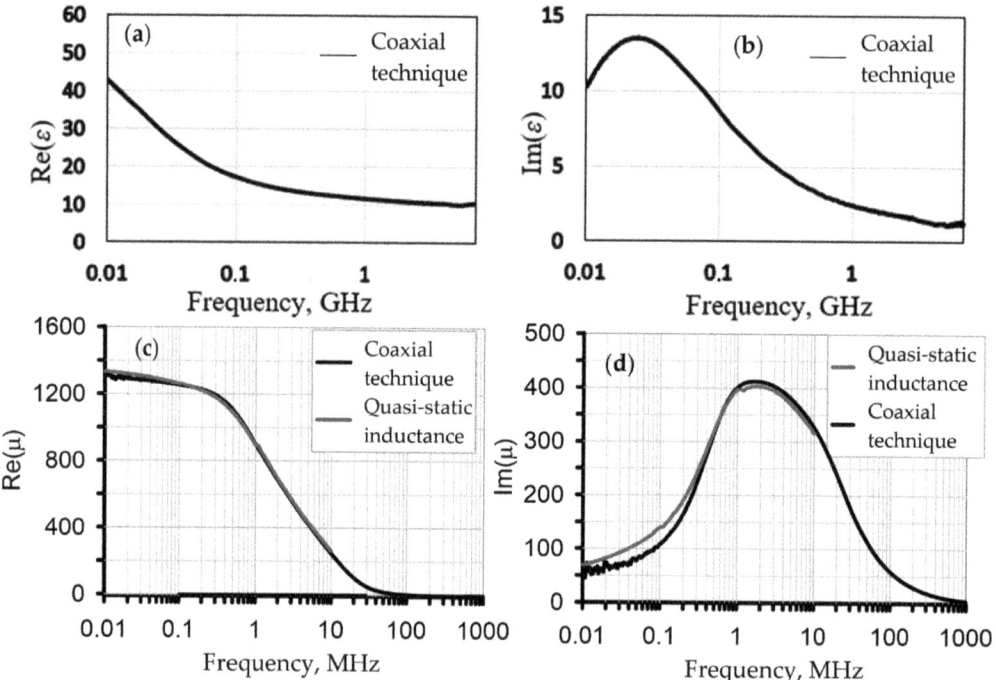

Figure 5. The measured frequency dependencies of permittivity and permeability of the bulk Ni-Zn ferrite sample. (**a**) is the real permittivity; (**b**) is the imaginary permittivity; (**c**) is the real permeability; (**d**) is the imaginary permeability.

4. Discussion

The inverse susceptibility η approach mentioned above is used to analyze the measured frequency dependencies of permittivity and permeability. The real and imaginary parts of inverse normalized susceptibility are derived from the measured data. The dependencies of inverse electric susceptibility on the volume fraction are shown in Figure 6a, and the data for the magnetic susceptibility are shown in Figure 6b. The dependencies are shown at frequencies of 0.5 to 5 GHz. The dependence of the real inverse magnetic susceptibility at a frequency of 0.5 GHz is close to that at a frequency of 1 GHz.

The dependence of the real inverse magnetic susceptibility on the concentration is linear. The imaginary part is constant. This suggests that the permeability obeys the MG mixing rule, see Equation (6). The effective form factor of inclusions is found from the coefficient of linear dependence. The form factor is the same for all studied frequencies and is approximately $N = 0.15 \pm 0.01$.

The electric susceptibility dependence behaves differently from the magnetic one. The real part of the inverse dielectric susceptibility has a linear dependence on concentration, which is typical for the Maxwell–Garnett mixing rule, see Figure 1. However, the imaginary part is not constant and has a systematic concentration dependence that is the same for all observed frequencies. Therefore, the MG mixing rule is not applicable to the permittivity. Moreover, the form factor found from the electric inverse susceptibility changes with the frequency, see Figure 7, and is different from the form factor obtained from the permeability data. Thus, it is experimentally shown that the permittivity and permeability obey different mixing rules.

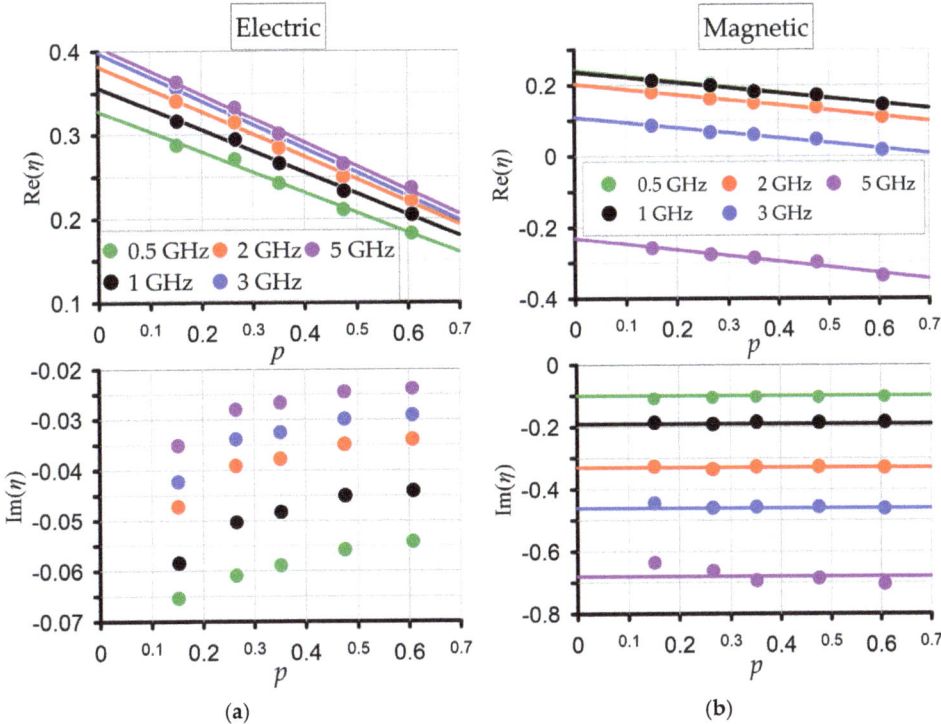

Figure 6. The dependencies of real and imaginary parts of inverse electric (**a**) and magnetic (**b**) susceptibilities on the volume fraction of inclusions. The data are given for frequencies of 0.5–5 GHz.

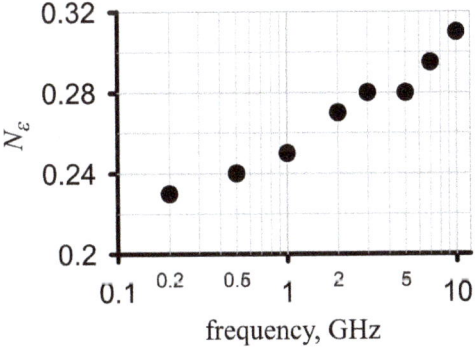

Figure 7. The frequency dependence of the effective form factor of inclusions found from the electric inverse susceptibility.

It is not clear which mixing rule to use to find the intrinsic permittivity of the ferrite. We retrieve it using the MG formula (1), but the obtained permittivity depends on the concentration and has an implausible frequency dependence. However, the measured and retrieved permittivities of the ferrite are close in order of magnitude.

The permittivity of composites depends not only on the geometric distribution of particle or cluster sizes but also on the effect of non-ideal electrical contacts. The resistance of the cluster should be determined mainly by the resistance of the contacts between the contiguous inclusions that make up the cluster. This resistance, as a rule, is significantly higher than the intrinsic resistance of the conductive particles. This effect is usually not taken into account within the percolation theory and mixing rules and may lead to their inapplicability.

The MG model (1) and the found form factor are applied to retrieve the intrinsic permeability of the ferrite. The retrieved permeability is the same with high accuracy for all studied volume fractions of inclusions, see Figure 8. Frequency dependencies at low frequencies have significant noise due to the peculiarities of measurement and calibration techniques, especially for samples with a low concentration of inclusions. Retrieving the intrinsic permeability from the measured data leads to an additional increase in the noise. Therefore, the curves are smoothed to better represent the results obtained. However, it does not coincide at all with the measured permeability of the bulk ferrite. The values of static permeability differ by two orders of magnitude, although the permeability of the composite is perfectly described by the mixing rule. Thus, it is experimentally shown that the retrieved permeability of the material and the measured one may not match.

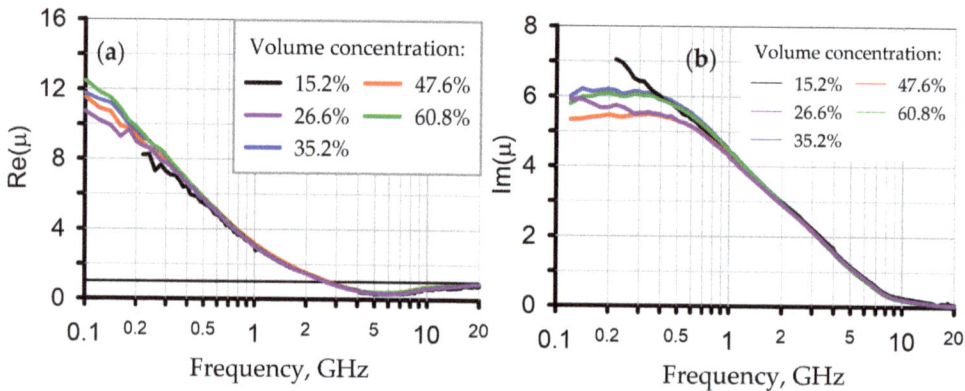

Figure 8. The frequency dependence of the real (**a**) and imaginary (**b**) intrinsic permeability of the Ni-Zn ferrite. The permeability retrieved from the measured data of the composite filled with stone-like ferrite powder by the MG mixing rule.

The reasons for such a strong discrepancy may be the influence of the skin effect, the errors in concentration calculations, the difference between the materials of the powdered and bulk ferrite, the poor retrieving accuracy and the different magnetic structures of particles and bulk ferrites.

We assume that the powder has the same density as the bulk ferrite. The densities may vary, for example, due to the presence of air pores in the bulk ferrite sample. This leads to an error in the determination of the volume concentration and, consequently, in the calculation of intrinsic properties. However, these errors cannot lead to a discrepancy in the permeability by two orders of magnitude. The found density coincides with the typical values for Ni-Zn ferrites.

The powder is obtained directly from the bulk ferrite sample to avoid material change. The appearance of an oxide shell layer can lead to the difference between powder and bulk materials, since the shell can occupy a significant part of a small particle. It is known that an oxide film appears on the surface of metals. The microstructure of Ni-Zn ferrites was studied in [36], and the presence of an oxide film was not found.

The composites filled with the spherical ferrite inclusions are additionally investigated. The SEM image of the particles is shown in Figure 3a. They are studied in the same way as the composites with stone-like particles. The frequency dependencies of the material parameters are measured, the inverse susceptibilities are calculated, and the intrinsic permeability of the inclusions is found. Its frequency dependence is shown in Figure 9. The intrinsic permeability is also retrieved by the MG model (1) with high accuracy but also does not coincide with that of the bulk ferrite, see Figure 10. Moreover, it does not match the permeability of the stone-like inclusions.

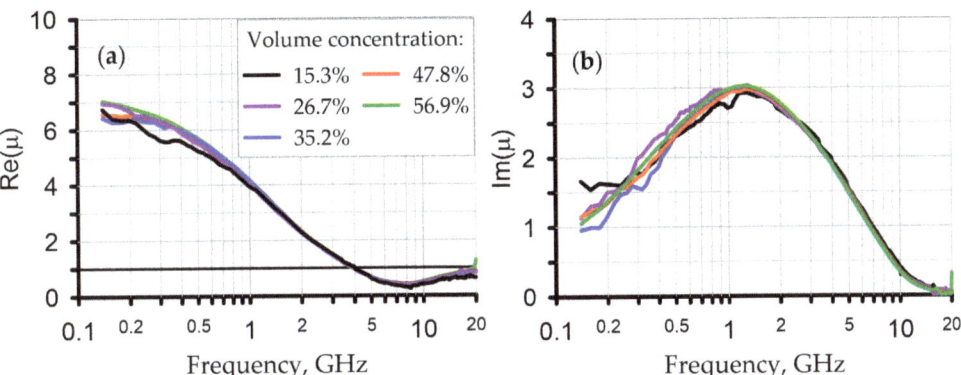

Figure 9. The frequency dependence of ferrite intrinsic permeability retrieved from the measured data of the composite filled with spherical powder by the MG mixing rule. (**a**) is the real permeability; (**b**) is the imaginary permeability.

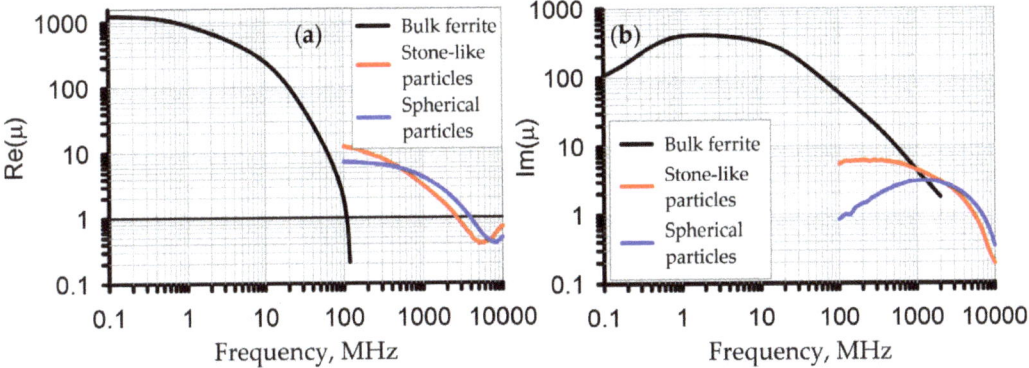

Figure 10. Comparison of the frequency dependencies of the measured and retrieved intrinsic permeability of the ferrite. (**a**) is the real permeability; (**b**) is the imaginary permeability.

It is known that the skin effect leads to a distortion in the frequency dependence of the microwave permeability [37]. The difference between the intrinsic permeability of the ferromagnetic inclusions of various shapes and sizes is usually attributed to the influence of the skin effect. The resistivity ρ of the studied ferrite is quite high, approximately 1000 Ohm × cm, and the skin effect should have a rather weak influence. The skin depth δ may be estimated using the formula:

$$\delta = \sqrt{2\rho/(2\pi f \mu)}. \qquad (7)$$

It is calculated to be approximately 5 cm at a frequency of 100 MHz, which is larger than the thickness of the sample and much larger than the particle size. Moreover, the permeability measured by the coaxial and the inductance techniques is the same. Therefore, the skin effect may be neglected.

Another reason for the discrepancy may be the poor retrieving accuracy if the intrinsic permeability is large. It is difficult to make a composite with a volume fraction above 70% using simple mixing. Figure 11 shows the theoretical dependencies of the static permeability of a composite on the volume fraction of inclusions at the permeability of inclusions of 1000, 50, and 10. The dependencies are calculated using the MG model. The greatest increase in permeability is observed at high concentrations. If the measurement error of the composite effective properties is approximately 10%, it is simply impossible to retrieve intrinsic permeability above 50. However, the permeability of the studied bulk ferrite reaches high values at low frequencies. In the region of 100 MHz, it is still rather small. Therefore, this cannot be the reason for the discrepancy.

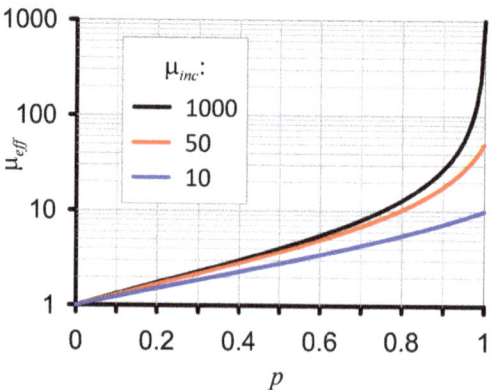

Figure 11. The theoretical dependencies of the static permeability of a composite on the volume fraction of inclusions calculated using the MG model.

The main reason for the discrepancy between permeabilities may be the different magnetic structure of the bulk ferrite and particles. The studied ferrite particles are rather small, due to which their domain structure may differ from that of bulk ferrite, which leads to a difference in magnetic properties. The inapplicability of the same mixing rule for permittivity and permeability may also be explained by the influence of the domain structure. Permittivity is determined by particles and permeability by domains.

The typical size of the domains in Ni-Zn ferrites is 1–4 µm [1,38,39]. Stone-like particles are quite strongly distributed in size, and among them should be present both single-domain and multi-domain particles. The interaction between inclusions in a composite with single-domain magnetic inclusions may have specific features. This problem was theoretically investigated in [26], where it was assumed that there are significant demagnetizing fields in single-domain particles. The interaction of these fields with the surrounding particles is not described by mixing rules. Note that demagnetizing fields may also be observed on the surface of multi-domain particles, which can cause differences in the effective magnetic properties of composites from the results of mixing rules. On the other hand, single-domain particles may form magnetic agglomerations during the preparation of composite samples, forming multi-domain clusters.

In any case, no experimental confirmation of the difference in the properties of magnetic composites containing single-domain and multi-domain inclusions was found in the literature. The differences in the microwave magnetic properties of composites containing micro- and nano-sized Fe_2O_3 powders are experimentally illustrated in [17]. However, it may be simply explained by the fact that these two types of powders contained different Fe_2O_3 phases, as can be concluded based on the difference in the color of the powders described in the paper.

According to [26], an increase in the concentration of magnetic inclusions in a composite significantly increases the interaction between these inclusions, and each particle should be considered as placed in some magnetic effective medium. The presence of this medium changes the demagnetization of the particle and, hence, the gyromagnetic spin resonance conditions. However, no change in the demagnetization factor up to volume concentrations of 60% is observed in our study. These effects likely appear at higher concentrations, which leads to a difference between the retrieved permeability and the measured one. The discrepancy should disappear for larger powders with a significant number of domains per particle.

The main methods for revealing the domain structure of the magnetic samples are magneto-optical techniques. In addition, to assess the influence of the domain structure on microwave magnetic properties of ferrite particles, the dependence of the effective and intrinsic permeability on the size of inclusions may be studied. In this case, the influence of the shape of the inclusions should be excluded. This can be performed by studying composites with spherical inclusions of various sizes. Solving these problems may be a topic for future research.

The experimental evidence for difference between mixing rules for permeability and permittivity for composites is obtained in this paper. This difference was also theoretically predicted in [26]. The magneto-dipole interaction in single-domain particles leads to the concentration dependence of the demagnetizing factor [27], while the depolarization factor is not affected by this interaction. However, the difference between the material parameters is observed at the studied concentrations, while the distortion of the demagnetization factor is not. Therefore, the reason for the difference is most likely the effect of non-ideal electrical contacts between neighboring inclusions, which is not taken into account by the mixing rules. This is supported by the fact that the permittivity increases more slowly with increasing concentration than the permeability.

The results obtained in the paper may be used in practice for the fabrication of composite materials for microwave absorbers [40], structured metasurfaces [41], infrared thin-film absorbers [42] and advanced optical composite materials [43].

5. Conclusions

The normalized inverse susceptibility approach is applied to analyze the microwave properties of composite materials filled with Ni-Zn ferrite powder. The applicability of the mixing rules to permittivity and permeability and the possibility of retrieving the intrinsic properties of inclusions are studied. The measured properties of the bulk ferrite are compared with those obtained by mixing rules from composite materials. The experimental evidence for difference between dielectric and magnetic mixing rules for composites, which was previously predicted only theoretically, is obtained. The reason for the difference in the mixing rules for the permittivity and permeability is considered to be the effect of non-ideal electrical contacts between neighboring inclusions.

It is also experimentally shown that the measured permeability of the bulk material may differ from the retrieved one. The influence of the skin effect, the errors in concentration calculations, the difference between the materials of powdered and bulk ferrite, the poor retrieving accuracy and the different magnetic structures of particles and bulk ferrites are considered as reasons for the discrepancy. The reason for the discrepancy is the difference between the domain structures and demagnetizing fields of the particles and bulk ferrite. There are significant demagnetizing fields in single-domain particles. The interaction of these fields with the surrounding particles is not described by the mixing rules.

Author Contributions: Conceptualization, A.S. and K.R.; material manufacturing, V.K. and I.I.; microwave measurements, A.S., D.P. and V.K.; software, D.P.; data analysis, A.S., K.R., D.P. and S.M.; writing—original draft preparation, A.S., writing—review and editing, K.R.; scanning electron microscopy and EDX analysis, A.D. and S.M. All authors have read and agreed to the published version of the manuscript.

Funding: Fabrication of the ferrite samples was funded under the Research and Development contract (Strategic Academic Leadership Program "Priority 2030", Project No. K7-2022-053, Contract No. M-2022-НП-ПЮ-053). The main part of the study was funded by government assignment FFUR-2021-0001, Project No. 075-00943-23-00.

Institutional Review Board Statement: Not applicable.

Informed Consent Statement: Not applicable.

Data Availability Statement: The data supporting this article are available from the corresponding author upon reasonable request.

Conflicts of Interest: The authors declare no conflict of interest.

References

1. Harris, V.G. Modern Microwave Ferrites. *IEEE Trans. Magn.* **2012**, *48*, 1075–1104. [CrossRef]
2. Liu, H.; Yu, Z.; Song, X.; Ran, M.; Jiang, X.; Lan, Z.; Sun, K. Effects of Substrates on Thin-Film Growth of Nickel Zinc Ferrite by Spin-Spray Deposition. *Coatings* **2023**, *13*, 690. [CrossRef]
3. Beatrice, C.; Tsakaloudi, V.; Dobák, S.; Zaspalis, V.; Fiorillo, F. Magnetic losses versus sintering treatment in Mn-Zn ferrites. *J. Magn. Magn. Mater.* **2017**, *429*, 129–137. [CrossRef]
4. Suarez, A.; Victoria, J.; Alcarria, A.; Torres, J.; Martinez, P.A.; Martos, J.; Soret, J.; Garcia-Olcina, R.; Muetsch, S. Characterization of Different Cable Ferrite Materials to Reduce the Electromagnetic Noise in the 2–150 kHz Frequency Range. *Materials* **2018**, *11*, 174. [CrossRef] [PubMed]
5. Shimada, K.; Ishizuka, K.; Tokuda, M. A Study of RF Absorber for Anechoic Chambers Used in the Frequency Range for Power Line Communication System. *PIERS Online* **2006**, *2*, 538–543. [CrossRef]
6. Randa, M.; Priyono. Ferrite phase of BaFe$_9$(MnCo)$_{1.5}$Ti$_{1.5}$O$_{19}$ as anti-radar coating material. In Proceedings of the 2015 International Conference on Radar, Antenna, Microwave, Electronics and Telecommunications (ICRAMET), Bandung, Indonesia, 5–7 October 2015; pp. 46–49. [CrossRef]
7. Qin, F.; Peng, M.; Estevez, D.; Brosseau, C. Electromagnetic composites: From effective medium theories to metamaterials. *J. Appl. Phys.* **2022**, *132*, 101101. [CrossRef]
8. Zhang, X.; Wu, Y. Effective medium theory for anisotropic metamaterials. *Sci. Rep.* **2015**, *5*, 7892. [CrossRef]
9. Rozanov, K.N.; Koledintseva, M.Y.; Yelsukov, E.P. Frequency-dependent effective material parameters of composites as a function of inclusion shape. In *Composites and Their Properties*; Hu, N., Ed.; InTech: New York, NY, USA, 2012; pp. 331–358. [CrossRef]
10. Kang, Y.; Tan, G.; Man, Q.; Ning, M.; Chen, S.; Pan, J.; Liu, X. A new low-density hydrogel-based matrix with hollow microsphere structure for weight reduction of microwave absorbing composites. *Mater. Chem. Phys.* **2021**, *266*, 124532. [CrossRef]
11. García-Valenzuela, A.; Acevedo-Barrera, A.; Vázquez-Estrada, O.; Nahmad-Rohen, A.; Barrera, R.G. Full dynamic corrections to the Maxwell Garnett mixing formula and corresponding extensions beyond the dipolar approximation. *J. Quant. Spectrosc. Radiat. Transf.* **2023**, *302*, 108578. [CrossRef]
12. Fung, T.H.; Veeken, T.; Payne, D.; Veettil, B.; Polman, A.; Abbott, M. Application and validity of the effective medium approximation to the optical properties of nano-textured silicon coated with a dielectric layer. *Opt. Express* **2019**, *27*, 38645–38660. [CrossRef]
13. Nazarov, R.; Zhang, T.; Khodzitsky, M. Effective Medium Theory for Multi-Component Materials Based on Iterative Method. *Photonics* **2020**, *7*, 113. [CrossRef]
14. Hasar, H.; Hasar, U.C.; Kaya, Y.; Ozturk, H.; Izginli, M.; Oztas, T.; Aslan, N.; Ertugrul, M.; Barroso, J.J.; Ramahi, O.M. Honey–Water Content Analysis by Mixing Models Using a Self-Calibrating Microwave Method. *IEEE Trans. Microw. Theory Tech.* **2023**, *71*, 691–697. [CrossRef]
15. Merrill, W.M.; Diaz, R.E.; LoRe, M.M.; Squires, M.C.; Alexopoulos, N.G. Effective medium theories for artificial materials composed of multiple sizes of spherical inclusions in a host continuum. *IEEE Trans. Antennas Propag.* **1999**, *47*, 142–148. [CrossRef]
16. Goyal, N.; Panwar, R. Dielectric Characterization of Electromagnetic Mixing Model Assisted Optimization Derived Heterogeneous Composites for Stealth Technology. *IEEE Trans. Dielectr. Electr. Insul.* **2023**, *30*, 690–699. [CrossRef]
17. Brosseau, C.; Talbot, P. Effective magnetic permeability of Ni and Co micro- and nanoparticles embedded in a ZnO matrix. *J. Appl. Phys.* **2005**, *97*, 104325. [CrossRef]
18. Bergman, D.J.; Stroud, D. Physical Properties of Macroscopically Inhomogeneous Media. *Solid State Phys.* **1992**, *46*, 147. [CrossRef]
19. Koledintseva, M.Y.; Xu, J.; De, S.; Drewniak, J.L.; He, Y.; Johnson, R. Systematic Analysis and Engineering of Absorbing Materials Containing Magnetic Inclusions for EMC Applications. *IEEE Trans. Magn.* **2011**, *47*, 317–323. [CrossRef]
20. Rozanov, K.N.; Bobrovskii, S.Y.; Lagarkov, A.N.; Mishin, A.D.; Osipov, A.V.; Petrov, D.A.; Shiryaev, A.O.; Starostenko, S.N. Revealing the effect of interaction between inclusions on the effective microwave permeability of composites. *Procedia Eng.* **2017**, *216*, 85–92. [CrossRef]
21. Hernandez-Cardoso, G.G.; Singh, A.K.; Castro-Camus, E. Empirical comparison between effective medium theory models for the dielectric response of biological tissue at terahertz frequencies. *Appl. Opt.* **2020**, *59*, D6–D11. [CrossRef]
22. Holcman, V.; Liedermann, K. New mixing rule of polymer composite systems. *WSEAS Trans. Electron.* **2007**, *4*, 181–185.
23. Tsutaoka, T.; Kasagi, T.; Hatakeyama, K.; Koledintseva, M.Y. Analysis of the permeability spectra of spinel ferrite composites using mixing rules. In Proceedings of the 2013 IEEE International Symposium on Electromagnetic Compatibility, Denver, CO, USA, 5–9 August 2013; pp. 545–550. [CrossRef]
24. Tsutaoka, T. Frequency dispersion of complex permeability in Mn-Zn and Ni-Zn spinel ferrites and their composite materials. *J. Appl. Phys.* **2003**, *93*, 2789–2796. [CrossRef]

25. Bellaredj, M.L.F.; Mueller, S.; Davis, A.K.; Kohl, P.; Swaminathan, M.; Mano, Y. Fabrication, Characterization and Comparison of FR4-Compatible Composite Magnetic Materials for High Efficiency Integrated Voltage Regulators with Embedded Magnetic Core Micro-Inductors. In Proceedings of the 2017 IEEE 67th Electronic Components and Technology Conference (ECTC), Orlando, FL, USA, 30 May–2 June 2017; pp. 2008–2014. [CrossRef]
26. Ramprasad, R.; Zurcher, P.; Petras, M.; Miller, M.; Renaud, P. Magnetic properties of metallic ferromagnetic nanoparticle composites. *J. Appl. Phys.* **2004**, *96*, 519–529. [CrossRef]
27. Mattei, J.; Le Floc'h, M. Percolative behaviour and demagnetizing effects in disordered heterostructures. *J. Magn. Magn. Mater.* **2003**, *257*, 335–345. [CrossRef]
28. Wei, X.; Pan, Y.; Chen, Z. 3D printing of NiZn ferrite architectures with high magnetic performance for efficient magnetic separation. *J. Eur. Ceram. Soc.* **2022**, *42*, 1522–1529. [CrossRef]
29. Kotru, S.; Paul, R.; Jaber, A.Q. Synthesis and magnetic studies of pure and doped NiZn ferrite films using Sol gel method. *Mater. Chem. Phys.* **2022**, *276*, 125357. [CrossRef]
30. Xiang, N.; Zhou, Z.; Ma, X.; Zhang, H.; Xu, X.; Chen, Y.; Guo, Z. The In Situ Preparation of Ni–Zn Ferrite Intercalated Expanded Graphite via Thermal Treatment for Improved Radar Attenuation Property. *Molecules* **2023**, *28*, 4128. [CrossRef]
31. Salem, M.M.; Morchenko, A.T.; Panina, L.V.; Kostishyn, V.G.; Andreev, V.G.; Bibikov, S.B.; Nikolaev, A.N. Dielectric and Magnetic Properties of Two-Phase Composite System: Mn-Zn or Ni-Zn ferrites in Dielectric Matrices. *Phys. Procedia* **2015**, *75*, 1360–1369. [CrossRef]
32. Starostenko, S.N.; Rozanov, K.; Shiryaev, A.O.; Lagarkov, A. A Technique to Retrieve High-Frequency Permeability of Metals from Constitutive Parameters of Composites with Metal Inclusions of Arbitrary Shape, Estimate of the Microwave Permeability of Nickel. *Prog. Electromagn. Res. M* **2018**, *76*, 143–155. [CrossRef]
33. Venkatarayalu, N.V.; Yuan, C.J. Eliminating errors due to position uncertainty in coaxial airline based measurement of material parameters. In Proceedings of the 2017 Progress in Electromagnetics Research Symposium—Fall (PIERS—FALL), Singapore, 19–22 November 2017; pp. 464–467. [CrossRef]
34. Costa, F.; Borgese, M.; Degiorgi, M.; Monorchio, A. Electromagnetic Characterisation of Materials by Using Transmission/Reflection (T/R) Devices. *Electronics* **2017**, *6*, 95. [CrossRef]
35. Angiulli, G.; Versaci, M. Extraction of the Electromagnetic Parameters of a Metamaterial Using the Nicolson–Ross–Weir Method: An Analysis Based on Global Analytic Functions and Riemann Surfaces. *Appl. Sci.* **2022**, *12*, 11121. [CrossRef]
36. Kostishin, V.G.; Vergazov, R.M.; Andreev, V.G.; Bibikov, S.B.; Podgornaya, S.V.; Morchenko, A.T. Effect of the microstructure on the properties of radio-absorbing nickel-zinc ferrites. *Russ. Microelectron.* **2011**, *40*, 574–577. [CrossRef]
37. Rozanov, K.N.; Koledintseva, M.Y. Application of generalized Snoek's law over a finite frequency range: A case study. *J. Appl. Phys.* **2016**, *119*, 073901. [CrossRef]
38. Aharoni, A.; Jakubovics, J.P. Theoretical single-domain size of NiZn ferrite. *J. Phys. IV Fr.* **1998**, *8*, Pr2-389–Pr2-392. [CrossRef]
39. Van der Zaag, P.J.; Van der Valk, P.J.; Rekveldt, M.T. A domain size effect in the magnetic hysteresis of NiZn-ferrites. *Appl. Phys. Lett.* **1996**, *69*, 2927–2929. [CrossRef]
40. Zhang, J.; Li, X.; Wang, X. Facile fabrication of urchin-like carbon nanotube–modified $Cu_{0.48}Ni_{0.16}Co_{2.36}O_4$/CuO with high optical–infrared–microwave attenuation. *Opt. Express* **2021**, *29*, 26004–26013. [CrossRef]
41. Liu, Y.; Ouyang, C.; Xu, Q.; Su, X.; Yang, Q.; Ma, J.; Li, Y.; Tian, Z.; Gu, J.; Liu, L.; et al. Moiré-driven electromagnetic responses and magic angles in a sandwiched hyperbolic metasurface. *Photon. Res.* **2022**, *10*, 2056–2065. [CrossRef]
42. Liu, Z.; Banar, B.; Butun, S.; Kocer, H.; Wang, K.; Scheuer, J.; Wu, J.; Aydin, K. Dynamic infrared thin-film absorbers with tunable absorption level based on VO_2 phase transition. *Opt. Mater. Express* **2018**, *8*, 2151–2158. [CrossRef]
43. Dong, T.; Luo, J.; Chu, H.; Xiong, X.; Peng, R.; Wang, M.; Lai, Y. Breakdown of Maxwell Garnett theory due to evanescent fields at deep-subwavelength scale. *Photon. Res.* **2021**, *9*, 848–855. [CrossRef]

Disclaimer/Publisher's Note: The statements, opinions and data contained in all publications are solely those of the individual author(s) and contributor(s) and not of MDPI and/or the editor(s). MDPI and/or the editor(s) disclaim responsibility for any injury to people or property resulting from any ideas, methods, instructions or products referred to in the content.

Review

Combining Non-Thermal Processing Techniques with Edible Coating Materials: An Innovative Approach to Food Preservation

Arezou Khezerlou [1], Hajar Zolfaghari [1], Samira Forghani [2], Reza Abedi-Firoozjah [3], Mahmood Alizadeh Sani [4], Babak Negahdari [5], Masumeh Jalalvand [5], Ali Ehsani [6,*] and David Julian McClements [7,*]

[1] Students Research Committee, Department of Food Sciences and Technology, Faculty of Nutrition and Food Sciences, Tabriz University of Medical Sciences, Tabriz 5166614711, Iran
[2] Department of Food Science and Technology, Faculty of Agriculture, Urmia University, Urmia 5166614711, Iran
[3] Student Research Committee, Department of Food Science and Technology, School of Nutrition Sciences and Food Technology, Kermanshah University of Medical Sciences, Kermanshah 5166614711, Iran
[4] Food Safety and Hygiene Division, Environmental Health Engineering Department, School of Public Health, Tehran University of Medical Sciences, Tehran 5166614711, Iran; saniam7670@gmail.com
[5] Department of Medical Biotechnology, School of Advanced Technologies in Medicine, Tehran University of Medical Sciences, Tehran 5166614711, Iran
[6] Nutrition Research Center, Department of Food Sciences and Technology, Faculty of Nutrition and Food Sciences, Tabriz University of Medical Sciences, Tabriz 5166614711, Iran
[7] Department of Food Science, University of Massachusetts Amherst, Amherst, MA 01003, USA
* Correspondence: ehsani@tbzmed.ac.ir (A.E.); mcclements@foodsci.umass.edu (D.J.M.)

Abstract: Innovative processing and packaging technologies are required to create the next generation of high-quality, healthy, safe, and sustainable food products. In this review, we overview the potential of combining edible coating materials with non-thermal processing technologies to improve the quality, increase the safety, extend the shelf life, and reduce the waste of foods and plastics. Edible coatings are typically assembled from food-grade structuring ingredients that can provide the required mechanical and barrier properties, such as proteins, polysaccharides, and/or lipids. These materials can be fortified with functional additives to further improve the quality, safety, and shelf life of coated foods by reducing ripening, gas exchange, and decay caused by bacteria and fungi. Non-thermal processing techniques include high hydrostatic pressure, pulsed light, ultrasound, and radiation technologies. These technologies can be used to inhibit the growth of pathogenic or spoilage microorganisms on packaged foods. Examples of the application of this combined approach to a range of highly perishable foods are given. In addition, the impact of these combined methods on the quality attributes of these food products is discussed.

Keywords: edible coatings; non-thermal processing; innovative technology; food safety; sustainable packaging

Citation: Khezerlou, A.; Zolfaghari, H.; Forghani, S.; Abedi-Firoozjah, R.; Alizadeh Sani, M.; Negahdari, B.; Jalalvand, M.; Ehsani, A.; McClements, D.J. Combining Non-Thermal Processing Techniques with Edible Coating Materials: An Innovative Approach to Food Preservation. *Coatings* **2023**, *13*, 830. https://doi.org/10.3390/coatings13050830

Academic Editor: Domingo Martínez-Romero

Received: 24 March 2023
Revised: 19 April 2023
Accepted: 23 April 2023
Published: 26 April 2023

Copyright: © 2023 by the authors. Licensee MDPI, Basel, Switzerland. This article is an open access article distributed under the terms and conditions of the Creative Commons Attribution (CC BY) license (https:// creativecommons.org/licenses/by/ 4.0/).

1. Introduction

Food safety is a key priority of the food industry [1]. Post-process contamination of food products due to inappropriate handling, packaging, and storage can lead to the spread of food-borne diseases and to increased food waste [2]. Hence, it is important to decontaminate foods before packaging and then ensure that microbial contamination does not occur after packaging [3]. The nature of the packaging materials used to protect foods is important because it affects their effectiveness as well as consumer perceptions. Ideally, any packaging material should not adversely affect the sensory appeal, quality, affordability, and health of a food product [4]. Moreover, it should ideally be produced and disposed of in an environmentally friendly manner. The use of synthetic compounds as film-forming substances (such as petroleum-based plastics) or as additives (such as sorbic acid, benzoic acid, propionic acid, and sulfur dioxide) can lead to packaging materials that can protect foods but that are often perceived negatively by consumers [5,6]. Consequently, there is interest in developing alternative kinds of food packaging materials that are more environmentally and consumer-friendly.

Microbial contamination can also be eliminated or reduced by using various thermal treatments of foods, such as pasteurization or sterilization. However, these processes often cause appreciable reductions in the sensory and nutritional profiles of foods. Consequently, there is interest in identifying alternatives to traditional thermal processing technologies that are able to improve product safety and shelf life without reducing product quality or nutrition [7]. For instance, natural antimicrobial compounds, such as essential oils (EOs), bacteriocins, and herbal extracts, are being explored for their potential application as additives in food packaging materials, such as films and coatings [8]. Non-thermal technologies, such as high hydrostatic pressure, pulsed light, ultrasound, and radiation technologies, are another food preservation method suitable for improving the safety and shelf life of foods [9,10]. Foods can also be preserved by using edible coatings, which consist of a thin and continuous layer of food-grade materials deposited around the food surfaces [11,12]. These coatings are often applied onto the surfaces of fresh produce by spraying, dipping, or brushing to enhance their safety, shelf life, and quality [13,14]. Edible coatings can be prepared from natural film-forming materials such as polysaccharides, proteins, lipids, and their blends [15]. Edible coatings are commonly applied to foodstuffs to inhibit their deterioration through oxidation, microbial spoilage, and gas exchange, as well as to improve their physical, tactile, and visual properties [16,17]. The functional performance of coatings can often be improved by incorporating active compounds into them, such as antimicrobial, antioxidant, or anti-browning agents [18]. However, the use of these any of these methods alone is typically unable to reduce pathogenic or spoilage microorganisms to a suitably low level.

A recent trend in food preservation has been the utilization of hurdle technologies, which use a combination of different approaches to increase the overall effect [19]. Indeed, combined treatments often exhibit synergistic effects, i.e., they lead to a greater effect than expected from the sum of the individual treatments. Previous studies have reported that coatings combined with non-thermal treatments have had an appreciable lethal effect on several microorganisms [20]. Consequently, there is interest in combining these two technologies together to improve their overall efficacy [9]. This review paper therefore describes the combined use of non-thermal technologies and edible coatings for the preservation of food products. It begins by describing different kinds of biodegradable packaging materials that can be used as edible coatings. It then discusses different types of non-thermal processing methods that can be used to treat foods. Finally, it provides examples of the use of combined methods to enhance the shelf life, safety, and quality of foods. Figure 1 presents a schematic of combining non-thermal methods with edible coatings as a new approach to food preservation.

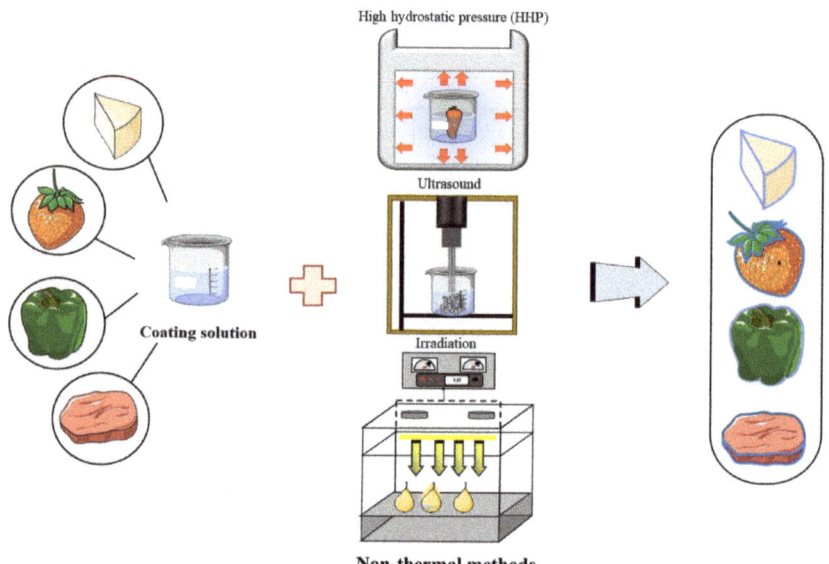

Figure 1. Schematic of combining non-thermal methods with edible coatings as a new approach to food preservation.

2. Food Packaging Materials

During recent decades, the expectations of consumers for food products with high quality, safety, and shelf life led to the emergence of many advancements in packaging systems. Active packaging materials are being designed that contain functional additives that can extend the shelf life of foods, including oxygen scavengers, moisture absorbers, antioxidants, and antimicrobials [21]. These active packaging systems often extend the shelf life of foodstuffs by decreasing their deterioration rates. In some cases, the active ingredients (such as antimicrobials or antioxidants) are designed to be released from the packaging materials into the food product or its head space during storage [22]. Smart packaging materials are also being produced that contain indicators such as temperature, pH, or gas indicators [21]. This type of packaging material is usually designed to provide visual information about the safety, quality, or freshness of packaged food in real time [23,24]. Indicators are typically attached to the interior or exterior of the packaging material to monitor, record, and communicate information to food producers, distributors, and consumers along the whole supply chain [25,26]. In the remainder of this review, we focus on the utilization of active packaging materials to form coatings that are designed to preserve foods, as well as the combination of non-thermal processing methods.

2.1. Biodegradable Packaging Materials

Due to the environmental problems associated with the production and disposal of petroleum-based plastic packaging materials, there has been growing interest in the development of more sustainable biodegradable packaging materials assembled from renewable natural resources. Edible film-forming substances, such as proteins, polysaccharides, and lipids, are commonly used to assemble this kind of packaging material.

2.2. Functional Additives for Active Packaging

The functional performance of biodegradable packaging materials can often be improved by including functional additives. Antioxidants, antimicrobials, light blockers, barrier enhancers, and mechanical modulators are often added to coatings for this purpose.

2.2.1. Preservatives: Antioxidants and Antimicrobials

Free radicals and reactive oxygen species (ROS) have deleterious effects on food quality and nutrition due to their ability to promote the oxidation of major food ingredients, such as lipids and proteins. Antioxidants can be incorporated into edible coatings to inhibit oxidation reactions in foods. Natural botanically derived antioxidants have received special attention recently because of consumer demand for greener labels [27]. The effectiveness of antioxidant compounds to scavenge free radicals is frequently determined using in vitro assays, such as the DPPH, FRAP, and ABTS assays [28]. These assays often measure the effectiveness of a coating material to scavenge free radicals.

Several kinds of antioxidants have been used to increase the antioxidant activity of edible films, including essential oils, phytochemicals, organic nanoparticles, and inorganic nanoparticles. For instance, López-Mata et al. (2018) reported that incorporating α-cinnamaldehyde into chitosan films increased their antioxidant properties. Qin et al. [29] showed that adding betacyanins to PVA/starch-based packaging films increased their radical scavenging activities in a dose-dependent manner. Moreover, Sholichah, Nugroho [30] reported that including quercetin in packaging films increased their antioxidant activity. The antioxidant activity of phytochemicals is closely attributed to the presence of numerous hydroxyl (-OH) groups in their structures and their electron-donating capacity to reactive free radicals during oxidation, which can neutralize free radical chain reactions [30].

The antimicrobial activity of food packaging materials is another important factor to consider when developing edible coatings. Natural substances that exhibit good antimicrobial activity, such as essential oils, phytochemicals, organic nanoparticles, and inorganic nanoparticles, can be incorporated into biodegradable packaging materials. For instance, Mohamad, Mazlan [31] showed that the antimicrobial activity of poly (lactic acid) films was increased by incorporating thymol, kesum, and curry essential oils. In another study, Chen, Zong [32] reported that the incorporation of cinnamaldehyde into PVA/starch films increased their ability to inhibit *Salmonella putrefaciens*. In addition, other kinds of active additives have also been incorporated into packaging materials to enhance their antioxidant activity, such as ZnO nanoparticles in chitosan/CMC films [33], ε-poly lysine in sodium lactate/whey protein films [34], anthocyanins in chitin/methylcellulose films [35], and silver nanoparticles in PVA/starch films [36].

2.2.2. Light Blockers

Many foods contain ingredients that are susceptible to degradation when exposed to light, especially electromagnetic radiation in the ultraviolet region. For instance, the chemical degradation of carotenoids, curcuminoids, or omega-3 fatty acids is accelerated in the presence of light [37]. Consequently, it is often important to design packaging materials that can block light from entering the food. Light absorbers and scatterers are substances that can block light, thereby protecting food components from photodegradation reactions [38]. Light scatterers are particulate materials with dimensions close to the wavelength of light, so they scatter light strongly, thereby blocking the ability of light to enter the packaging material and damage the food. However, these materials also make the packaging material appear cloudy or opaque. Light absorbers are chromophores that selectively absorb light waves over certain wavelength ranges and that can also be included in packaging materials to protect packaged foods from photodegradation. A wide variety of UV-protective chromophores have been studied for this purpose, including proteins, natural pigments, anthraquinone, lignin, flavonoids, tannin, curcuminoids, chalcones, and bixin [37–39]. As an example, lignin has chromophore functional groups (e.g., aromatic rings, conjugated carbonyl groups, and C=C bonds) that can absorb a broad spectrum of UV light (250–400 nm) [37]. Consequently, they can be used as light blockers to protect photo-labile substances from degradation when exposed to light.

2.2.3. Barrier Enhancers

The safety, quality, and shelf life of packaged foods are mainly influenced by the transfer of certain molecules, such as gases (such as O_2, CO_2, water vapor, or organic vapor) or liquids (such as water or oil), between the packaging materials and the surrounding environment as well as by the diffusion of other ingredients through the packaging film, including nanoparticles [40]. Consequently, additives are required to control the movement of different substances through packaging materials and to control the rate of oxidation reactions, microbial growth, enzymatic browning, and other processes responsible for changes in the look, feel, taste, and nutrition of foods. Controlling the oxygen and water vapor permeability (WVP) of films is critical for many applications due to the important role oxygen and water play in various chemical reactions and in microbial growth. Therefore, low oxygen and water vapor permeability are generally required for food packaging materials to minimize oxygen and moisture transfer between the food and the surrounding environment [41,42].

The permeability of natural or synthetic polymer-based films depends on their thickness, porosity, integrity, and rheology. Therefore, it can be modified and controlled by incorporating various kinds of additives into the films, such as blockers, plasticizers, or crosslinking agents [40]. These substances may either decrease or increase a film's permeability depending on their effects on the polymer chain interactions, the ratio between any crystalline and amorphous zones, the degree of porosity, and the hydrophilic/hydrophobic ratio. Recently, Tanwar, Gupta [43] reported that the addition of coconut shell extract increased the WVP of PVA/starch films, which may have been because of the hydrophilic nature of the components of the prepared films. In contrast, Ceballos, Ochoa-Yepes [44] reported that incorporating yerba mate extract into starch films decreased their permeability to water vapor and oxygen. Consequently, an appropriate additive must be selected for the required application.

2.2.4. Mechanical Modulators

A food packaging material is expected to possess certain mechanical properties including flexibility, stretchability, integrity, and strength to protect the food throughout the distribution chain. The mechanical properties of packaging materials can be assessed by various parameters such as their tensile strength (TS), elastic modulus (EM), and elongation at break (EAB) [40,45]. Various kinds of additives incorporated into packaging materials may either positively or negatively influence their mechanical properties. In addition, several factors including the type or nature of biopolymer, as well as the number and strength of the interactions between the polymer molecules, can impact the mechanical properties of packaging systems [45]. As an example, it was reported that adding grapefruit seed extract and TiO_2 nanoparticles reduced the TS and EM of corn starch–chitosan films, while the EAB increased significantly ($p < 0.05$) [46]. In another study, the TS of CMC films was reported to decrease from 37 to 23 MPa, the EM to decrease from 114 to 41 MPa, and the EAB to increase from 32 to 53% after adding α-tocopherol nanocapsules [47]. Chen, Zong [32] reported that incorporating cinnamaldehyde into PVA/starch-based films decreased the TS and increased the EAB of the films. This change in TS can be partially related to the heterogeneous film structure with a discontinuous phase created after adding the cinnamaldehyde. The increase in EAB can be partially attributed to the plasticizing effect of this essential oil [32].

3. Non-Thermal Methods in Combination with Food Coating Materials

3.1. High Hydrostatic Pressure (HHP)

HHP is a non-thermal process in which a pressure of 100 to 1000 MPa is applied to a food, which can be either liquid or solid food [48]. Typically, the temperatures used are below those normally utilized in traditional thermal processing operations. However, the temperature does increase as the pressure increases, by almost 3 °C per 100 MPa, which has to be taken into account [49]. A commercial-scale high-pressure processing time is around

20 min. HHP is a simple, flexible, and reliable process that does not require the use of additives; as a result, it has been successfully applied to food products [50,51]. This process has been used commercially in various products such as fruit juice, jam, jelly, sauces, meat, fish, ready-to-eat products, and yogurts [52]. In addition, new applications of HHP are being developed in the pharmaceutical and medical fields [53]. HHP can be applied to products that are packaged into flexible containers (high-pressure-resistant packaging). The pressure is applied to a chamber containing a liquid medium (commonly water), causing it to be uniformly and instantly transmitted all over the sample, independent of its shape, size, and composition [54].

HHP can be used for various purposes in food applications, including inhibiting bacterial growth, inactivating enzymes, prolonging shelf life, maintaining natural nutrients, improving sensory attributes, and increasing desirable properties (digestibility) [55,56]. It has been reported that HHP can inactivate microorganisms by breaking non-covalent bonds and damaging cell membranes [57]. The pressure applied in this process has a very small effect on covalent bonds. The combined use of HHP and an edible coating can be applied as a two-hurdle factor approach to reduce the survival of microorganisms, inactivate enzymes, and enhance the quality of food products [58].

Table 1 summarizes recent studies related to the combined use of HHP and edible coating on various products. Gómez-Estaca, López-Caballero [59] applied high-pressure processing (250 MPa) and an edible film composed of gelatin, chitosan, and clove essential oil on vacuum-packed salmon carpaccio. The combined approach reduced the total viable bacteria (TVC), *pseudomonads*, H_2S-producing organism, and enterobacteria content. Donsì, Marchese [60] studied the effects of combining a modified chitosan coating with an HHP treatment on the color, firmness, and microbial (*Listeria innocua*) count of green beans during storage for 14 days at 4 °C. The green beans were coated by spraying a modified chitosan solution containing a nanoemulsion of mandarin essential oil for 10 s, and then they were inoculated with 10^7 cfu/g of *L. innocua*. The coated green beans were packed in multilayer polymer/aluminum/polymer film and then exposed to HHP treatment at pressure levels of 200, 300, or 400 MPa for 5 min at 25 °C. According to the results, combining the coating with 200, 300, and 400 MPa pressure declined the population of *L. innocua* by 1.6 to 3.5 logs. This combined treatment improved the firmness of the green beans due to the ability of the pressure to thicken the cell walls. This treatment also led to a significant color change: the L* (darkness) and b* (yellowness) values of the green beans decreased, while the a* (greenness) values increased. This may be due to the disruption of the chloroplasts and leakage of chlorophyll, which was indicated by a bright green color on the surface. Gonçalves, Gouveia [61] produced cellulose acetate films with oregano essential oils using a casting method and then subjected the films to an HHP treatment at pressures of 300 or 400 MPa for 5 or 10 min. The ability of the films to inhibit *L. monocytogenes*, *S. aureus*, and *E. coli* on Coalho cheese was measured during 3 weeks of storage at 4 °C. At the end of storage, the microbial count for the three types of microorganisms was reduced by using the combined treatment.

Table 1. The effects of non-thermal processes combined with edible coating on food product quality and safety.

Product	Type of Food	Type of Process	Process Conditions	Polymer	Concentration of Polymer (%w/v)	Active Packaging Materials	Significant Results	Ref
Meat	Rainbow trout fillets	High hydrostatic pressure (HHP)	220 MPa, 15 °C, 5 min	Chitosan	1.5	-	Slight change in major bond of sarcoplasmic and myofibrillar muscle fractions	[62]
	Rainbow trout fillets	HHP	220 MPa, 15 °C, 5 min	Chitosan	1.5	-	Extend the shelf life by about 24 days	[63]
	Trout fillets	HHP	300 MPa, 12 °C, 10 min	Chitosan	1.5	Clove EO	Strong additive antimicrobial effect against mesophilic aerobic and coliform bacteria	[64]
	Cured Iberian ham	HHP	600 MPa, 8 min	Chitosan	2	Nisin, Rice bran extract	6 Log CFU/g of *L. monocytogenes* reduction	[65]
	Fermented sausages	HHP	600 MPa, 12 °C, 5 min	PVOH	13	Nisin	No extra protection on *L. monocytogenes*	[66]
	Chicken	γ-irradiation	2.5 kGy	Chitosan	2 0.1	Grape seed extract	Reduction of bacterial growth Increasing shelf life	[67]
	Minced chicken thigh	γ-irradiation	0, 2, 4, and 6 kGy	Pectin	~3	Papaya leaf extract	Improving the quality and safety of minced chicken thigh meat Reduced the initial total bacterial count, psychrophilic bacteria, and LAB Prolonged shelf life	[68]
	Carp fillets	Irradiation	3 kGy	Chitosan	2	Rose polyphenols	Extending the shelf life of fish Preserving sensory quality Preventing bacterial growth, oxidation, and changes in color	[69]
	Carp fillets	γ-irradiation	0, 1, 3, and 5 kGy	Calcium caseinate	4.7	Rosemary Oil	Increasing in the bacterial inhibitory effect Improving the quality and safety Extending the refrigerated shelf life	[70]
	Minced meat	γ-irradiation	3 kGy	CMC Chitosan PC	3 0.5 3	ZnO	Improving microbiological, chemical, and sensory quality Increasing the chilling life of minced meat	[71]
	Carp fillets	Ultrasound	40 KHz	Chitooligoaccharides	1	-	High score of sensory properties for coating and ultrasound Increased shelf life by 11 days 1.40 Log CFU/g of TVC reduction Applying coating with ultrasound led to reduction of TVB-N by 37%	[72]

Table 1. Cont.

Product	Type of Food	Type of Process	Process Conditions	Polymer	Concentration of Polymer (%w/v)	Active Packaging Materials	Significant Results	Ref
Fruit & vegetable	Fresh-cut Apple	Pulsed light (PL)	12 J/cm^2	Gellan	0.5	Ascorbic acid	Delayed the microbiological spoilage Preserved the sensory quality Decreased softening and browning of apple slices	[73]
	Fresh-cut Apple	PL	0.4 J/cm^2 per pulse	Pectin	2	Ascorbic acid	Reduced browning and softening of apple slices Led to 2 log CFU/g decline of microbial papulation Preserved sensory characteristics	[74]
	Fresh-cut cantaloupe	PL	0.9 J/cm^2 every 48 h up to 26 days	Sodium alginate	1.86	-	Compared with PL, alginate coating revealed more effectiveness in preserving high pectin content in cantaloupe slices. PL treatment was more effective than alginate coating in maintaining hemicellulose The combination of PL treatment with alginate manifested a synergistic effect on maintaining the overall cell wall fractions and cell wall integrity of cantaloupes	[75]
	Fresh-cut cucumber slices	PL	4, 8, and 12 J/cm^2	Chitosan	2	Carvacrol EO	Coating was less effective on *E. coli ATCC 26* reductions. PL treatments showed more effectiveness on microbial inactivation The inactivation of *E. coli ATCC 26* increased by increasing PL fluences Applying chitosan coating containing 0.08% carvacrol in combination with PL treatment (12 J/cm^2) led to reduction of more than 5 log cycles in the *E. coli* population	[76]
	Tomatoes	PL	2, 4, and 8 J/cm^2	Sodium alginate	0.5	Oregano EO	Applying coating containing 0.17% Oregano EO in combination with PL treatment (4 J/cm^2) led to reduction in the TVC, yeast, and mold	[77]
	Apple cubes	HHP	400 MPa, 35 °C, 5 min	Alginate	2	Vanillin	Reduction of *E. coli* by >5 log Reduced color changes Maintain firmness Increased phloridzin concentration (17%)	[78]
	Fresh-Cut Kiwifruit	Ultrasound	40 KHz, 350 W, 10 min	Chitosan	1	ZnO	Reduced ethylene, carbon dioxide production, and water loss with combination treatment with 1.2 g/L ZnO	[79]
	Fresh-Cut Cucumber	Ultrasound	20 kHz, 400 W, 10 min	Chitosan	1	Carbon dots	5.18 log CFU/g of microbial papulation reduction 3.45 log CFU/g of mold and yeast reduction Reduced respiration rate and weight loss Increased TSS, brix, and ascorbic acid amount Maintain flavor and taste	[80]

Table 1. Cont.

Product	Type of Food	Type of Process	Process Conditions	Polymer	Concentration of Polymer (%w/v)	Active Packaging Materials	Significant Results	Ref
	Pumpkin	Ultrasound	40 KHz, 150 W	Sodium alginate	3	-	Reduced processing time and solid uptake Increased water removal rate Improved texture	[81]
	Bell pepper	UV-C irradiation	254 nm, at 8 ± 1 °C, 24 days, 80%–85% RH	Aloe gel cinnamon oil chitosan	(1.5 and 2.5) (0.30 and 0.40) (1 and 1.5)	Cinnamon oil	Improving the quality of fruit Reduction in softening, weight loss, and electrolyte leakage	[82]
	Plum	γ-irradiation	1.5 kGy, 25 ± 2 °C, RH 70% and 3 ± 1 °C, RH 80%	CMC	0.5–1.0	-	Maintaining the storage quality Delaying the decaying Reduction in yeast and mold count	[83]
	Cherry	γ-irradiation	1.2 kGy, 25 ± 2 °C, RH 70% and 3 ± 1 °C, RH 80%, at 28 days	CMC	0.5–1.0	-	Maintaining the storage quality Delaying the decaying Delaying the onset of mold growth	[84]
	Jujube	Ultraviolet irradiation	253.7 nm, 4, 6, 8, and 10 min	Chitosan	1, 1.5, 2, and 2.5	-	Reduction of decay incidence Restraining increase in respiration rate, weight loss, malonaldehyde content, and electrolyte leakage Maintaining the activities of superoxide dismutase, peroxidase, and catalase at higher level Restraining decrease in ascorbic acid and chlorophyll	[85]
	Green beans	γ-irradiation	0.25 kGy	Chitosan	3	Mandarin EO	Reduction in microbial population and controlling their growth	[86]
	Carrot	γ-irradiation	0.5 kGy	Calcium caseinate	5	Cinnamon, citronella, lemongrass, and oregano EOs	NO significant effect on weight loss, color, or firmness Decreased the TMF and yeast and mold count after 7 days	[87]
	Peanut	Ultrasound	25, 40, and 80 kHz	WPI Zein CMC	11 15 0.5	-	Delayed hexanal formation (11% for CMC, 48% for WPI)	[88]

3.2. Ultrasound

Ultrasonic technologies involve the use of oscillating pressure waves with frequencies typically in the range from about 20 kHz to 10 MHz in most industrial applications [89]. Based on the magnitude of the intensities employed, ultrasound can be classified as high intensity (destructive), which is used to change the properties of foods, or low intensity (non-destructive), which is used to measure the properties of foods [90]. Two essential requirements of this method are a source of ultrasound and a condensed medium [91]. Ultrasonic waves are typically applied to liquid, semi-solid, or solid systems [92]. Samples can be treated with ultrasound irradiation by immersing them within an ultrasonic bath or by directing the pressure waves generated by an ultrasonic probe onto them [93]. In some cases, ultrasonic waves are directly applied to the surfaces of samples, whereas in other cases, they may pass through the air first [94].

Ultrasonic treatments can inactivate bacteria and enzymes, which is useful for improving the shelf life and safety of foods [95]. The high-intensity ultrasound technologies used in food processing can cause physical and/or chemical changes in foods through cavitation, which involves the formation and rapid collapse of gas bubbles in fluids in the presence of fluctuating pressure waves [96]. Cavitational forces can break up structures within foods as well as accelerate mass transfer processes. Coating foods with edible films alters the effects of ultrasound on mass transfer processes [97].

Edible coatings and ultrasound can be used in combination to minimize quality deterioration in foods. For instance, the peanut samples were first subjected to ultrasonic treatments (25, 40, and 80 kHz/10 min) and subsequently dipped in carboxymethyl cellulose (CMC) solution containing α-tocopherol, rosemary, and tea extracts, after which there was a striking increase in the oxidative stability of peanuts stored for 12 weeks at 35 °C [98]. This effect was attributed to the ability of sonication to remove some of the surface lipids from the peanuts, as well as to the barrier properties provided by the coatings. Reducing the amount of surface lipids available to react with oxygen reduced lipid oxidation. A combination of sonication and a CMC coating has also been shown to improve the quality and nutritional profile of banana slices [99]. In this case, the banana slices were first coated by immersion in CMC solutions, and then they were sonicated.

Researchers have evaluated the effects of combining sonication with chito-oligosaccharide (COS) coatings on the microbial and chemical properties of grass carp fillets during 12 days of storage at 4 °C [72]. The combined treatment was shown to reduce the chemical and microbial deterioration of the fish, thereby extending its shelf life considerably. Moreover, no deleterious effects of the combined treatment on the sensory properties of the fish were observed.

3.3. Pulsed Light

Pulsed light (PL) treatment is a non-thermal processing method that can be used for the rapid inactivation of microorganisms on food surfaces and packaging materials [100]. PL technology involves the use of intense light pulses of short duration and a broad wavelength spectrum [101]. The PL generation system comprises one or more inert-gas flash lamps (e.g., xenon lamps), a power unit, and a high-voltage connection. When a high-current electric pulse passes through the gas chamber of the lamp, the inert gas molecules are excited and collide with each other, leading to the emission of short intense pulses of light with wavelengths ranging from around 200 to 1100 nm [101–103]. This range includes ultraviolet (200–400 nm), visible (400–700 nm), and infrared (700–1100 nm) light [102]. In food applications, PL usually involves applying 1 to 20 flashes per second with an energy density ranging from 0.01 to 50 J/cm^2 at the surface [100].

Microbial decontamination by PL treatments has mainly been attributed to UV light [103]. Conjugated carbon–carbon double bonds in proteins and nucleic acids absorb ultraviolet radiation, which leads to structural changes in enzymes, receptors, transporters, membranes, and genetic materials, thereby causing disruption of key biochemical pathways that lead to cell death [100,103]. Moreover, applying ultraviolet light on the target surface stimulates the generation of reactive oxygen species (ROS) such as H_2O_2, single oxygen, and hydroxyl radicals that affect the cell membranes and cell walls [104,105].

The application of PL technology for food preservation has some benefits over conventional methods, such as efficient inactivation of microorganisms, no need for chemical disinfectants or preservatives, low operation costs, the capability of either continuous or batch operation, short processing times, and high throughputs [100]. Nevertheless, it does have some limitations. Foods with smooth surfaces, such as many fresh fruits and vegetables, cheeses, and meat slices are suitable for PL treatment, while foods with uneven or porous surfaces are unsuitable because shadow effects reduce the ability of the light waves to interact with all of the surfaces [106]. Because PL technology is a surface decontamination technique, it is affected by the light scattering and absorption properties of foods, which means that it is unsuitable for the treatment of grains, cereals, and spices due to their opaque nature [100,106]. Other potential drawbacks of this technology are the high initial investment costs, the short lifetime of lamps, the potential for changes in pH and color at high intensities, and overheating [22].

Combining edible packaging and PL treatments has been shown to have synergistic benefits on food preservation by increasing microbial decontamination [23]. Studies on the combination of PL and edible packaging are summarized in Table 1, and a few examples are provided here.

Researchers have evaluated the effects of various combinations of alginate coating, malic acid dipping, and PL treatment on the quality of fresh-cut mango during 14 days of storage at 4 °C [24]. Fresh mango slices inoculated with *L. innocua* were dipped in sodium alginate solution (2% w/v) and then dipped in a calcium chloride solution (2% w/v) containing malic acid (2% w/v) or in a malic acid solution (2% w/v). The PL treatment involved applying 20 pulses with a fluence of 0.4 J.cm^2/pulse. This study showed that combined treatments led to around a 4 log reduction in *L. innocua* in the mango. Coating the mango pieces prior to the PL treatment helped to avoid tissue softening during storage.

Koh, Noranizan [26] assessed the effects of an alginate coating followed by a PL treatment on the sensory properties of fresh-cut cantaloupes during 36 days of storage at 4 °C. The fresh-cut cantaloupes were coated by dipping them in an alginate solution containing glycerol and sunflower oil. The coated cantaloupes were then packed in polypropylene bags and exposed to the PL at a fluence of 0.9 J/cm^2 every 48 h for up to 26 days. Combining the alginate coating with the PL treatment reduced the decrease in the sugar content of the cantaloupes during storage. The alginate coating was more effective than the PL treatment when they were used alone in preventing changes in the organic acid content of the cantaloupes. However, the combination of the alginate coating and PL treatment reduced the formation of lactic acid and helped preserve the desirable aroma profile of the cantaloupes.

Researchers have assessed the effects of combining a PL treatment with starch films containing preservatives (sodium benzoate and/or citric acid) on microbial growth and the quality of Cheddar cheese slices during refrigerated storage [2]. The surfaces of the cheese slices were first inoculated with *L. innocua* at a level of 7 log CFU/cheese slice, which were then coated or not coated before being exposed to the PL treatment. The results showed that combining the coatings and PL treatments greatly reduced the number of *L. innocua* on the cheese surfaces during storage. However, there were some undesirable changes in the quality attributes of the cheese caused by the treatments. After 7 days of storage, the pH value of cheese reduced to 4.0, which resulted in increased cheese hardness.

3.4. Irradiation

Food irradiation involves exposing foods to a controlled level of ionizing radiation, which has the ability to break chemical bonds and deactivate microorganisms [107]. Irradiation has been used to kill harmful microorganisms in poultry, meat, seafood, and spices; extend the storage time of fresh vegetables and fruits; and control the sprouting of tubers, onions, and potatoes. The dose of radiation used depends on the application: low dose (1 kGy) to delay ripening and prevent germination, medium dose (1–10 kGy) to kill pathogens, and high dose (>10 kGy) for disinfection and sterilization [108]. According to the World Health Organization (WHO), there is no risk of applying irradiation to foodstuffs at the levels normally used, and it may even help maintain the nutritional content [109]. Irradiated food products must comply with strict international regulations with regard to safety [110]. Three types of ionizing radiation are commonly used for this purpose: ultraviolet light, gamma rays, and electron beams.

UV light, which has a wavelength ranging from 100 to 400 nm, is that part of the electromagnetic spectrum that falls between visible light and ionizing radiation. The nucleic acids of microorganisms absorb UV light strongly between 250 and 260 nm. Microorganisms are destroyed when they are exposed to sufficiently high intensities of UV light due to changes in the molecular structure of nucleic acids and proteins that disrupt their metabolism [87]. Ultraviolet light can also directly damage the ester bonds in key molecules in microorganisms, either by directly absorbing UV energy or by generating reactive species, such as oxygen-free or hydroxyl radicals, which react with them [111]. For instance, it has been reported that the antimicrobial activity of UV light towards various microorganisms is due to the formation of pyrimidine dimers in the DNA strands [20,112]. Researchers have reported that applying low doses of UV-C light (254 nm) to fruits and vegetables can reduce their tendency to rot during storage, thereby increasing their quality and shelf life [113].

Electron irradiation involves creating an electron beam using a cathode, which is then directed at the sample to be treated. At sufficiently high energy, the electron beam is capable of breaking molecular bonds or releasing electrons from atoms, which can lead to the deactivation of microorganisms [114]. The radiation dose required to have a beneficial effect depends on the nature of the food being treated, so it must be optimized for each product. A major benefit of electron irradiation is that no pretreatment of the samples is required and the processing times are relatively short [108].

Gamma rays are electromagnetic radiation with a relatively short wavelength and high frequency that can easily penetrate foods with little or no heat generation [115]. This method is already used commercially to sterilize a variety of foods. Cesium (137) or cobalt (60) radionuclides are gamma ray sources that have been used in biological applications for decades [116]. A commonly used gamma ray supply consists of cobalt 60 rods contained within rustproof steel tubes. These tubes are raised within a concrete irradiation crate containing the food. Studies have shown that irradiating foods with bioactive coatings or in modified atmospheric packaging helps to enhance the radiation sensitivity of food pathogens without negatively impacting the sensory properties of the food products [117].

Several studies have examined the combined impact of irradiation and coatings on the quality attributes of food products. For instance, combining alginate coatings (containing essential oils, sodium diacetate, and natamycin) with γ-radiation (0.4 and 0.8 kGy) was shown to be effective in decreasing the viability of several spoilage and pathogenic microorganisms (*A. niger*, *E. coli*, *L. monocytogenes*, and *S. Typhimurium*) on broccoli florets under refrigeration conditions, thereby increasing their shelf life [117]. In another study, it was shown that a combination of an edible coating and γ-irradiation (0.5 kGy) was effective in reducing *E. coli*, *Salmonella enteric*, and *L. innocua* on green peppers without adversely affecting their quality attributes [10]. Other researchers have shown that combining a chitosan coating (loaded with a mandarin essential oil nanoemulsion) with a UV-C irradiation treatment decreased the levels of *L. innocua* contamination on green beans while also improving their firmness and color retention [20]. Similarly, combining CMC coatings with UV-C or γ-irradiation inhibited the growth of *L. innocua* in pears, thereby extending their shelf life and quality attributes [118]. A combination of an alginate coating (loaded with essential oils and citrus extract), ozonation, and irradiation has also been shown to increase the shelf life of fish fillets (Figure 2) [119]. Similarly, combining a chitosan coating (loaded with cumin essential oil) and γ-irradiation (2.5 kGy) was shown to reduce the growth of *L. monocytogenes*, *E. coli* O157:H7, and *Salmonella Typhimurium* on beef [120]. Salem, Naweto [121] showed that a combination of γ-irradiation (0.5 and 1.0 kGy) and a paraffin oil coating reduced the levels of blue mold (*Penicillium expansum*) on apples during cold storage. The coated and irradiated apples had the lowest weight loss, highest firmness, highest calcium levels, and longest shelf life. Similarly, a combination of electron beam radiation (0, 0.5, and 1 kGy) and a shellac coating was shown to reduce changes in the color, chlorophyll levels, and chlorophyllase activity of pears during storage at 13 °C for 30 days while increasing the rate of respiration and vitamin C concentration (Figure 3) [108]. Other researchers showed that combining an ultraviolet light treatment with a chitosan coating improved the quality and nutritional content of strawberries during storage of 15 days at 1 °C and 90% relative humidity [114]. Although irradiation methods (pulsed light or UV) in combination with food coatings have successful effects on the preservation of coated food, some bioactive compounds or nanomaterials such as anthocyanins, quercetin, some essential oils, nanoparticles, etc., have the property of blocking the irradiated rays. Therefore, there may be a need to use a higher dose of antimicrobial or irradiated radiation, which should be considered in future studies.

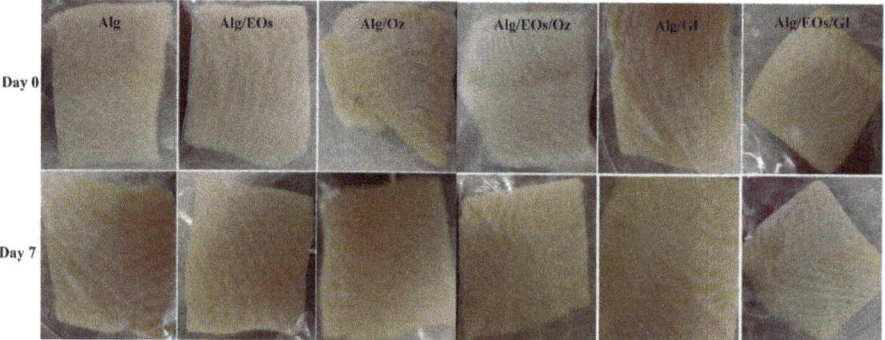

Figure 2. Effect of alginate coating with ozonation or gamma irradiation on *Merluccius* sp. fillets. Reprinted from [96], copyright 2019, with permission from Elsevier.

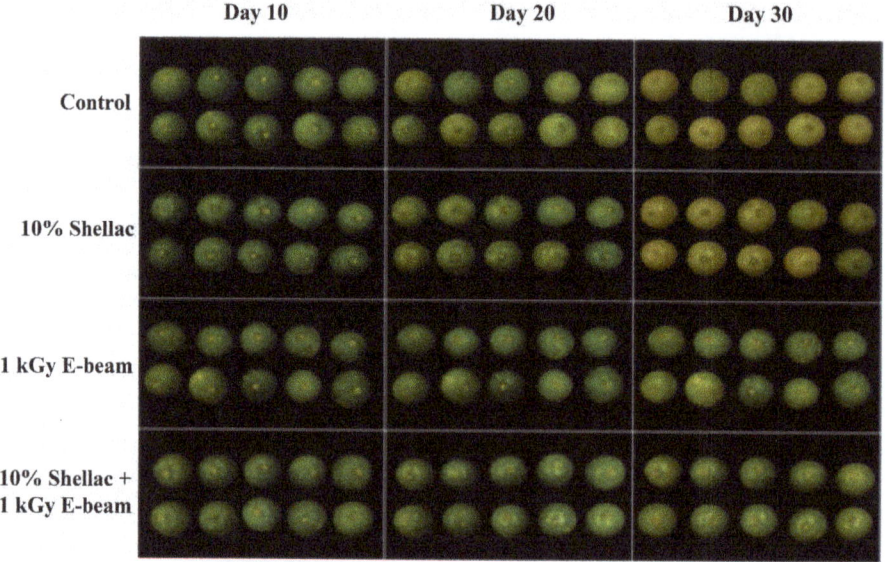

Figure 3. Effect of shellac coating with E-beam irradiation on lime. Reprinted from [83], copyright 2021, with permission from Elsevier.

4. Conclusions

In this review article and according to the reported results, it was concluded that a combination of biodegradable coatings and non-thermal processing methods can be used to improve the quality, safety, and shelf life of various kinds of food products. The utilization of this approach may reduce the need for plastic packaging materials and synthetic chemicals, which can adversely affect human health and the environment. However, further research is required to ensure that these technologies are safe and efficacious to employ under realistic usage conditions and that they can be performed economically at the large scale required for industrial applications. If these hurdles can be overcome, then combining biodegradable coatings and non-thermal processing methods may be a means of improving the sustainability and reducing the negative environmental impact of the food supply chain.

Author Contributions: Conceptualization, A.E., D.J.M., A.K. and M.A.S.; methodology and software, A.K., R.A.-F. and H.Z.; validation, S.F., A.E. and M.A.S.; formal analysis, A.E., B.N. and M.J.; investigation, A.E., M.A.S. and D.J.M.; resources, A.E., M.A.S., A.K. and D.J.M.; data curation, A.K., M.J., H.Z. and S.F; writing—original draft preparation, A.K., R.A.-F.; writing—review and editing, D.J.M., A.E. and B.N.; visualization, A.E. and D.J.M.; supervision, A.E., B.N. and D.J.M.; project administration, D.J.M. and A.E.; funding acquisition, D.J.M. All authors have read and agreed to the published version of the manuscript.

Funding: This research received no external funding.

Institutional Review Board Statement: Not applicable.

Informed Consent Statement: Not applicable.

Data Availability Statement: Not applicable.

Conflicts of Interest: The authors declare no conflict of interest.

Sample Availability: Samples of the compounds are not available from the authors.

References

1. Flynn, K.; Villarreal, B.P.; Barranco, A.; Belc, N.; Björnsdóttir, B.; Fusco, V.; Rainieri, S.; Smaradóttir, S.E.; Smeu, I.; Teixeira, P.; et al. An introduction to current food safety needs. *Trends Food Sci. Technol.* **2018**, *84*, 1–3. [CrossRef]
2. De Moraes, J.O.; Hilton, S.T.; Moraru, C.I. The effect of Pulsed Light and starch films with antimicrobials on Listeria innocua and the quality of sliced cheddar cheese during refrigerated storage. *Food Control* **2020**, *112*, 107134. [CrossRef]
3. Yong, H.I.; Kim, H.-J.; Park, S.; Kim, K.; Choe, W.; Yoo, S.J.; Jo, C. Pathogen inactivation and quality changes in sliced cheddar cheese treated using flexible thin-layer dielectric barrier discharge plasma. *Food Res. Int.* **2015**, *69*, 57–63. [CrossRef]
4. Khezerlou, A.; Jafari, S.M. Nanoencapsulated bioactive components for active food packaging. In *Handbook of Food Nanotechnology*; Elsevier: Amsterdam, The Netherlands, 2020; pp. 493–532.
5. Silva, M.M.; Lidon, F. Food preservatives—An overview on applications and side effects. *Emir. J. Food Agric.* **2016**, *28*, 366–373. [CrossRef]
6. Rangan, C.; Barceloux, D.G. Food Additives and Sensitivities. *Disease-A-Month* **2009**, *55*, 292–311. [CrossRef]
7. De Corato, U. Improving the shelf-life and quality of fresh and minimally-processed fruits and vegetables for a modern food industry: A comprehensive critical review from the traditional technologies into the most promising advancements. *Crit. Rev. Food Sci. Nutr.* **2019**, *60*, 940–975. [CrossRef] [PubMed]
8. Al-Maqtari, Q.A.; Rehman, A.; Mahdi, A.A.; Al-Ansi, W.; Wei, M.; Yanyu, Z.; Phyo, H.M.; Galeboe, O.; Yao, W. Application of essential oils as preservatives in food systems: Challenges and future prospectives—A review. *Phytochem. Rev.* **2021**, *21*, 1209–1246. [CrossRef]
9. Chauhan, O. Combination of Non thermal Processes and Their Hurdle Effect. In *Non-Thermal Processing of Foods*, CRC Press: Boca Raton, FL, USA, 2019; pp. 329–372.
10. Maherani, B.; Harich, M.; Salmieri, S.; Lacroix, M. Antibacterial properties of combined non-thermal treatments based on bioactive edible coating, ozonation, and gamma irradiation on ready-to-eat frozen green peppers: Evaluation of their freshness and sensory qualities. *Eur. Food Res. Technol.* **2018**, *245*, 1095–1111. [CrossRef]
11. Pop, O.L.; Pop, C.R.; Dufrechou, M.; Vodnar, D.C.; Socaci, S.A.; Dulf, F.V.; Minervini, F.; Suharoschi, R. Edible Films and Coatings Functionalization by Probiotic Incorporation: A Review. *Polymers* **2019**, *12*, 12. [CrossRef]
12. Dhall, R.K. Advances in Edible Coatings for Fresh Fruits and Vegetables: A Review. *Crit. Rev. Food Sci. Nutr.* **2013**, *53*, 435–450. [CrossRef]

13. Poonia, A.; Mishra, A. Edible nanocoatings: Potential food applications, challenges and safety regulations. *Nutr. Food Sci.* **2021**, *52*, 497–514. [CrossRef]
14. Ansorena, M.R.; Ponce, A.G. Coatings in the Postharvest. In *Polymers for Agri-Food Applications*; Springer: Berlin/Heidelberg, Germany, 2019; pp. 339–354.
15. Khezerlou, A.; Zolfaghari, H.; Banihashemi, S.A.; Forghani, S.; Ehsani, A. Plant gums as the functional compounds for edible films and coatings in the food industry: A review. *Polym. Adv. Technol.* **2021**, *32*, 2306–2326. [CrossRef]
16. Tkaczewska, J. Peptides and protein hydrolysates as food preservatives and bioactive components of edible films and coatings—A review. *Trends Food Sci. Technol.* **2020**, *106*, 298–311. [CrossRef]
17. Ribeiro, A.M.; Estevinho, B.N.; Rocha, F. Preparation and Incorporation of Functional Ingredients in Edible Films and Coatings. *Food Bioprocess Technol.* **2020**, *14*, 209–231. [CrossRef]
18. Khezerlou, A.; Azizi-Lalabadi, M.; Mousavi, M.M.; Ehsani, A. Incorporation of essential oils with antibiotic properties in edible packaging films. *J. Food Bioprocess Eng.* **2019**, *2*, 77–84.
19. Padhan, S. Hurdle technology: A review article. *Trends Biosci.* **2018**, *11*, 3457–3462.
20. Severino, R.; Vu, K.D.; Donsì, F.; Salmieri, S.; Ferrari, G.; Lacroix, M. Antibacterial and physical effects of modified chitosan based-coating containing nanoemulsion of mandarin essential oil and three non-thermal treatments against Listeria innocua in green beans. *Int. J. Food Microbiol.* **2014**, *191*, 82–88. [CrossRef]
21. Han, J.-W.; Ruiz-Garcia, L.; Qian, J.-P.; Yang, X.-T. Food Packaging: A Comprehensive Review and Future Trends. *Compr. Rev. Food Sci. Food Saf.* **2018**, *17*, 860–877. [CrossRef]
22. Heinrich, V.; Zunabovic, M.; Varzakas, T.; Bergmair, J.; Kneifel, W. Pulsed Light Treatment of Different Food Types with a Special Focus on Meat: A Critical Review. *Crit. Rev. Food Sci. Nutr.* **2015**, *56*, 591–613. [CrossRef]
23. Pirozzi, A.; Pataro, G.; Donsì, F.; Ferrari, G. Edible Coating and Pulsed Light to Increase the Shelf Life of Food Products. *Food Eng. Rev.* **2020**, *13*, 544–569. [CrossRef]
24. Salinas-Roca, B.; Soliva-Fortuny, R.; Welti-Chanes, J.; Martín-Belloso, O. Combined effect of pulsed light, edible coating and malic acid dipping to improve fresh-cut mango safety and quality. *Food Control* **2016**, *66*, 190–197. [CrossRef]
25. Forghani, S.; Almasi, H.; Moradi, M. Electrospun nanofibers as food freshness and time-temperature indicators: A new approach in food intelligent packaging. *Innov. Food Sci. Emerg. Technol.* **2021**, *73*, 102804. [CrossRef]
26. Koh, P.C.; Noranizan, M.A.; Karim, R.; Nur Hanani, Z.A. Sensory quality and flavour of alginate coated and repetitive pulsed light treated fresh-cut cantaloupes (*Cucumis melo* L. Var. *Reticulatus* Cv. *Glamour*) during storage. *J. Food Sci. Technol.* **2019**, *56*, 2563–2575. [PubMed]
27. Lu, W.; Shi, Y.; Wang, R.; Su, D.; Tang, M.; Liu, Y.; Li, Z. Antioxidant Activity and Healthy Benefits of Natural Pigments in Fruits: A Review. *Int. J. Mol. Sci.* **2021**, *22*, 4945. [CrossRef]
28. Cai, L.; Wang, Y. Physicochemical and Antioxidant Properties Based on Fish Sarcoplasmic Protein/Chitosan Composite Films Containing Ginger Essential Oil Nanoemulsion. *Food Bioprocess Technol.* **2021**, *14*, 151–163. [CrossRef]
29. Qin, Y.; Xu, F.; Yuan, L.; Hu, H.; Yao, X.; Liu, J. Comparison of the physical and functional properties of starch/polyvinyl alcohol films containing anthocyanins and/or betacyanins. *Int. J. Biol. Macromol.* **2020**, *163*, 898–909. [CrossRef] [PubMed]
30. Sholichah, E.; Nugroho, P.; Purwono, B. Preparation and characterization of active film made from arrowroot starch/PVA film and isolated quercetin from shallot (*Allium cepa* L. var. *aggregatum*). In *AIP Conference Proceedings*; AIP Publishing LLC: Melville, NY, USA, 2018; Volume 2024, p. 020013.
31. Mohamad, N.; Mazlan, M.M.; Tawakkal, I.S.M.A.; Talib, R.A.; Kian, L.K.; Fouad, H.; Jawaid, M. Development of active agents filled polylactic acid films for food packaging application. *Int. J. Biol. Macromol.* **2020**, *163*, 1451–1457. [CrossRef]
32. Chen, C.; Zong, L.; Wang, J.; Xie, J. Microfibrillated cellulose reinforced starch/polyvinyl alcohol antimicrobial active films with controlled release behavior of cinnamaldehyde. *Carbohydr. Polym.* **2021**, *272*, 118448. [CrossRef]
33. Lukic, I.; Vulic, J.; Ivanovic, J. Antioxidant activity of PLA/PCL films loaded with thymol and/or carvacrol using scCO2 for active food packaging. *Food Packag. Shelf Life* **2020**, *26*, 100578. [CrossRef]
34. Zinoviadou, K.G.; Koutsoumanis, K.P.; Biliaderis, C.G. Physical and thermo-mechanical properties of whey protein isolate films containing antimicrobials, and their effect against spoilage flora of fresh beef. *Food Hydrocoll.* **2010**, *24*, 49–59. [CrossRef]
35. Sani, M.A.; Tavassoli, M.; Hamishehkar, H.; McClements, D.J. Carbohydrate-based films containing pH-sensitive red barberry anthocyanins: Application as biodegradable smart food packaging materials. *Carbohydr. Polym.* **2021**, *255*, 117488. [CrossRef] [PubMed]
36. Cano, A.; Cháfer, M.; Chiralt, A.; González-Martínez, C. Development and characterization of active films based on starch-PVA, containing silver nanoparticles. *Food Packag. Shelf Life* **2016**, *10*, 16–24. [CrossRef]
37. Sadeghifar, H.; Ragauskas, A. Lignin as a UV light blocker—A review. *Polymers* **2020**, *12*, 1134. [CrossRef]
38. Kwon, S.; Orsuwan, A.; Bumbudsanpharoke, N.; Yoon, C.; Choi, J.; Ko, S. A Short Review of Light Barrier Materials for Food and Beverage Packaging. *Korean J. Packag. Sci. Technol.* **2018**, *24*, 141–148. [CrossRef]
39. Islam, M.T.; Repon, R.; Liman, L.R.; Hossain, M.; Al Mamun, A. Functional modification of cellulose by chitosan and gamma radiation for higher grafting of UV protective natural chromophores. *Radiat. Phys. Chem.* **2021**, *183*, 109426. [CrossRef]
40. Abedi-Firoozjah, R.; Yousefi, S.; Heydari, M.; Seyedfatehi, F.; Jafarzadeh, S.; Mohammadi, R.; Rouhi, M.; Garavand, F. Application of Red Cabbage Anthocyanins as pH-Sensitive Pigments in Smart Food Packaging and Sensors. *Polymers* **2022**, *14*, 1629. [CrossRef]

41. Yekta, R.; Mirmoghtadaie, L.; Hosseini, H.; Norouzbeigi, S.; Hosseini, S.M.; Shojaee-Aliabadi, S. Development and characterization of a novel edible film based on Althaea rosea flower gum: Investigating the reinforcing effects of bacterial nanocrystalline cellulose. *Int. J. Biol. Macromol.* 2020, 158, 327–337. [CrossRef]
42. Yong, H.; Liu, J. Recent advances in the preparation, physical and functional properties, and applications of anthocyanins-based active and intelligent packaging films. *Food Packag. Shelf Life* 2020, 26, 100550. [CrossRef]
43. Tanwar, R.; Gupta, V.; Kumar, P.; Kumar, A.; Singh, S.; Gaikwad, K.K. Development and characterization of PVA-starch incorporated with coconut shell extract and sepiolite clay as an antioxidant film for active food packaging applications. *Int. J. Biol. Macromol.* 2021, 185, 451–461. [CrossRef]
44. Ceballos, R.L.; Ochoa-Yepes, O.; Goyanes, S.; Bernal, C.; Famá, L. Effect of yerba mate extract on the performance of starch films obtained by extrusion and compression molding as active and smart packaging. *Carbohydr. Polym.* 2020, 244, 116495. [CrossRef] [PubMed]
45. KKuorwel, K.K.; Cran, M.J.; Orbell, J.D.; Buddhadasa, S.; Bigger, S. Review of Mechanical Properties, Migration, and Potential Applications in Active Food Packaging Systems Containing Nanoclays and Nanosilver. *Compr. Rev. Food Sci. Food Saf.* 2015, 14, 411–430. [CrossRef]
46. Jha, P. Effect of grapefruit seed extract ratios on functional properties of corn starch-chitosan bionanocomposite films for active packaging. *Int. J. Biol. Macromol.* 2020, 163, 1546–1556. [CrossRef] [PubMed]
47. Mirzaei-Mohkam, A.; Garavand, F.; Dehnad, D.; Keramat, J.; Nasirpour, A. Physical, mechanical, thermal and structural characteristics of nanoencapsulated vitamin E loaded carboxymethyl cellulose films. *Prog. Org. Coatings* 2019, 138, 105383. [CrossRef]
48. Khaliq, A.; Chughtai MF, J.; Mehmood, T.; Ahsan, S.; Liaqat, A.; Nadeem, M.; Sameed, N.; Saeed, K.; Ur Rehman, J.; Ali, A. High-Pressure Processing; Principle, Applications, Impact, and Future Prospective. In *Sustainable Food Processing and Engineering Challenges*; Elsevier: Amsterdam, The Netherlands, 2021; pp. 75–108.
49. Picart-Palmade, L.; Cunault, C.; Chevalier-Lucia, D.; Belleville, M.-P.; Marchesseau, S. Potentialities and Limits of Some Non-thermal Technologies to Improve Sustainability of Food Processing. *Front. Nutr.* 2019, 5, 130. [CrossRef]
50. Wang, C.-Y.; Huang, H.-W.; Hsu, C.-P.; Yang, B.B. Recent Advances in Food Processing Using High Hydrostatic Pressure Technology. *Crit. Rev. Food Sci. Nutr.* 2015, 56, 527–540. [CrossRef]
51. Liepa, M.; Zagorska, J.; Galoburda, R. High-Pressure processing as novel technology in dairy industry: A Review. *Res. Rural. Dev.* 2016, 1, 76–83.
52. Rathnakumar, K.; Martínez-Monteagudo, S.I. High-Pressure Processing: Fundamentals, Misconceptions, and Advances. *Ref. Modul. Food Sci.* 2019.
53. Diehl, P.; Schauwecker, J.; Mittelmeier, W.; Schmitt, M. High hydrostatic pressure, a novel approach in orthopedic surgical oncology to disinfect bone, tendons and cartilage. *Anticancer. Res.* 2008, 28, 3877–3883.
54. Knorr, D.; Jäger, H.; Reineke, K.; Schlüter, O.; Schössler, K. *Emerging and New Technologies in Food Science and Technology*; International Union of Food Science and Technology (IUFoST): Oakville, ON, Canada, 2010.
55. Hogan, E.; Kelly, A.L.; Sun, D.-W. 1—High Pressure Processing of Foods: An Overview. In *Emerging Technologies for Food Processing*; Sun, D.-W., Ed.; Academic Press: London, UK, 2005; pp. 3–32.
56. Wgiorgis, G.A.; Yildiz, F. Review on high-pressure processing of foods. *Cogent Food Agric.* 2019, 5, 1568725.
57. Marcos, B.; Aymerich, T.; Garriga, M. Evaluation of High Pressure Processing as an Additional Hurdle to Control Listeria monocytogenes and Salmonella enterica in Low-Acid Fermented Sausages. *J. Food Sci.* 2005, 70, m339–m344. [CrossRef]
58. Morris, C.; Brody, A.L.; Wicker, L. Non-thermal food processing/preservation technologies: A review with packaging implications. *Packag. Technol. Sci. Int. J.* 2007, 20, 275–286. [CrossRef]
59. Gómez-Estaca, J.; López-Caballero, M.E.; Martínez-Bartolomé, M.; de Lacey, A.M.L.; Gómez-Guillen, M.C.; Montero, M.P. The effect of the combined use of high pressure treatment and antimicrobial edible film on the quality of salmon carpaccio. *Int. J. Food Microbiol.* 2018, 283, 28–36. [CrossRef] [PubMed]
60. Donsì, F.; Marchese, E.; Maresca, P.; Pataro, G.; Vu, K.D.; Salmieri, S.; Lacroix, M.; Ferrari, G. Green beans preservation by combination of a modified chitosan based-coating containing nanoemulsion of mandarin essential oil with high pressure or pulsed light processing. *Postharvest Biol. Technol.* 2015, 106, 21–32. [CrossRef]
61. Günlü, A.; Sipahioğlu, S.; Alpas, H. The effect of high hydrostatic pressure on the muscle proteins of rainbow trout (Oncorhynchus mykiss Walbaum) fillets wrapped with chitosan-based edible film during cold storage (4 ± 1° C). *High Pressure Res.* 2014, 34, 122–132. [CrossRef]
62. Günlü, A.; Sipahioğlu, S.; Alpas, H. The effect of chitosan-based edible film and high hydrostatic pressure process on the microbiological and chemical quality of rainbow trout (Oncorhynchus mykiss Walbaum) fillets during cold storage (4 ± 1 °C). *High Press. Res.* 2014, 34, 110–121. [CrossRef]
63. Albertos, I.; Rico, D.; Diez, A.M.; González-Arnáiz, L.; García-Casas, M.J.; Jaime, I. Effect of edible chitosan/clove oil films and high-pressure processing on the microbiological shelf life of trout fillets. *J. Sci. Food Agric.* 2014, 95, 2858–2865. [CrossRef] [PubMed]
64. Martillanes, S.; Rocha-Pimienta, J.; Llera-Oyola, J.; Gil, M.V.; Ayuso-Yuste, M.C.; García-Parra, J.; Delgado-Adámez, J. Control of Listeria monocytogenes in sliced dry-cured Iberian ham by high pressure processing in combination with an eco-friendly packaging based on chitosan, nisin and phytochemicals from rice bran. *Food Control* 2021, 124, 107933. [CrossRef]

65. Marcos, B.; Aymerich, T.; Garriga, M.; Arnau, J. Active packaging containing nisin and high pressure processing as post-processing listericidal treatments for convenience fermented sausages. *Food Control* **2012**, *30*, 325–330. [CrossRef]
66. Hassanzadeh, P.; Tajik, H.; Rohani, S.M.R.; Moradi, M.; Hashemi, M.; Aliakbarlu, J. Effect of functional chitosan coating and gamma irradiation on the shelf-life of chicken meat during refrigerated storage. *Radiat. Phys. Chem.* **2017**, *141*, 103–109. [CrossRef]
67. Abdeldaiem, M. Using of combined treatment between edible coatings containing ethanolic extract of papaya (*carica papaya* L.) leaves and gamma irradiation for extending shelf-life of minced chicken meat. *Am. J. Food Sci. Technol.* **2014**, *2*, 6–16.
68. Zhang, Q.Q.; Rui, X.; Guo, Y.; He, M.; Xu, X.L.; Dong, M.S. Combined Effect of Polyphenol-Chitosan Coating and Irradiation on the Microbial and Sensory Quality of Carp Fillets. *J. Food Sci.* **2017**, *82*, 2121–2127. [CrossRef] [PubMed]
69. Abdeldaiem, M.H.; Mohammad, H.G.; Ramadan, M.F. Improving the Quality of Silver Carp Fish Fillets by Gamma Irradiation and Coatings Containing Rosemary Oil. *J. Aquat. Food Prod. Technol.* **2018**, *27*, 568–579. [CrossRef]
70. Sayed, W.; El-Banna, M.; Ibrahim, M. Improving Minced Meat Quality by Edible Antimicrobial Polymers and Gamma Radiation. *Egypt. J. Radiat. Sci. Appl.* **2019**, *32*, 245–253. [CrossRef]
71. Yu, D.; Zhao, W.; Yang, F.; Jiang, Q.; Xu, Y.; Xia, W. A strategy of ultrasound-assisted processing to improve the performance of bio-based coating preservation for refrigerated carp fillets (Ctenopharyngodon idellus). *Food Chem.* **2020**, *345*, 128862. [CrossRef] [PubMed]
72. Moreira, M.R.; Tomadoni, B.; Martín-Belloso, O.; Soliva-Fortuny, R. Preservation of fresh-cut apple quality attributes by pulsed light in combination with gellan gum-based prebiotic edible coatings. *LWT-Food Sci. Technol.* **2015**, *64*, 1130–1137. [CrossRef]
73. Moreira, M.R.; Álvarez, M.V.; Martín-Belloso, O.; Soliva-Fortuny, R. Effects of pulsed light treatments and pectin edible coatings on the quality of fresh-cut apples: A hurdle technology approach. *J. Sci. Food Agric.* **2017**, *97*, 261–268. [CrossRef]
74. Koh, P.C.; Noranizan, M.A.; Hanani, Z.A.N.; Karim, R.; Rosli, S.Z. Application of edible coatings and repetitive pulsed light for shelf life extension of fresh-cut cantaloupe (*Cucumis melo* L. reticulatus cv. Glamour). *Postharvest Biol. Technol.* **2017**, *129*, 64–78. [CrossRef]
75. Taştan, Ö.; Pataro, G.; Donsì, F.; Ferrari, G.; Baysal, T. Decontamination of fresh-cut cucumber slices by a combination of a modified chitosan coating containing carvacrol nanoemulsions and pulsed light. *Int. J. Food Microbiol.* **2017**, *260*, 75–80. [CrossRef] [PubMed]
76. Pirozzi, A.; Del Grosso, V.; Ferrari, G.; Pataro, G.; Donsì, F. Combination of edible coatings containing oregano essential oil nanoemulsion and pulsed light treatments for improving the shelf life of tomatoes. *Chem. Eng. Trans.* **2021**, *87*, 61–66.
77. Bambace, M.F.; Moreira, M.R.; Sánchez-Moreno, C.; De Ancos, B. Effects of combined application of high-pressure processing and active coatings on phenolic compounds and microbiological and physicochemical quality of apple cubes. *J. Sci. Food Agric.* **2021**, *101*, 4256–4265. [CrossRef]
78. Meng, X.; Zhang, M.; Adhikari, B. The Effects of Ultrasound Treatment and Nano-zinc Oxide Coating on the Physiological Activities of Fresh-Cut Kiwifruit. *Food Bioprocess Technol.* **2013**, *7*, 126–132. [CrossRef]
79. Fan, K.; Zhang, M.; Chen, H. Effect of Ultrasound Treatment Combined with Carbon Dots Coating on the Microbial and Physicochemical Quality of Fresh-Cut Cucumber. *Food Bioprocess Technol.* **2020**, *13*, 648–660. [CrossRef]
80. Jansrimanee, S.; Lertworasirikul, S. Synergetic effects of ultrasound and sodium alginate coating on mass transfer and qualities of osmotic dehydrated pumpkin. *Ultrason. Sonochemistry* **2020**, *69*, 105256. [CrossRef] [PubMed]
81. Abbasi, N.A.; Ashraf, S.; Ali, I.; Butt, S.J. Enhancing storage life of bell pepper by UV-C irradiation and edible coatings. *Pak. J. Agric. Sci.* **2015**, *52*, 405–413.
82. Hussain, P.R.; Suradkar, P.; Wani, A.M.; Dar, M.A. Retention of storage quality and post-refrigeration shelf-life extension of plum (*Prunus domestica* L.) cv. Santa Rosa using combination of carboxymethyl cellulose (CMC) coating and gamma irradiation. *Radiat. Phys. Chem.* **2015**, *107*, 136–148. [CrossRef]
83. Hussain, P.R.; Rather, S.A.; Suradkar, P.; Parveen, S.; Mir, M.A.; Shafi, F. Potential of carboxymethyl cellulose coating and low dose gamma irradiation to maintain storage quality, inhibit fungal growth and extend shelf-life of cherry fruit. *J. Food Sci. Technol.* **2016**, *53*, 2966–2986. [CrossRef]
84. Zhang, S.; Yu, Y.; Xiao, C.; Wang, X.; Lei, Y. Effect of ultraviolet irradiation combined with chitosan coating on preservation of jujube under ambient temperature. *LWT* **2014**, *57*, 749–754. [CrossRef]
85. Severino, R.; Ferrari, G.; Vu, K.D.; Donsì, F.; Salmieri, S.; Lacroix, M. Antimicrobial effects of modified chitosan based coating containing nanoemulsion of essential oils, modified atmosphere packaging and gamma irradiation against Escherichia coli O157:H7 and Salmonella Typhimurium on green beans. *Food Control* **2015**, *50*, 215–222. [CrossRef]
86. Ben-Fadhel, Y.; Cingolani, M.C.; Li, L.; Chazot, G.; Salmieri, S.; Horak, C.; Lacroix, M. Effect of γ-irradiation and the use of combined treatments with edible bioactive coating on carrot preservation. *Food Packag. Shelf Life* **2021**, *28*, 100635. [CrossRef]
87. Wambura, P.; Yang, W.W. Ultrasonication and Edible Coating Effects on Lipid Oxidation of Roasted Peanuts. *Food Bioprocess Technol.* **2009**, *3*, 620–628. [CrossRef]
88. Gonçalves, S.M.; de Melo, N.R.; da Silva, J.P.; Chávez, D.W.H.; Gouveia, F.S.; Rosenthal, A. Antimicrobial packaging and high hydrostatic pressure: Combined effect in improving the safety of coalho cheese. *Food Sci. Technol. Int.* **2020**, *27*, 301–312. [CrossRef] [PubMed]
89. Ahari, H.; Nasiri, M. Ultrasonic Technique for Production of Nanoemulsions for Food Packaging Purposes: A Review Study. *Coatings* **2021**, *11*, 847. [CrossRef]
90. Clark, J.P. Commercial Applications of Ultrasound in Foods. *Food Technol.* **2010**, *64*, 78.

91. Patist, A.; Bates, D. Ultrasonic innovations in the food industry: From the laboratory to commercial production. *Innov. Food Sci. Emerg. Technol.* 2008, 9, 147–154. [CrossRef]
92. Cárcel, J.; García-Pérez, J.; Benedito, J.; Mulet, A. Food process innovation through new technologies: Use of ultrasound. *J. Food Eng.* 2012, 110, 200–207. [CrossRef]
93. Bendicho, C.; Lavilla, I. Ultrasound extractions. *Encycl. Sep. Sci.* 2000, 1448–1454.
94. García-Pérez, J.V.; Carcel, J.A.; Mulet, A.; Riera, E.; Gallego-Juarez, J.A. Ultrasonic drying for food preservation. In *Power Ultrasonics*; Elsevier: Amsterdam, The Netherlands, 2015; pp. 875–910.
95. Huang, G.; Chen, S.; Dai, C.; Sun, L.; Sun, W.; Tang, Y.; Xiong, F.; He, R.; Ma, H. Effects of ultrasound on microbial growth and enzyme activity. *Ultrason. Sonochemistry* 2017, 37, 144–149. [CrossRef] [PubMed]
96. Izadifar, Z.; Babyn, P.; Chapman, D. Ultrasound Cavitation/Microbubble Detection and Medical Applications. *J. Med. Biol. Eng.* 2018, 39, 259–276. [CrossRef]
97. Khin, M.M.; Zhou, W.; Perera, C.O. A study of the mass transfer in osmotic dehydration of coated potato cubes. *J. Food Eng.* 2006, 77, 84–95. [CrossRef]
98. Wambura, P.; Yang, W.; Mwakatage, N.R. Effects of Sonication and Edible Coating Containing Rosemary and Tea Extracts on Reduction of Peanut Lipid Oxidative Rancidity. *Food Bioprocess Technol.* 2008, 4, 107–115. [CrossRef]
99. Dehsheikh, F.N.; Dinani, S.T. Coating pretreatment of banana slices using carboxymethyl cellulose in an ultrasonic system before convective drying. *Ultrason. Sonochemistry* 2018, 52, 401–413. [CrossRef]
100. Mahendran, R.; Ramanan, K.R.; Barba, F.J.; Lorenzo, J.M.; López-Fernández, O.; Munekata, P.E.; Roohinejad, S.; Sant'Ana, A.S.; Tiwari, B.K. Recent advances in the application of pulsed light processing for improving food safety and increasing shelf life. *Trends Food Sci. Technol.* 2019, 88, 67–79. [CrossRef]
101. Barba, F.J.; Ahrné, L.; Xanthakis, E.; Landerslev, M.G.; Orlien, V. *Innovative Technologies for Food Preservation: Inactivation of Spoilage and Pathogenic Microorganisms*; Academic Press: Cambridge, MA, USA, 2017.
102. Mandal, R.; Mohammadi, X.; Wiktor, A.; Singh, A.; Singh, A.P. Applications of Pulsed Light Decontamination Technology in Food Processing: An Overview. *Appl. Sci.* 2020, 10, 3606. [CrossRef]
103. John, D.; Ramaswamy, H.S. Pulsed light technology to enhance food safety and quality: A mini-review. *Curr. Opin. Food Sci.* 2018, 23, 70–79. [CrossRef]
104. Koh, P.C.; Noranizan, M.A.; Karim, R.; Nur Hanani, Z.A.; Yusof, N.L. Cell wall composition of alginate coated and pulsed light treated fresh-cut cantaloupes (*Cucumis melo* L. Var. *Reticulatus* Cv. Glamour) during chilled storage. *J. Food Sci. Technol.* 2020, 57, 2206–2221. [CrossRef]
105. Abedi-Firoozjah, R.; Ghasempour, Z.; Khorram, S.; Khezerlou, A.; Ehsani, A. Non-thermal techniques: A new approach to removing pesticide residues from fresh products and water. *Toxin Rev.* 2020, 40, 562–575. [CrossRef]
106. Oliu, G.O.; Martín-Belloso, O.; Soliva-Fortuny, R. Pulsed Light Treatments for Food Preservation. A Review. *Food Bioprocess Technol.* 2008, 3, 13–23. [CrossRef]
107. Pongsri, R.; Aiamla-Or, S.; Srilaong, V.; Uthairatanakij, A.; Jitareerat, P. Impact of electron-beam irradiation combined with shellac coating on the suppression of chlorophyll degradation and water loss of lime fruit during storage. *Postharvest Biol. Technol.* 2020, 172, 111364. [CrossRef]
108. Ravindran, R.; Jaiswal, A.K. Wholesomeness and safety aspects of irradiated foods. *Food Chem.* 2019, 285, 363–368. [CrossRef]
109. Ehlermann, D.A. The early history of food irradiation. *Radiat. Phys. Chem.* 2016, 129, 10–12. [CrossRef]
110. Roberts, P.B. Food irradiation: Standards, regulations and world-wide trade. *Radiat. Phys. Chem.* 2016, 129, 30–34. [CrossRef]
111. Gu, J.-D.; Wang, Y. Microbial transformation of phthalate esters: Diversity of hydrolytic esterases. *Environ. Contam.–Health Risks Bioavailab. Bioremediat.* 2013, 313–346.
112. Wright, J.R.; Sumner, S.S.; Hackney, C.R.; Pierson, M.D.; Zoecklein, B.W. Efficacy of ultraviolet light for reducing *Escherichia coli* O157: H7 in unpasteurized apple cider. *J. Food Prot.* 2000, 63, 563–567. [CrossRef] [PubMed]
113. Bal, E. Influence of chitosan-based coatings with UV irradiation on quality of strawberry fruit during cold storage. *Turk. J. Agric. -Food Sci. Technol.* 2019, 7, 275–281. [CrossRef]
114. Lung, H.-M.; Cheng, Y.-C.; Chang, Y.-H.; Huang, H.-W.; Yang, B.B.; Wang, C.-Y. Microbial decontamination of food by electron beam irradiation. *Trends Food Sci. Technol.* 2015, 44, 66–78. [CrossRef]
115. Ajibola, O.J. An overview of irradiation as a food preservation technique. *Nov. Res. Microbiol. J.* 2020, 4, 779–789.
116. Lester, G.E.; Hallman, G.J.; Pérez, J.A. γ-Irradiation dose: Effects on baby-leaf spinach ascorbic acid, carotenoids, folate, α-tocopherol, and phylloquinone concentrations. *J. Agric. Food Chem.* 2010, 58, 4901–4906. [CrossRef]
117. Ben-Fadhel, Y.; Saltaji, S.; Khlifi, M.A.; Salmieri, S.; Vu, K.D.; Lacroix, M. Active edible coating and γ-irradiation as cold combined treatments to assure the safety of broccoli florets (*Brassica oleracea* L.). *Int. J. Food Microbiol.* 2017, 241, 30–38. [CrossRef]
118. Hussain, P.R.; Meena, R.S.; Dar, M.A.; Wani, A.M. Carboxymethyl Cellulose Coating and Low-Dose Gamma Irradiation Improves Storage Quality and Shelf Life of Pear (*Pyrus Communis* L., Cv. Bartlett/William). *J. Food Sci.* 2010, 75, M586–M596. [CrossRef]
119. Shankar, S.; Danneels, F.; Lacroix, M. Coating with alginate containing a mixture of essential oils and citrus extract in combination with ozonation or gamma irradiation increased the shelf life of *Merluccius* sp. fillets. *Food Packag. Shelf Life* 2019, 22, 100434. [CrossRef]

120. Dini, H.; Fallah, A.A.; Bonyadian, M.; Abbasvali, M.; Soleimani, M. Effect of edible composite film based on chitosan and cumin essential oil-loaded nanoemulsion combined with low-dose gamma irradiation on microbiological safety and quality of beef loins during refrigerated storage. *Int. J. Biol. Macromol.* **2020**, *164*, 1501–1509. [CrossRef] [PubMed]
121. Salem, E.A.; Naweto, M.A.R.; Mahmoud, M.M. Effect of Irradiation and Edible Coating as Safe Environmental Treatments on the Quality and The Marketability of "Anna" Apples During Cold Storage. *Arab. J. Nucl. Sci. Appl.* **2019**, *52*, 193–202. [CrossRef]

Disclaimer/Publisher's Note: The statements, opinions and data contained in all publications are solely those of the individual author(s) and contributor(s) and not of MDPI and/or the editor(s). MDPI and/or the editor(s) disclaim responsibility for any injury to people or property resulting from any ideas, methods, instructions or products referred to in the content.

www.ingramcontent.com/pod-product-compliance
Lightning Source LLC
LaVergne TN
LVHW070731100526
838202LV00013B/1210

9 783725 818556